OBSERVATIONAL PLASMA ASTROPHYSICS: FIVE YEARS OF YOHKOH AND BEYOND

ASTROPHYSICS AND SPACE SCIENCE LIBRARY

VOLUME 229

OBSERVATIONAL PLASMA ASTROPHYSICS: FIVE YEARS OF YOHKOH AND BEYOND

Edited by

TETSUYA WATANABE

National Astronomical Observatory, Mitaka, Tokyo, Japan

TAKEO KOSUGI

National Astronomical Observatory, Mitaka, Tokyo, Japan

and

ALPHONSE C. STERLING

Computational Physics Inc., Fairfax, VA, U.S.A.

KLUWER ACADEMIC PUBLISHERS

DORDRECHT / BOSTON / LONDON

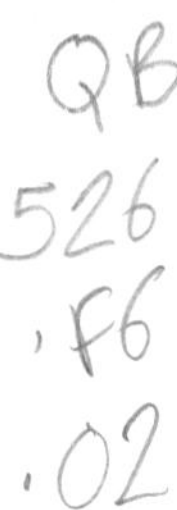

A C.I.P. Catalogue record for this book is available from the Library of Congress.

ISBN 0-7923-4985-7

Published by Kluwer Academic Publishers,
P.O. Box 17, 3300 AA Dordrecht, The Netherlands.

Sold and distributed in the U.S.A. and Canada
by Kluwer Academic Publishers,
101 Philip Drive, Norwell, MA 02061, U.S.A.

In all other countries, sold and distributed
by Kluwer Academic Publishers,
P.O. Box 322, 3300 AH Dordrecht, The Netherlands.

Printed on acid-free paper

Printed in the Netherlands

Table of Contents

<Joint Observation of Yohkoh and SoHO>

IV. High-Energy Particle Acceleration in Cosmic Plasmas

<Posters>

PREFACE

The Yohkoh Fifth Anniversary Symposium was held in the National Olympics Memorial Youth Center in the Yoyogi district of Tokyo, Japan, from 6 to 8 November, 1996. It was attended by more than a hundred participants from around the world, and featured invited oral presentations, along with oral and poster contributions covering various aspects of solar physics and closely related fields. This volume is a collection of the material presented at that conference.

Yohkoh was launched on August 30, 1991, from the Kagoshima Space Center on the island of Kyushu, southern Japan. It carries an array of instruments for observing the Sun, including a Hard X-ray Telescope (HXT), Soft X-ray Telescope (SXT), Bragg Crystal Spectrometer (BCS), and a Wide-Band Spectrometer (WBS). At the time of launch, the Sun was near the maximum of the solar cycle. SXT soft X-ray images from the earliest years of Yohkoh show an incredibly complex and dynamic solar corona, often with numerous active regions present simultaneously and a high frequency of flaring activity. Five years later, at the time of the conference which is the subject of this volume, the Sun was near the minimum of the solar cycle, and the SXT images were markedly different. Gone were the large-scale bright, nebulous features that filled out the active-Sun corona. Instead, the SXT corona was largely dark, punctuated by small-scale X-ray features and occasional active regions and activity complexes. Of course, flaring activity was also drastically reduced.

With Yohkoh, we have had the opportunity to observe this amazing transition from solar maximum to solar minimum. In these pages are discussions and analyses dealing with numerous aspects of this dynamic, changing solar atmosphere, including many aspects of solar flares, long-term coronal evolution, coronal mass ejections (CMEs), magnetic field evolution, active region studies, and several other topics. In addition to Yohkoh results, some preliminary findings from the solar satellite SOHO and their connection to Yohkoh observations are also presented. There are also contributions dealing with astrophysical and plasma physics topics, reflecting the richness and diversity of physical processes of mutual interest to solar scientists and their brethren in other areas of the scientific community.

This meeting would not have been possible without the dedicated assistance of our co-participants on the Scientific Organizing Committee, Y. Ogawara, L. W. Acton, J. L. Culhane, R. A. Harrison, H. Inoue, K. Makishima, and Y. Uchida, and our co-participants on the Local Organizing Committee, H. Hara, H. Hudson, F. Iwata, K. Mitsuda, T. Murakami, T. Sakao, T. Sakurai, K. Shibasaki, K. Shibata, T. Shimizu, and S. Tsuneta. We thank the Institute of Space and Astronautical Science (ISAS), who sponsored the meeting as one in a series of symposia that it organizes annually. This Symposium was also partly supported by the National Astronomical Observatory (NAOJ), who supported the meeting with a Grant-in-Aid for Science Research from the Ministry of Education, Science, Sports and Culture (Grant-in-Aid for Co-operative Research No.08304021). Finally, we thank all the participants, whose contributions made the meeting such a success.

Tokyo, 30-August-1997, the sixth anniversary of Yohkoh's launch

Te. Watanabe, T. Kosugi, and A. C. Sterling

Yohkoh Fifth Anniversary Symposium on Observational Plasma Astrophysics: Five Years of Yohkoh and Beyond
November 6 – 8 1996, National Olympics Memorial Youth Center

I. Basic Properties of Cosmic Plasmas

EVOLUTION OF THE SOLAR CORONA FROM THE MAXIMUM TO MINIMUM

H. HARA
National Astronomical Observatory
2-21-1 Osawa, Mitaka, Tokyo 181, Japan

Abstract

The evolution of the solar corona from the maximum to minimum phase in the cycle 22 is reported from the soft X-ray observation with *Yohkoh* soft X-ray telescope (SXT). The variation of X-ray flux is investigated for two coronal components; active regions and other dark features consisting of quiet regions and coronal holes. The X-ray flux from active regions is strongly dependent on the size of active regions and changes with the solar activity cycle in phase. We find that the X-ray flux from the other dark features also changes with the solar cycle, and that there is a strong correlation between the X-ray flux from the dark features and the corresponding photospheric magnetic flux outside active regions. These show that the solar corona is heated in conjunction with the magnetic field, and that there is no possibility of being due to the acoustic heating. The soft X-ray flux does not monotonically decrease. Rather, there is a period of enhancement with about a one-year interval in the whole-Sun X-ray flux. The activity appears as bright clusters in the butterfly diagram of the soft X-ray intensity and corresponds to the emergence of *complexes of activity* in the sunspot zones. The high-latitude activity is also studied, and we find the X-ray intensity of the high-latitude regions fluctuates with a time scale of about one year.

1. Introduction

The solar soft X-ray flux changes with the relative sunspot number and total photospheric magnetic flux in phase. Although the X-ray variability has long been monitored by simple photometric data (*e.g.* GOES soft X-ray data), the soft X-ray telescope (SXT) aboard *Yohkoh* allows us for the

T. Watanabe et al. (eds.), Observational Plasma Astrophysics: Five Years of Yohkoh and Beyond, 3–12.

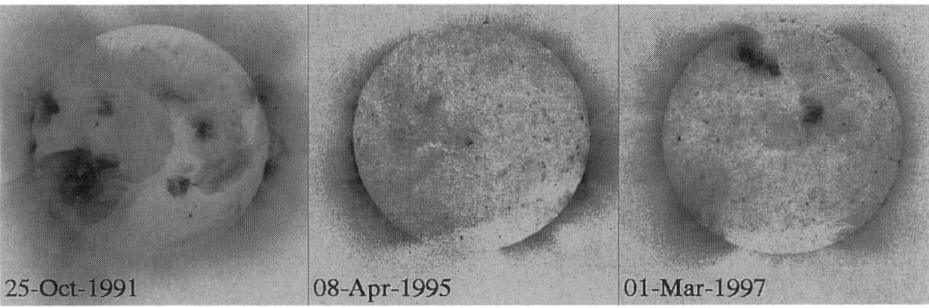

Figure 1. Soft X-ray images; (left) near solar maximum, and (middle/right) near the minimum phase. An active region of cycle 23 is seen at mid-latitudes in the right panel, while there is an active region of cycle 22 near the equator.

first time to study the long-term soft X-ray variability of individual coronal structures in detail. *Yohkoh* has been observing the Sun since September 1991 during a period in which the solar activity was drastically changing (Figure 1). A remarkable change is the long-term variation of X-ray flux coming from dark region outside active region, as shown later. Active regions belonging to cycle 23 have appeared since 1996, and they affect the global coronal structures of the old cycle.

In the present paper, we study the evolution of the solar corona from the solar maximum of cycle 22 to the next minimum, based upon the *Yohkoh* soft X-ray observations. *Yohkoh* soft X-ray full-disk images taken with the thin-Al filter are used, and the pixel size of the images is $5''$. We use soft X-ray intensity histograms made from the X-ray images for the purpose of investigating the long-term variability of the solar corona. Coronal synoptic charts are used to see the relationship between the temporal variation of the soft X-rays and the corresponding spatial structures.

2. Long-Term Soft X-ray Variability of the Sun

2.1. X-RAY INTENSITY HISTOGRAMS OF *YOHKOH* IMAGES

Figure 2 shows an intensity histogram $-dN(>I)/dI$ of a soft X-ray image as a function of X-ray intensity I, where $N(>I)$ is the number of pixels with X-ray intensities larger than I. The unit of soft X-ray intensity I is DN/s/$5''$pixel, where 1 DN is equivalent to an energy deposition of approximately 365 eV onto a detector cell of CCD. The ordinate of the histogram is proportional to the projected area in a given intensity range. At an X-ray intensity of $I_{th} = 100$ DN/s/$5''$pixel, the histogram slope typically changes. This X-ray intensity level discriminates active regions from

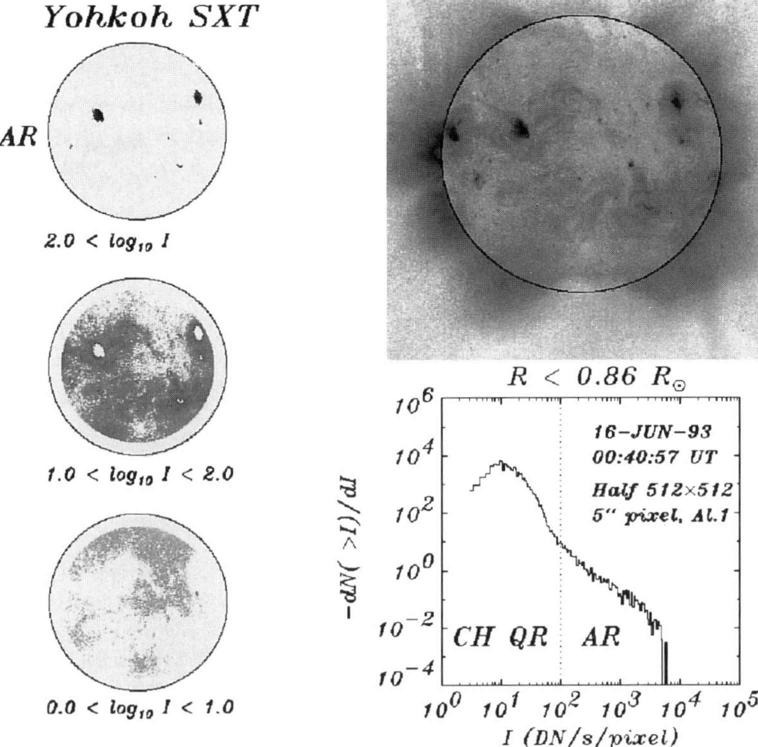

Figure 2. A soft X-ray histogram (bottom–right) on 1993 June 16 made from an SXT composite image (top–right) within 0.86 $R_\odot$. AR, QR, and CH indicate active regions, quiet regions, and coronal holes, respectively. The left panels show the same images in three different ranges of flux. Circles in the X-ray images show the limb of the Sun.

other darker regions as seen in Figure 2. Active regions correspond to a less steep power-law portion in the X-ray intensity histogram, and darker regions (hereafter background components), which consist of quiet regions and coronal holes, show a steep slope of intensity distribution in the histogram (Hara 1994).

2.2. TEMPORAL VARIATION OF THE WHOLE-SUN SOFT X-RAY FLUX

Figure 3 shows the temporal variation of the soft X-ray flux. The sinusoidal curve in the bottom–left of the figure reflects the change of the area due to the seasonal variation of the distance between the Sun and Earth. The total X-ray flux (top–left in Figure 3) gradually decreases with time, oscillating

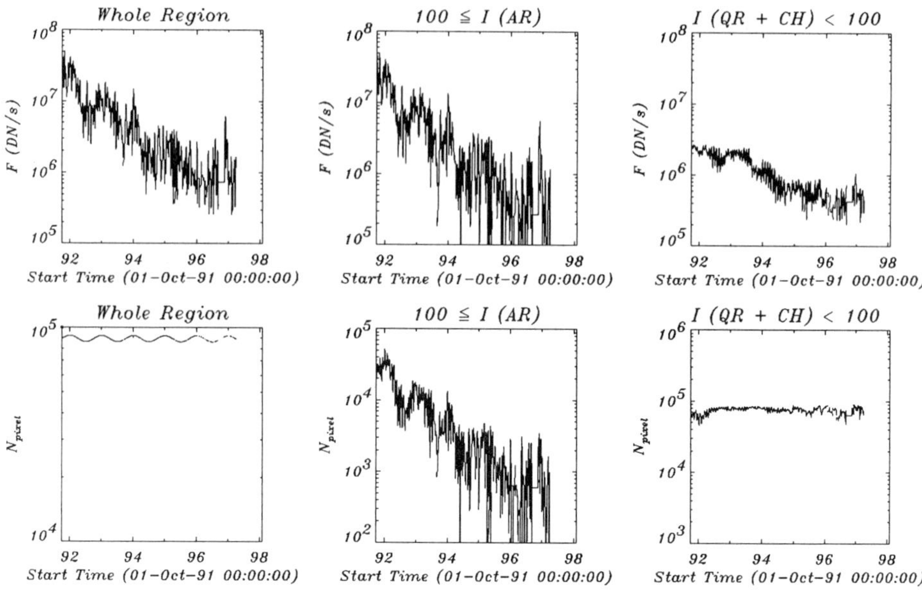

Figure 3. Temporal variation of the soft X-rays in the region within 0.86 $R_\odot$. The 10-days' running mean of the integrated X-ray flux F (DN/s)through the thin-Al filter (top) and the number of image pixels N_{pixel} (bottom) are shown. Integrations over the whole region (left), active regions (middle), and dark regions that consist of the quiet regions and coronal holes (right) are indicated. The X-ray flux in the top panels is corrected for the annual variation of distance between the Sun and Earth.

with time scales of ~27 days and ~1 year. The period of ~27 days is due to solar rotation and implies the nonuniformity of occurrence of the bright regions in longitude. A one-year period modulation in the X-ray flux can be also seen in the temporal variation of the total magnetic flux (Harvey 1993). In the previous cycle, 10.7 cm radio flux, Ca K plage index, and Lα flux also show a one-year period modulation in the declining phase (Pap *et al.* 1991).

2.3. VARIABILITY OF THE ACTIVE-REGION CORONA

The temporal variation of total soft X-ray intensity of all regions (top–left in Figure 3) shows a similar change to that of active regions (top–middle in Figure 3). The correlation coefficient (hereafter CC) between the two temporal variations is 0.998. Thus the variation of the total X-ray intensity is mostly due to changes of the X-ray intensity from active regions. The bottom–middle panel in Figure 3 shows the temporal variation of the projected area occupied by active regions. The overall variation of the area is similar to the variation of the soft X-ray flux coming from active regions (CC = 0.992), suggesting that the variation is due to variation in fractional area of the disk covered with active regions, and not to a remarkable change of the soft X-ray intensity per unit area in active regions (Hara 1996a).

2.4. VARIABILITY OF BACKGROUND COMPONENTS

Although X-ray flux monitors such as those aboard the GOES satellites have been mostly monitoring the variation of bright regions, *i.e.* active regions, such observations cannot discriminate between the background components and active regions because of lack of spatial resolution. We show the variation of the X-ray flux from the background components based upon full-disk observations in the top-right portion of Figure 3. Since the area occupied by the background components is almost constant over the observing period as seen in the bottom-right of Figure 3, the background components has changed its brightness in the declining phase of cycle 22. The X-ray intensity from the background component around 1995 March is smaller than that around 1991 October by a factor of three. The decrease of X-ray intensity from the background component is well seen as a shift of the peak to the darker side in the X-ray intensity histogram (Figure 4; see also Hara 1996b).

The shift of the peak in X-ray histograms does not imply that the area occupied by coronal holes substantially expands in the declining phase. Figure 5 shows two X-ray synoptic charts; one of them is in the maximum phase, and the other is near the minimum phase. The X-ray intensity profiles across a Carrington longitude in these synoptic charts are shown in Figure 6. These profiles indicate that the X-ray intensity in quiet regions decreases by a factor of three.

2.5. SOFT X-RAY VARIABILITY AND CORONAL HEATING MECHANISM

It is reported by Harvey (1993, 1994) that the photospheric magnetic flux in the quiet sun, which corresponds to the photospheric region under quiet regions and coronal holes in soft X-ray structures, also varies in phase with the sunspot cycle, as shown in the left panel of Figure 7. A correlation plot between total magnetic flux and total soft X-ray flux for any regions outside active regions is indicated in the right panel of Figure 7. In order to remove any effects regarding short-term variability, each flux is averaged for three rotation periods. Since the area covered by active regions may affect the relationship between the two flux parameters, we corrected each flux by considering the area occupied by active regions. That is why the

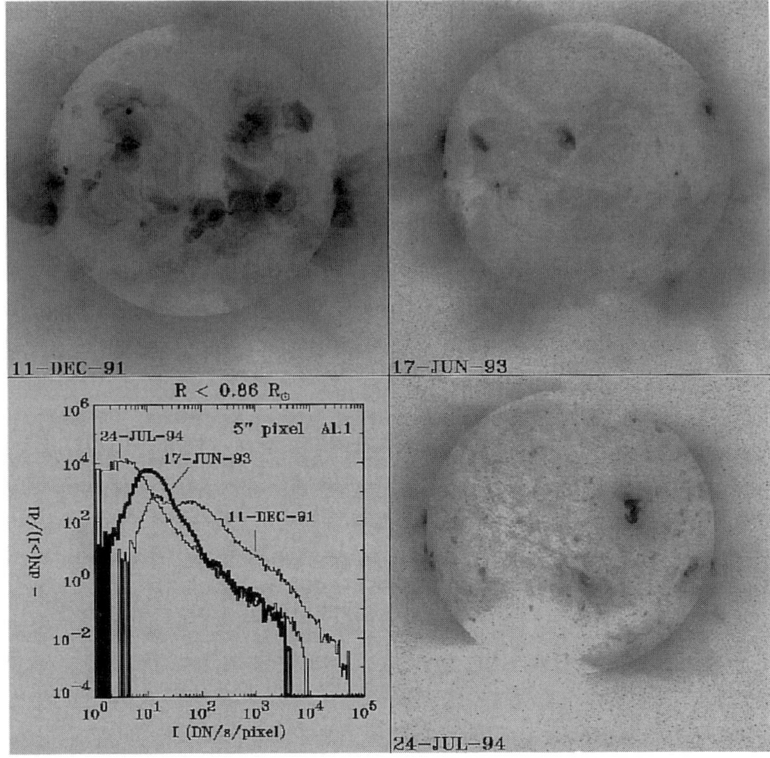

Figure 4. Soft X-ray histograms at several phases during the declining phase of the solar activity cycle are shown in the bottom-left panel together with the corresponding X-ray images. An inverse color scale is used. The peak of the histogram on 24 July 1994 shifts to the darker side compared with the histogram on 11 December 1991 and 17 June 1993.

level of quiet-sun magnetic flux in the right panel of Figure 7 is inconsistent with that in the left panel. From this correlation plot we find that there is a strong power-law relationship between the two flux parameters. A similar relationship was found by Golub *et al.* (1980) for active regions. Since there has not been a report that the photospheric granular motion, which is believed to be the energy source of heating of the solar corona, changes with the sunspot cycle, the correlation suggests that the presence of a magnetic field is required for the coronal heating of the Sun, and that the heating rate in the solar corona is determined by the number of magnetic field lines in a given area. This implies that the magnetic field is the path in which the energy for coronal heating is transferred from the photosphere to the corona. Although the acoustic heating theory has survived by introducing a small filling factor to explain the small acoustic flux estimated from ob-

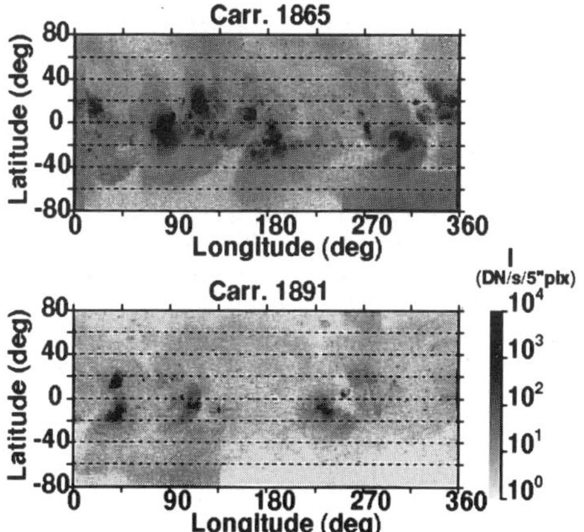

Figure 5. Soft X-ray synoptic charts for Carrington number 1865 (21 January 1993 – 17 February 1993) and 1891 (31 December 1994 – 27 January 1995).

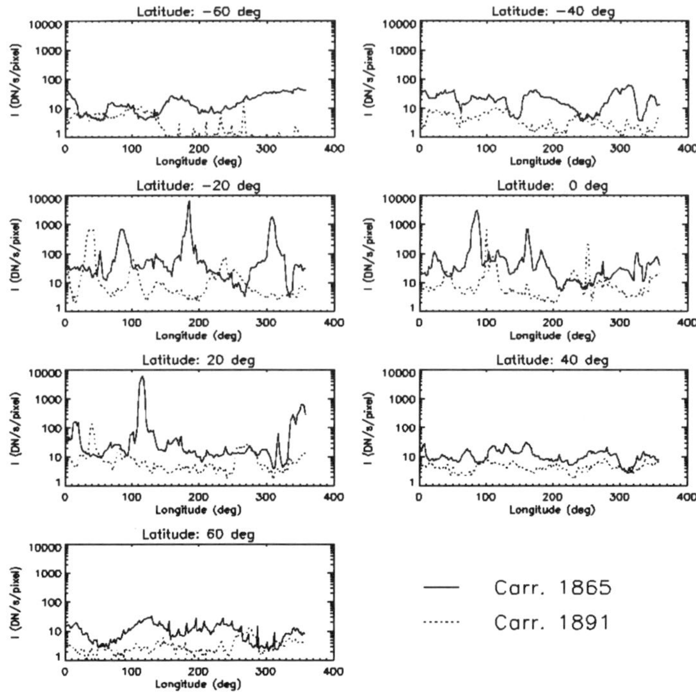

Figure 6. One-dimensional intensity profiles for Carrington number 1865 (1891) by solid (dotted) line at a given latitude as a function of the Carrington longitude.

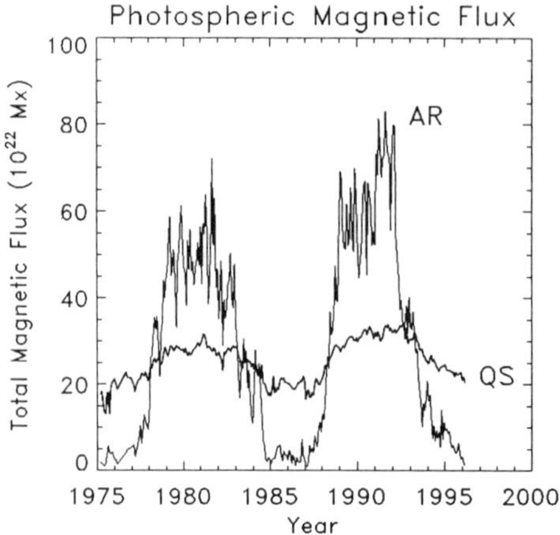

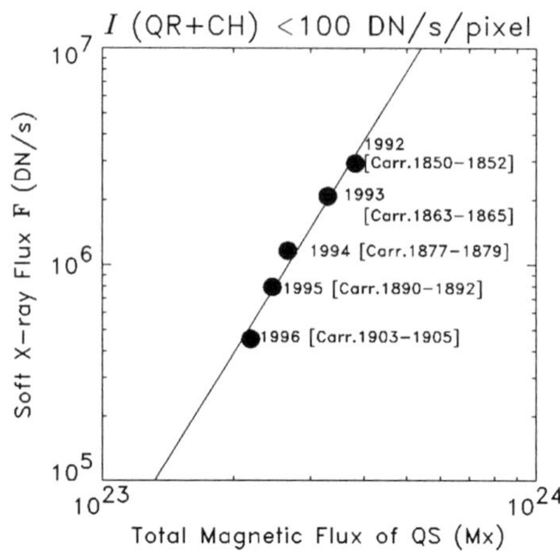

Figure 7. (left) Integrated photospheric magnetic flux of active regions (AR; solid line) and quiet sun (QS; thick solid line) over two sunspot cycles (Courtesy K. L. Harvey). The data were taken at Kitt Peak National Observatory. (right) Correlation plot between total magnetic flux and soft X-ray flux for the region outside active regions. The averaged flux over three rotations is compared. Flux change due to the areal change by emergence of active region is corrected. The data points are fitted by a power law, and the power index is 3.3 ± 0.3.

servations in the 1970s (Ulmschneider 1979), our result effectively rules out the possibility of coronal heating by acoustic waves.

3. Variation of Global Coronal Structures

3.1. LOW-LATITUDE ACTIVITIES

The time–latitude diagram of the soft X-ray intensity, similar to the butterfly diagram for sunspots, is presented in Figure 8. This figure is made from the soft X-ray synoptic charts by averaging the X-ray intensity in the direction of the Carrington longitude, therefore structures in the longitudinal direction are smeared out. In low-latitude regions ($\theta < 30°$) it is clear that bright structures appear intermittently, and an active phase roughly comes almost every one-year interval, though there are also cases of shorter intervals. These bright structures correspond to the bumps in the soft X-ray temporal variation as seen in Figure 3. The bright structures do not, of course, consist of a single active region with a long lifetime of nearly a year, but rather they comprise many active regions, as is apparent from several local maxima inside of individual structures in both the

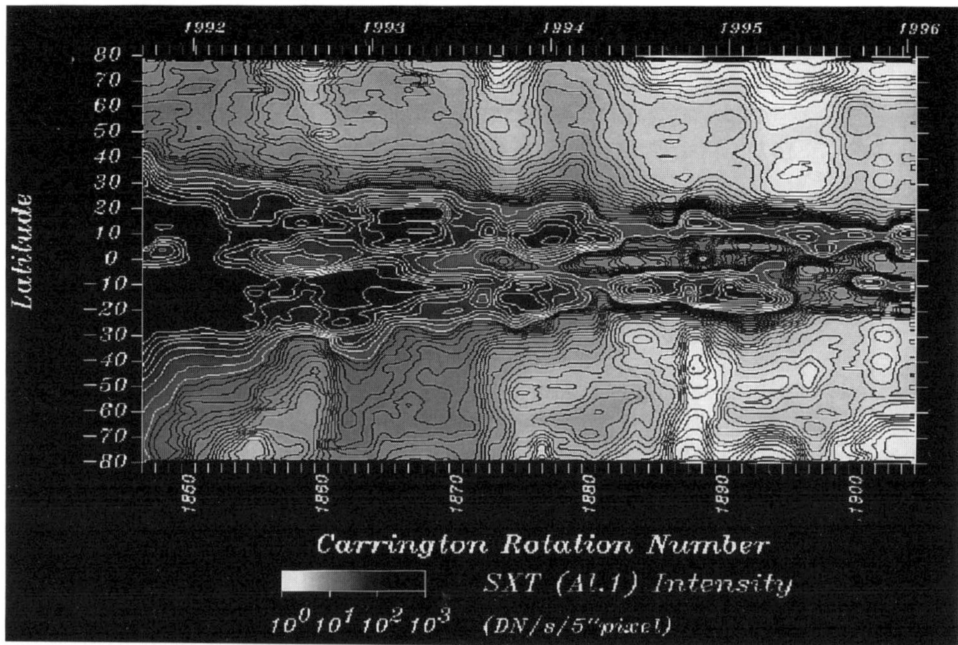

Figure 8. Time–latitude diagram of soft X-ray intensity through the thin Al filter covering a period between Oct 1991 and Jan 1996. The annual change of the distance between the Sun and Earth is corrected for. Iso-intensity contours are indicated by white (black) lines for low-latitude (high-latitude) regions. The contour level steps are constant on a logarithmic scale, and the intensity at a certain contour level is 1.31 (1.11) times brighter or darker than the intensities of the neighboring contours in white (black) lines.

latitudinal and temporal directions. The bright regions in the low latitudes of the time–latitude diagram consist of clusters of active regions, that is, *complexes of activity* found in the magnetic-field observations (Bumba & Howard 1965, 1969; Gaizauskas *et al.* 1983).

3.2. HIGH-LATITUDE ACTIVITIES

We can also see activity at high latitudes as shown in Figure 8. The X-ray intensity in high-latitude regions fluctuates with time. Periods when high-latitude regions become bright are Carrington rotations 1862–1865, 1877–1879, and 1889–1891 in the northern hemisphere, and 1863–1869, 1877–1882, and 1895–1899 in the southern hemisphere. Existence of the high-latitude activity zones in the corona has been previously known (Waldmeier 1957; Trellis 1957; Bretz & Billings 1959; Hansen *et al.* 1969; Leroy & Noens 1983; Altrock 1988, 1997). However, the long-term intensity fluctuation of this high-latitude structures has been found for the first time in the present study. This phenomenon might not be related to the activity of *complexes of activity*, because low-latitude activity maxima do not coincide with the high-latitude activity maxima.

REFERENCES

Altrock, R. C. 1988, in *Solar and Stellar Coronal Structure and Dynamics*: Proc. of Sacramento Peak 9th Summer Workshop, ed. R. C. Altrock, p.414.

Altrock, R. C. 1997, *Solar Phys.*, **170**, 411.

Bretz, M. C., and Billings, D. E. 1959, *Astrophys. J.*, **129**, 134.

Bumba, V., and Howard, R. 1965, *Astrophys. J.*, **141**, 1502.

Bumba, V., and Howard, R. 1969, *Solar Phys.*, **7**, 28.

Gaizauskas, V., Harvey, K. L., Harvey, J. W., and Zwaan, C. 1983, *Astrophys. J.*, **265**, 1056.

Golub, L., Maxson, C., Rosner, R., Serio, S., and Vaiana, G. S., 1980, *Astrophys. J.*, **238**, 343.

Hansen, R. T., Garcia, C. J., Hansen, S. F., and Loomis, H. G. 1969, *Solar Phys.*, **7**, 417.

Hara, H. 1994, in *New Look at the Sun with Emphasis on Advanced Observations of Coronal Dynamics and Flares, NRO Report*, **360**, eds. S. Enome and T. Hirayama, p.57.

Hara, H. 1996a, in *Magnetic Phenomena in the Solar Atmosphere – Prototypes of Stellar Magnetic Activity*, Proc. of IAU Colloq. 153, eds Y. Uchida *et al.*, p.321.

Hara, H. 1996b, Ph.D. thesis, University of Tokyo.

Harvey, K. L. 1993, Ph.D. thesis, University of Utrecht.

Harvey, K. L. 1994, in *The Sun as a Variable Star*, Proc. of IAU Colloq. 143, eds. J. M. Pap, C. Fröhlich, H. S. Hudson, and S. K. Solanki, p.217.

Leroy, J. L., and Noens, J. C. 1983, *Astron. Astrophys.*, **120**, L1.

Pap, J. M., London, J., and Rottmann, G. J. 1991, *Astron. Astrophys.*, **245**, 648.

Trellis, M. 1957, *Ann. d'Astrophys. Suppl.*, **5**, 1.

Ulmschneider, P. 1979, *Space Sci. Rev.*, **24**, 71.

Waldmeier, M. 1957, *Die Sonnenkorona, Vol. 2*, Verlag Birkhauser, Basel.

DENSITY, TEMPERATURE AND MAGNETIC FIELD STRUCTURE OF HIGH LATITUDE CORONA

MADHULIKA GUHATHAKURTA
Catholic Unversity of America/GSFC
Code 682, Greenbelt, MD 20771, USA

RICHARD FISHER
GSFC
Code 682, Greenbelt, MD 20771, USA

AND

KEITH STRONG
Lockheed/Martin Solar & Astrophysics Laboratory
LPARL, Palo Alto, CA 94304, USA

Abstract.

This paper explores the physical and morphological characteristics of the large-scale coronal structures such as polar coronal rays and high latitude streamers as seen in white-light and relates these structures to observations in soft X-ray (3-45 Å), red (6374 Å Fe X) and green (5303 Å Fe XIV) line emissions to estimate their densities and temperatures. Analysis shows that polar rays can be characterized by at least two temperature classes. Cool (0.7-1.3 10^6 K) rays are a dominant feature of the polar corona during the quiescent phase of the solar cycle. The hot (1.8-2.6 10^6 K) rays when present form a small subset of the array of rays seen in white-light. Hot rays seem to emerge from the boundary of the polar coronal hole (polar crown filament belt). The location of the cool rays on the other hand can be on the boundary or inside the coronal hole. We do not always find a one to one correspondence between the polar rays observed in white-light versus those observed in XUV and visible emission lines. We find the emission line-ratio temperature to be high in the high latitude ($> 45^o$ N,S) coronal streamers with enhanced white-light emission. These streamers are located along a neutral line which separates the weak old cycle polar field from the weak new cycle high-latitude magnetic field of opposite polarity (Ap J (Letters), 471, 1, L69). This paper is an extended abstract for the Ap J Letters paper.

T. Watanabe et al. (eds.), Observational Plasma Astrophysics: Five Years of Yohkoh and Beyond, 13.

STUDIES OF CORONAL TEMPERATURE

L. W. ACTON
Montana State University
Bozeman, Montana, 59717, USA

AND

J. R. LEMEN
Solar and Astrophysics Laboratory
Lockheed Martin Advanced Technology Center
Palo Alto, California, 94304, USA

1. Extended abstract

The temperature of the solar corona is high, variable, and inhomogenous. As a result, the large scale temperature structure of the entire corona is poorly known. There are few, if any, full sun temperature maps of the solar corona in the literature. The soft x-ray telescope (SXT) on *Yohkoh* is able to provide useful temperature maps for regions where the x-ray signal is sufficiently strong. Previous SXT studies have shown that the coronal temperature decreases with height immediately over active regions (Klimchuk and Gary, 1995). Within coronal holes and beneath coronal streamers the mean temperature is observed to increase with height (Sturrock, Wheatland and Acton, 1996, Foley, Culhane and Acton, 1997). Here we present a full-sun temperature map to illustrate these and other features of the temperature structure of the corona.

For sufficiently long exposures SXT images through 2 different analysis filters may be used to prepare a temperature map of the entire corona (Tsuneta, *et al.* 1991). The isothermal temperatures derived by this technique represent averages along each line of sight through the optically-thin corona. The average is biased towards the higher temperatures because of the spectral response of the SXT. However, apart from experimental error, an SXT temperature gives a reliable lower limit to the temperature of the hottest plasma along the line of sight. That is, plasma must exist which is at least as hot as the derived value.

T. Watanabe et al. (eds.), Observational Plasma Astrophysics: Five Years of Yohkoh and Beyond, 15–17.

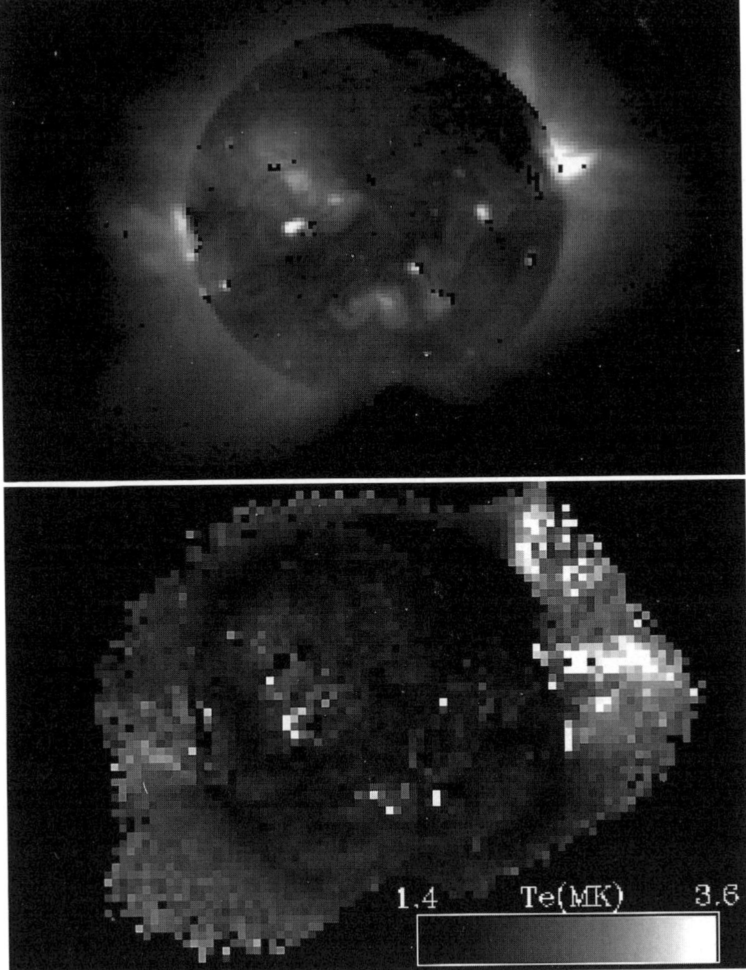

Figure 1. A long-exposure x-ray image of the corona (top) and the corresponding coronal temperature map (bottom). The pixel size of the log-scaled intensity image is 5" and that of the temperature map is 40". North is at the top and east to the left. Blacked-out pixels are not included in the analysis.

On 1992 Augsust 26 *Yohkoh* was offpointed from the center of the sun in order to allow SXT to observe the high corona with long (1 minute) exposures. Interleaved shorter exposures recorded the brighter coronal features. Figure 1 (top) is a composite image prepared from 8 co-aligned SXT exposures taken through the SXT thin-aluminum analysis filter. The images were acquired over a period of 138 minutes (one orbit offpointed west followed by one orbit offpointed east). The effective exposure is 2 minutes.

Pixels with insufficient signal for this temperature analysis, or a too-great time variation, are blacked out in the figure. The data have been corrected for off-axis vignetting and x-ray scattering by the SXT mirror. For the temperature analysis the intensity data have been binned into 40" pixels to further improve statistics and to emphasize the large scale variations of coronal temperature. These temperature results should be quite reliable except for the region just to the right of the non-radial cusped structure in the northwest. This region was affected by saturation bleed from the very bright feature at 3 o'clock and will be blacked out in the final analysis.

Note the following features of the coronal temperature structure:

- The lowest mean temperatures are at the limb, reflecting the fact that coronal temperatures are lower in the legs of the magnetic loops.

- High temperatures on the disk are associated with complexes of activity.

- The large regions of high temperature, above the west and northwest limbs, are associated with transient coronal activity. The distinctly non-radial cusped structure in the northwest appears to be the aftermath of a coronal mass ejection.

- The temperature above active regions (e.g., near equator on west limb) appears to decrease with height as noted by Klimchuk and Gary (1995).

- The mean temperature of the quiet closed corona generally increases with height. Note especially the large coronal streamer in the southwest.

- The SXT signal levels from coronal holes were too faint to pass the intensity criterion of this analysis.

2. Acknowledgements

This work was supported by NASA contract NAS8-40801 with the Lockheed Martin Advanced Technology Center.

References

Foley, C. A., Culhane, J. L. and Acton, L. W. (1997) *YOHKOH* soft X-ray determination of plasma parameters in a polar coronal hole, Submitted to *ApJ.*

Klimchuk, J. A. and Gary, D. E. (1995) A comparison of active region temperatures and emission measures observed in soft X-rays and microwaves and implications for coronal heating, *ApJ* **448**, 925–937.

Sturrock, P. A., Wheatland, M. S. and Acton, L. W. (1996) *YOHKOH* soft x-ray telescope images of the diffuse solar corona, *ApJ Letters* **461**, L115–L117.

Tsuneta, S., Acton, L., Bruner, M., Lemen, J., Brown, W., Caravalho, R., Catura, R., Freeland, S., Jurcevich, B., Morrison, M., Ogawara, Y., Hirayama, T., and Owens, J., (1991) The Soft X-Ray Telescope for the SOLAR-A Mission, *Solar Physics* **136**, 37–67.

THE BEHAVIOR OF CORONAL HOLES

SANCHEZ-IBARRA, A. AND BARRAZA-PAREDES, M.
Area de Astronomía/ CIF-US, Universidad de Sonora
Hermosillo Sonora, México.

Abstract. Five years of soft X-ray images from Yohkoh have provided us with a unique view of high-energy solar phenomena, as well as a new way of observing the lowest activity levels of the solar corona. For the first time, the Sun has been observed at soft X-ray wavelengtsHs during the minimum phase of solar cycle 22, showing an almost turned-off X-ray Sun. The most evident phenomenon, is that related to low-level coronal activity, namely Coronal Holes, (CHs), have been observed within superior detail compared with observations made through He I images in former times. In contrast with the situation near solar maximum, when CHs are easy to identify, CHs are a feature difficult to discern from the quiet regions and "dark lanes" dominant at times near the minimum of the cycle. In this paper we discuss the role of fine structures in the solar corona with densities comparable to CHs, their relations with photospheric fields, and the connection with the global behavior of CHs through the solar cycle.

1. Introduction

Coronal Holes CHs are thought to be regions of low temperature and density, and for many years have been considered to be the lowest level of activity in the solar corona. Deduced first as "M" regions of periodic magnetic disturbances (Broun; Chapman *et al.*; Bartels), they were really discerned through the observations of the OSO satellites, and mainly with the Skylab observations at short wavelengts (Bohlin *et al.*, 1975). A global view of CH activity was obtained with several studies made with the material from Skylab.

T. Watanabe et al. (eds.), Observational Plasma Astrophysics: Five Years of Yohkoh and Beyond, 19–26.

¿From 1975 to current times, CHs were observed in He I 1083 nm images taken at the National Solar Observatory of Kitt Peak. These data were extensively analyzed and a Catalog of Coronal Holes (Sanchez-Ibarra *et al.*, 1992) was produced. The work with the Catalog data allowed us to determine several characteristics of CHs in connection with the eleven year solar cycle, such as frequency, sizes, frequency of polarity predominance in average latitudes, and rotation.

The global evolution of CHs over the cycle (Sanchez-Ibarra, 1996) permited to see a possible relation of CHs with the new model proposed of the Extended Solar Cycle, where evolution of CHs would be through a period of 18-22 years, overlaped on the sunspot cycle of eleven years. Moreover, available data in soft x-ray wavelengtsHs from Yohkoh since 1991, let us explore in detail that possibility, and also to find the presence of small structures similar to CHs extending through the solar disk over the last 12 months where the minimum of the sunspot solar cycle is ocurring.

The relation of CHs with the Extended Solar Cycle pattern, as well as the presence of small structures, are discussed here as a possible way to establish a new empirical magnetic model for solar activity.

2. Coronal Holes and the Extended Solar Cycle Pattern

In the Catalog of Coronal Holes (Sanchez-Ibarra *et al.*, 1992), CHs were divided into two types:

– Polar CHs: Those that are extensions beyond $\pm 60^o$ of the main polar holes to lower latitudes.

– Equatorial CHs: Those that are isolated from the main polar holes, appearing at latitudes between 0 and $\pm 60^o$.

These types were selected arbitrarily from their appearence and from the analysis of data from the Catalog of Coronal Holes. Also it was possible to establish (Sanchez-Ibarra, 1996) that:

a) Polar extensions reach the lowest latitudes at times right after the polarity reversal of the sunspot cycle, and become contracted during the minimum and before the inversion of polarity.

b) Equatorial or isolated CHs migrate during the cycle from high to low latitudes in two bands with different polarities.

c) The total number of CHs of the same polarity diminishes from the time of the polarity reversal until the minimum of the next solar cycle.

Both the migration of CHs from polar to equatorial latitudes, and also the decay of the frequency of CHs with one polarity, suggested a possible relation between the evolution of CHs and the Extended Solar Cycle (ESC) pattern.

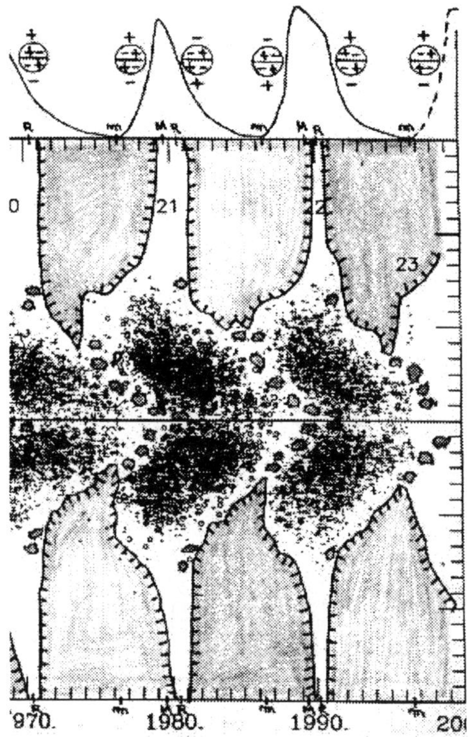

Fig 1. Average latitudes of polar extensions and isolates CHs on the butterfly diagram for solar cycles 21 and 22. Positive predominant polarity and negative predominant polarity correspond to those indicated at the upper diagram.

Plotting the average annual latitudes of both polar extensions and isolated CHs on the butterfly diagram (Harvey, 1996) for the last two sunspot cycles (Fig. 1), made it possible to find that:

a) A few isolated CHs with dominant polarity corresponding to the next polarity reversal appear at high latitudes one or two years ahead of that polarity reversal.

b) After polarity reversal, the polar extensions with the new polarity become dominate, extending to the lowest latitudes from both poles and decreasing the number of isolated holes.

c) Before the solar minimum, the polar extensions contract to average latitudes of $\pm 50^o$, and isolated holes reappear at lower latitudes.

d) Meanwhile the polar extensions remain at latitudes close to $\pm 45^o$ until the next polarity reversal period, and the isolated holes migrate to lower latitudes until the maximum of the sunspot cycle when they finally disappear.

On the same butterfly diagram, we ploted the mean photospheric magnetic field and the bands of faster and slower than average rotational speed following the rule that the main activity belt is located at the shear zone between a fast and a slow band (Fig 2).

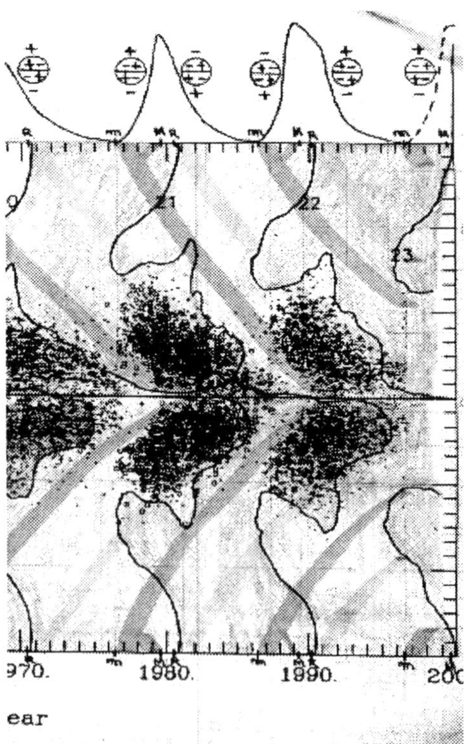

Fig 2. Bands of faster and slower than average rotational speed and mean photospheric field are plotte on the butterfly diagram for solar cycles 21 and 22.

In figure 3, adding mean CH latitudes in the evolution model of CHs through several cycles is reveled that:

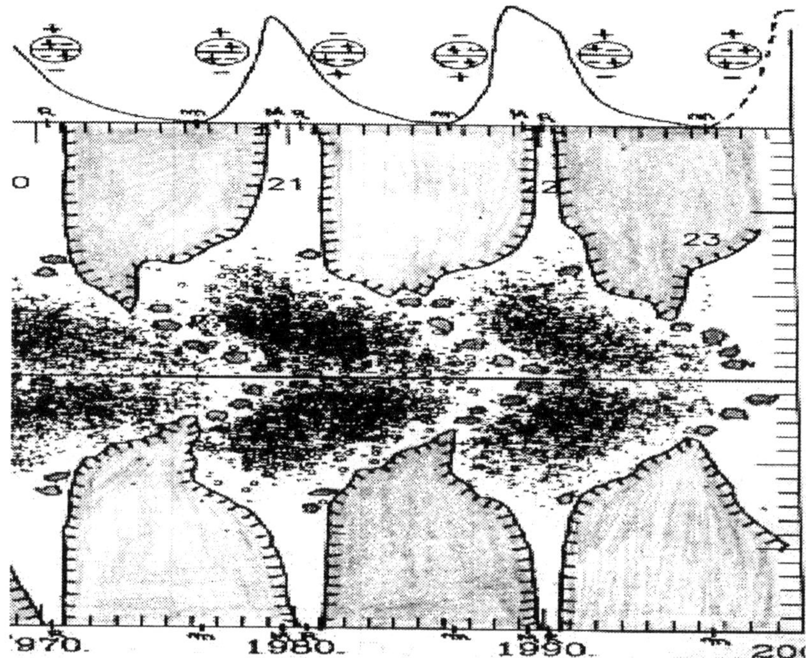

Fig 3. Simulation of evolution of CHs through several solar cycles. Polar extensions and isolated CHs contours in the butterfly diagram. Also bands of faster and slower than average rotadional speed bands are plotted.

a) The first group of isolated CHs appearing before the reversal of polarity at high latitudes are located in the shear band between a slow and a fast band of Torsional Osccilations (TO).

b) The second group of isolated CHs appearing after the contraction of polar extensions before the minimum of the solar cycle, are also present with average latitudes in the shear band between a slow and a fast band of TO.

Also plotted in Fig. 3 other features that support the ESC pattern, such as:

a) The mean latitudes of Ephemeral Regions (ER's) which migrate from polar to equatorial latitudes in connection with the mean latitude of active regions.

b) The filament crown that migrates (Makarov et al., 1989; McIntosh, 1992) from medium ($\pm 45^o$) latitudes to the pole from one reversal of polarity to the next.

Althought the filament crown seems to be crossing the main latitude zones during CHs polar extensions, we must remember that at the time of maximum extension of polar holes through the equator, they appear and remain at opposite solar longitudes. So, the filament crowns also migrates in opposite solar longitudes to polar latitudes at this time.

3. Small Structures similar to CHs

The Yohkoh images have permitted higher resolution observations of CHs than those obtained with the He I 1083 nm line. Also, for the first time we have had the opportunity to observe the Sun during the minimum of the sunspot cycle in soft X-rays.

In soft X-ray images from 1991, it is easy to recognize the presence both polar extension as isolated CHs. Determining the average brightness of pixels right at the polar holes, it is possible to create a contour plot over the solar disk to identify CHs.

Soft X-ray images allow us to see a wider extension of CHs at positions near the central meridian than can be seen in the He I images. Also in some cases (and increasing so near the limb) the CHs are contained by streamers from the active regions.

However, since February of 1996, as the Sun has been aproaching the minimum phase of the sunspot solar cycle, although the polar holes continue to be very well defined, isolated CHs have been appearing as very small structures below the average CH.

Although some of these structures appear to be related to filament channels, and the so-called "dark lanes" (Moses *et al.*, 1994), about 80% measured to date have same low-level coronal brightness that corresponds to the polar CHs.

In a previous work (Sanchez-Ibarra *et al.*, 1996a), we discussed the possibility that we were detecting "mini" CHs, which could make an important contribution to the flux of the low and the medium-speed solar wind.

Looking for a correspondence, the clusters of "mini" CHs were plotted on synoptic charts for the last six Carrington rotations, to compare with the solar wind flux. We found that there was a significant increase of the solar wind with the presence of CHs clusters in three of fourteen cases.

A possible way to explain the detection of "mini" CHs could be the low level of X-ray irradiance during the minimum phase of the cycle. At such times the brightness of some sectors of the solar disk could be as low as that of the polar CHs regions.

However, if we return to the plot of Graph 3, we can see that the presence of "mini" CHs happens at the time of contraction of polar extensions and increases the presence of isolated CHs at low latitudes for the rest of the solar cycle. During these times, the increase of very small bipoles and new large-scale structures of the photospheric magnetic field prevent the polar extensions from forming complete Coronal Holes. Rather, the large scale holes appear fragmented.

4. Discussion and conclusions

A precise determination of the differences between the so-called quiet Sun and Coronal Holes at times of the minimum of the cycle must be done. An important contribution to this problem woul be the project of Solanki et al. who plans to determine several features of Coronal Holes versus the Normal Quiet Sun with SOHO data.

Meanwhile, we may concluded that:

a) CHs of one polarity appear during the minimum of the sunspot cycle, or right after it at high latitudes as isolated CHs, dominating the corona right after the polarity reversal as polar extensions until 1-2 years before the next minimum of the cycle. In this last period, isolated holes appear again migrating to the equator until the maximum of the next cycle.

b) CHs of one polarity spend about 16-18 years in a clear migration from poles to the equator.

c) At least the isolated CHs appear in the shear zone between a slow and a fast band of Torsional Oscillation pattern.

d) The polar crown of filaments and surge of active regions at the begining of the cycle prevent polar extensions from reaching low latitudes, and produce a slow contraction to the poles from the minimum of the cycle until the reversal of polarity.

e) The presence of "mini" CHs near the minimum of the sunspot cycle could be related to the transition from big polar extensions to the appearence of isolated CHs at low latitudes which could be due to a rearrangement of the photospheric magnetic fields.

References

Bartels, J.: (1932), *Terr. Magn. and Atmos. Elect.*, 37, 1.

Bohlin, J.D. and Rubenstein, D.M.: (1975), Report UAG-51, *World Data Center for Solar-Terrestrial Physics*, NOAA, Boulder, CO, USA.

Broun, J.A.: (1858), *Philos Magazine*, 16, 81.

Chapman, S. and Bartels, J.: (1940), Geomagnetism, *The Clarendon Press*, Oxford.

Harvey, K.L.: (1996), personal communication.

Makarov, V.I. and Sivaraman, K.R.: (1989), *Solar Phys.*, 119, 35.

McIntosh, P.S.: (1992), *Proceeding of the National Solar Observatory/Sacramento Peak 12th Summer Workshop on the Solar Cycle*, ed. K.L. Harvey, ASP Conference Series 27, 14.

Moses, D., Cook, J.W., Bartoe, J.-D- F-, Brueckner, G.E., Dere, K.P., Webb, D.F., Davis, J.M., Harvey, J.W., Recely, F., Martin, S.F., and Zirin, H.: (1994), Ap. J., 430, 913.

Sanchez-Ibarra, A. and Barraza-Paredes, M.: (1992), *UAG Report 102, World Data Center for Solar-Terrestrial Physics*, NOAA, Boulder, CO, USA.

Sanchez-Ibarra, A.: (1996), Proceedings of "*The Solar Cycle: Recent Progress and Future Research*," in press.

Sanchez-Ibarra, A. and Barraza-Paredes, M.: (1996), Proceedings of "*The Solar Cycle: Recent Progress and Future Research*," in press.

DEEP SURVEY OF SOLAR NANO-FLARES WITH YOHKOH

T. SHIMIZU AND S. TSUNETA
National Astronomical Observatory
2-21-1 Osawa, Mitaka, Tokyo 181, JAPAN

Extended Abstract

The *Yohkoh* Soft X-Ray Telescope has observed numerous transient brightenings of compact active-region coronal loops (microflares). Shimizu (1995) found that the distribution function of the total energy of transient brightenings is a decreasing power-law with an index ($1.5 \sim 1.6$) in the energy range greater than 10^{27} erg, suggesting that the large number of small brightenings is energetically less important than the few large ones. However, the distribution function for smaller brightenings ($< 10^{27}$ erg) is still unknown. This paper performs a deep search for smaller brightenings (termed nanoflares, Parker 1988) using the soft X-ray position-dependent time profiles. For details, please refer to the full-length paper published in the 1997 September issue of the *Astrophysical Journal* with the same title (Shimizu and Tsuneta 1997).

Short time-scale variability fainter than transient brightenings is found in the soft X-ray position-dependent light curves. The time variability is found almost everywhere in active regions and X-ray bright points, while no significant variability is found in quiet regions (Figure 1). An intensity correlation is found between the magnitudes of the time variability and the intensities of the persistent corona (Figure 2). The time variability is apparently related to the heating mechanism of the persistent active-region corona. The intensity correlation can be explained with the idea that the persistent corona is made of extremely numerous nanoflares, larger ones of which are observed as the time variability. The alternative explanation is that a common parameter controls both the persistent corona and the time variability.

References

Parker, E.N. (1988) Nanoflares and the Solar X-Ray Corona, *Astrophys. J.*, **330**, 474-479.

T. Watanabe et al. (eds.), Observational Plasma Astrophysics: Five Years of Yohkoh and Beyond, 27–28.

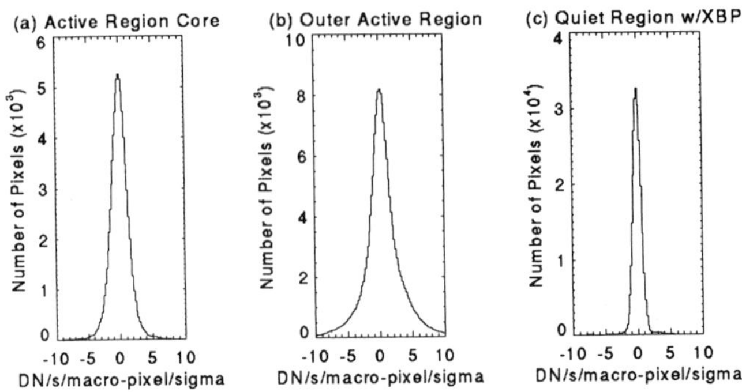

Figure 1. Histograms of the deviation of observed X-ray signals from the averages of macro-pixel (4×4 pixels) light curves: (a) the core part of the active region NOAA 7558, (b) the outer part of the active region, and (c) a quiet region. The horizontal axis is normalized with the 1σ noise of the Poisson statistics of the incident photons.

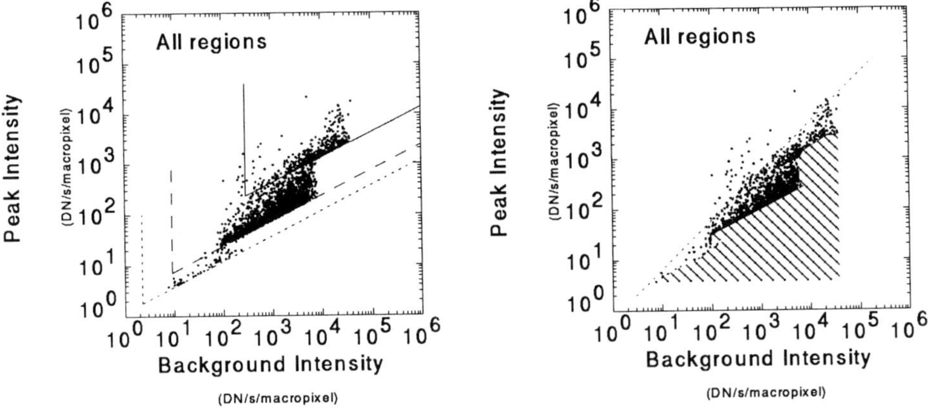

Figure 2. The magnitude of the enhancements is plotted against the corresponding persistent-coronal (background) intensity. Each point corresponds to each X-ray enhancement detected with the position-dependent light-curve analysis. The three regions with different X-ray emission level (the active-region bright core, the active-region outer region, and a quiet region) are examined. Each data set has its own lower detection limit, depending on the number of incident photons (solid, dashed, and dotted lines in the left panel). Smaller events below the photon noise would exist in the hatched region of the right panel. The upper envelope of the distribution says that the intensity of the largest time variability observed is correlated with the intensity of the persistent corona (background intensity) over a wide intensity range ($10 \sim 5 \times 10^4$ DN s^{-1} macro-pixel^{-1}).

Shimizu, T. (1995) Energetics and Occurrence Rate of Active-Region Transient Brightenings and Implications for the Heating of the Active Region Corona, *Publ. Astron. Soc. Japan*, **47**, 251-263.

Shimizu, T. and Tsuneta, S. (1997) Deep Survey of Solar Nano-Flares with Yohkoh, *Astrophys. J.*, in press.

THE SUN AS AN X-RAY STAR: OVERVIEW OF THE METHOD

G. PERES, S. ORLANDO AND F. REALE
Istituto ed Osservatorio Astronomico
Piazza del Parlamento 1, 90134 Palermo, Italy

R. ROSNER
Department of Astronomy and Astrophysics,
University of Chicago; Enrico Fermi Institute, Chicago, Illinois, USA

AND

H. HUDSON
Solar Physics Research Corp.
Tucson, Arizona, USA

Abstract. We present a method to study the solar-stellar connection, i.e., the close similarity of the physical phenomena occurring on the Sun and on late-type active stars, by taking advantage of Yohkoh/SXT X-ray images. From such images, we first find the distributions of the whole disk differential emission measure, and from them we then synthesize spectra analogous to those collected by X-ray instruments aimed at stars other than the Sun. Here we illustrate the application of this method to ROSAT/PSPC and ASCA/SIS, and discuss test cases as well as future applications.

Key words: X-ray - Corona - Solar stellar connection

1. Introduction

Solar observations give us a unique opportunity to study in detail an example of a late type star and, in particular, to examine the spatial structure of the source regions of its X-ray emission. Such observations are not possible for the other late-type stars; it is therefore very important to be able to construct a variety of tests of the widely–accepted proposition (cf. papers

T. Watanabe et al. (eds.), Observational Plasma Astrophysics: Five Years of Yohkoh and Beyond, 29–37.

in Linsky and Serio 1993; Noyes, 1985; Noyes, 1986; Wilson, 1994) that the Sun is in fact a reasonable prototype for coronal activity on such stars (cf. Vaiana et al. 1981). Quite aside from the fact that solar observations allow for spatial resolution (and the stellar observations do not), it is also the case that instruments used to observe the solar corona in general differ substantially from those used for stellar coronal observations; this results in the additional obstacle that measurements of coronal physical properties (e.g., temperature and density) for the solar corona are not easily compared with those of stellar coronae.

This paper addresses these problems by describing a method for exploring the coronal solar-stellar analogy, using the observations from the *Yohkoh* soft X-ray telescope (SXT), as well as providing initial illustrative results. More specifically, by using full disk images, we generate data that a non-solar X-ray instrument, such as *ROSAT/PSPC* or *ASCA/SIS* would collect when observing a star identical to the Sun, thus providing a general framework as well as a tool to compare directly solar and stellar data. The scope of the project includes to trace how the various kinds of structures, regions and phenomena resolved on the Sun's disk influence the stellar-like synthesized X-ray spectra, thus linking some stellar analogues of the phenomena we observe on the Sun with the stellar spectra. For instance, we could fit the derived solar spectra with the standard one- or two-temperature plasma model routinely applied to stellar X-ray spectra and then see how the fitting compares to the factual distribution of temperatures on the Sun's disk.

We present our method in Section 2, and the results in Section 3; our present findings are summarized, and future developments are discussed, in Section 4.

2. The Method

The Yohkoh/SXT data are X-ray wide-band photometric images with 2.5 arcsec angular resolution, at best (Tsuneta *et al.* 1991). These data are not directly comparable to the stellar spectra collected with ROSAT/PSPC ($\Delta E/E \approx 0.45$ at 1 keV) and ASCA/SIS ($\Delta E/E \approx 0.05$ at 1 keV): while these instruments do not resolve the star's disk, Yohkoh/SXT yields only broad-band spectral information.

Our approach is based on deriving the Differential Emission Measure vs. Temperature (DEMT) of the whole solar corona from Yohkoh/SXT images and on synthesizing the stellar-like spectra from the DEMT. From two practically simultaneous images taken with Yohkoh/SXT through two different filters, properly selected among the six available (Fig.1), the plasma temperature (T) and emission measure (EM) in each pixel of the field of view are derived from the ratio of flux measured in the bands determined by the two filters (Tsuneta *et al.* 1991), using the standard Yohkoh data processing. We then obtain an EM vs. T distribution on the Sun's disk by coupling the temperature and emission measure values in each pixel and then generating a histogram of all the emission measure values vs. the corresponding temperature. After dividing the range of temperature detectable by the instrument (nominally from around 10^6 K to a few 10^7 K) into several intervals (or bins) we sum all the EM values within each "bin" and finally divide by the temperature bin width to obtain a graph of differential emission measure (Fig. 1).

From the emission measure distribution, taking into account the plasma emissivity vs. temperature and the instrument spectral response, it is straightforward to synthesize spectra of the Sun as if it were a star observed by other instruments. More in detail, we synthesize the spectrum for each temperature bin of the DEMT, then sum the resulting spectra, taking into account the emission measure of each bin, into a single global spectrum that we finally filter through the instrument spectral response. The synthesis is accomplished with the Analysis System for Astrophysical Plasmas (Maggio *et al.* 1994), developed in IDL for presenting and interpreting the results of models of astrophysical plasmas.

3. Results

The distribution of emission measure, per se, provides interesting information, such as the relative role of plasma at different temperatures in determining the whole Sun X-ray emission and the relative contribution of the various regions on the disk. In this respect it is worth noting, for instance, how the emission measure distribution can be significantly different at different times, given the Sun's variability on various time scales. Fig. 2 shows the DEMT at two phases of solar activity: the very quiet Sun of the present solar minimum (1 June 94, 07:32) and the "active" Sun (27 February 1992, 21:14) with a few active regions. The latter are responsible for the hump in the DEMT around $5 - 7 \times 10^6$ K (cf. also Watanabe *et al.* 1995) as we have verified generating an analogous DEMT distribution after removing the active regions from the image.

Fig. 3 shows the corresponding X-ray spectra, folded through the ROSAT/PSPC spectral response, derived from the two DEMT distributions of Fig. 2. The simulation assumes that the star is located at a distance of 1 pc and the exposure time is 1000 s for the active-Sun-like conditions and 2000 s for the quiet-Sun-like conditions. We also take into account the number randomization due to the photon counting statistics. The number of counts is amply sufficient to perform a statistically sound analysis on

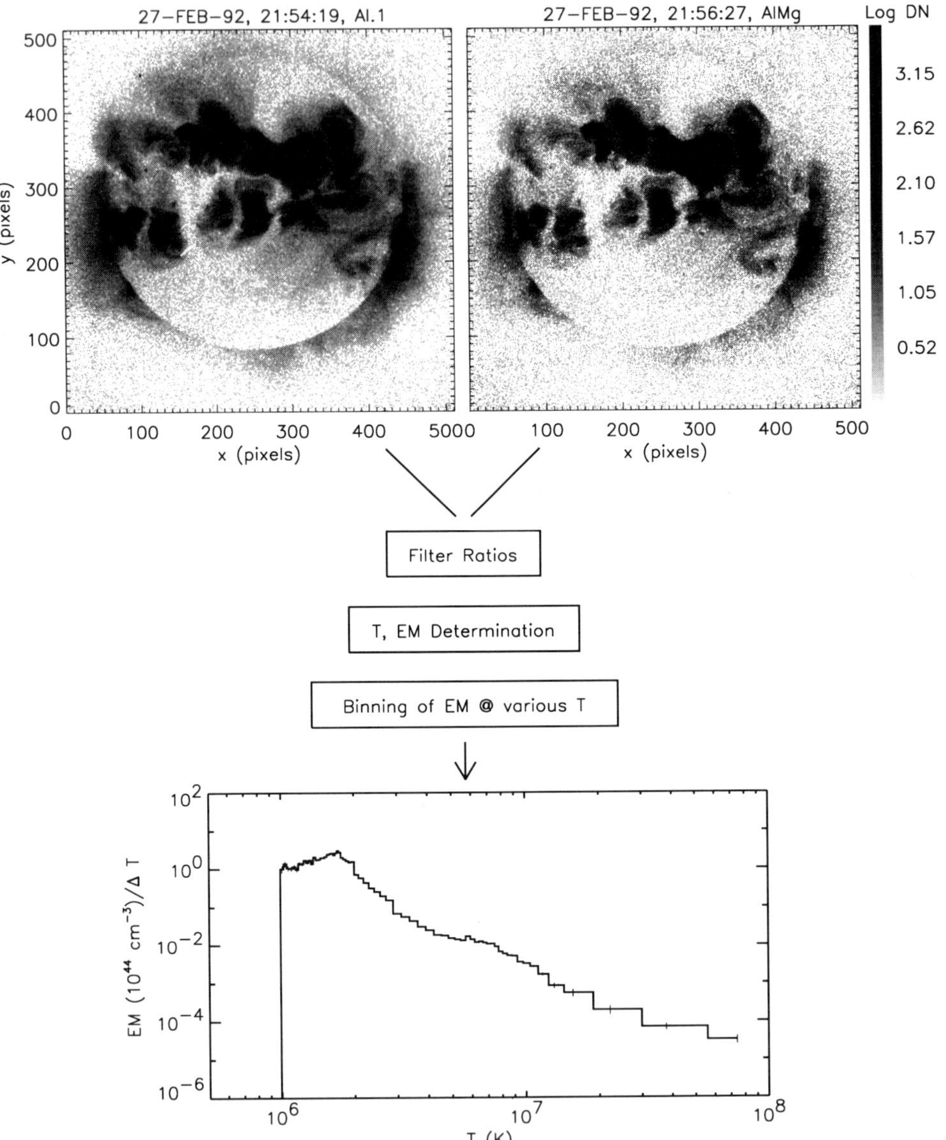

Figure 1. This figure outlines the method to obtain the whole-disk emission measure vs. temperature distribution. Two quasi simultaneous X-ray images obtained by Yohkoh/SXT through two different filters (Al 1265 A, upper left panel; Al 2930 A + Mg 2070 A + Mn 562 A, upper right panel), allow to derive the temperature and the emission measure in each pixel (middle box), on the basis of the calibration of filtered flux ratio. Coupling the T and EM values of each pixel and making an histogram of the emission measure vs. temperature values, we generate a differential emission measure distribution of the kind reported in the bottom panel.

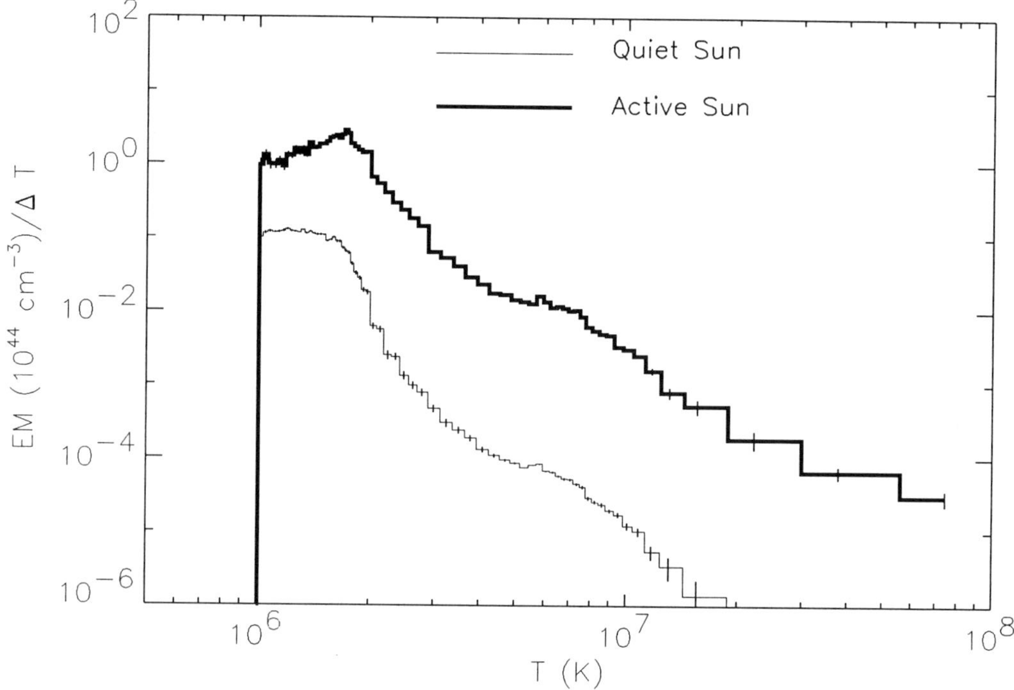

Figure 2. Two markedly different examples of differential emission measure pertaining to different times: a phase with significant activity (27 February 1992, 21:14 - solid line) and a phase without any sign of significant activity on the Sun (1 June 94, 07:32 - thin line).

ROSAT spectra; the higher number of photons obtained for the active Sun is typical of a good stellar observation accomplished with ROSAT. These spectra have then been fitted with the one- and the two-component thermal models within the XSPEC environment for data analysis, a method routinely used for ROSAT/PSPC data interpretation. In the same figure the fitted temperature(s) in keV, the relevant fitted emission measure(s), the reduced χ^2 value of the best fitting model, and the number of degrees of freedom in the fitting. While the quiet Sun ROSAT spectrum can be well fitted with only one temperature, the active Sun cannot be fitted with a single temperature component but is well fitted with two thermal components. The first fitting is in good agreement with the findings typical of low activity stellar coronae, the other with moderately active coronae.

Fig. 4 shows the analogous X-ray spectra folded through the ASCA/SIS spectral response. The observed spectra would be obtained with 10^4 s of exposure, for the active Sun, obtaining 3364 photon counts, and 1.5×10^5 s

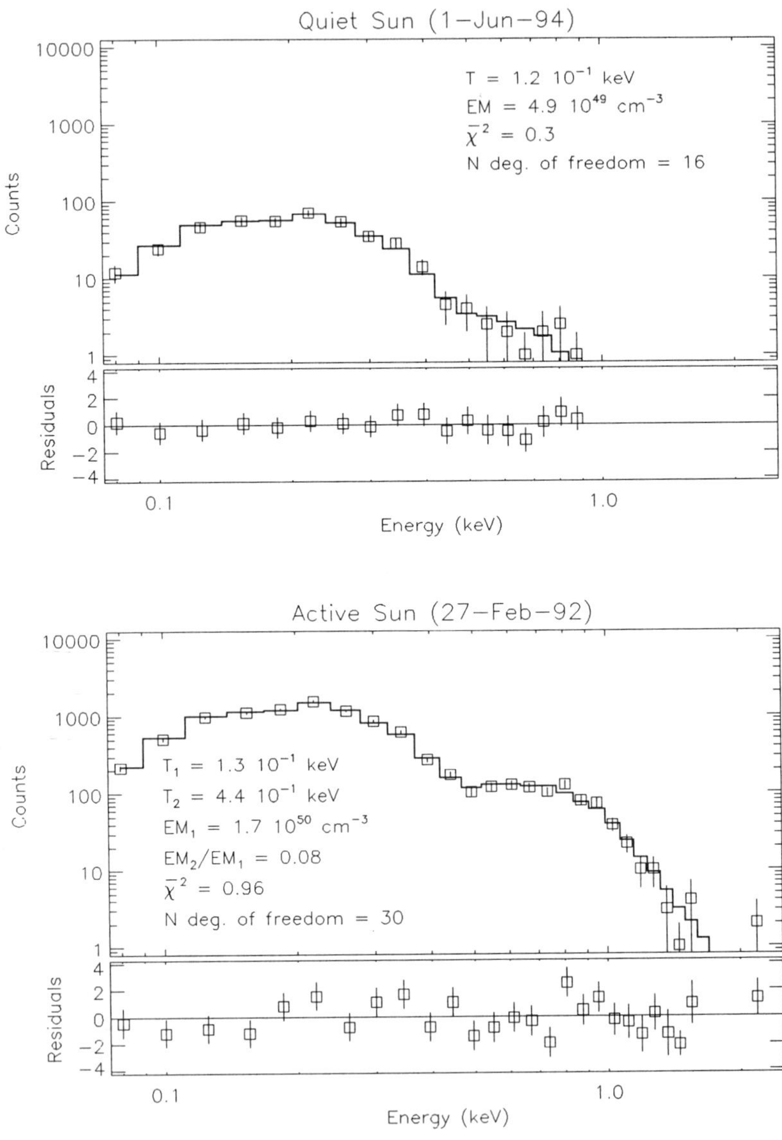

Figure 3. X-ray spectra of a star identical to the Sun observed with the ROSAT/PSPC, derived from the two DEMT's shown in Fig.2 and folded through the instrument spectral response. The star is located at a distance of 1 pc and the observing time is 2000 s for the quiet conditions and 1000 s for the active conditions. The quiet conditions are well fitted with one thermal component, the active conditions with two components. The relevant parameters, described in the text, are shown in the graph.

for the quiet Sun obtaining 965 counts. The difference between the spectra of the two phases of the Sun's activity is now even more evident because of the larger sensitivity of ASCA/SIS to high energy photons. The synthesized spectra also show the ASCA/SIS spectral resolution capable to discriminate some lines or groups of lines.

4. Final considerations and future prospects

This method, already at its present level, allows us to characterize the distribution of plasma temperature on the Sun's disk and to identify the effect of active regions and other phenomena.

There is considerable room for refinements and we are presently working on expanding and improving our approach.

On one hand, the derivation of DEMT from Yohkoh images can be refined in several respects. It could be worthwhile to remove the influence of the radiation scattered by the telescope mirrors from the very bright regions of the image onto the very faint ones.

It is also worth exploring the effect of pixel summation, i.e. computing the DEMT after summing signal over "big pixels" made of 2×2 pixels, or 4×4 pixels, or bigger. This would help in reducing the noise which, along with the scattered light, may affect mainly the pixels with low signal and give rise to unrealistic flux ratios and, therefore, temperatures. Such a noise influence can, most likely, be present both in the high and in the low temperature tails of the distribution, respectively $\geq 2 \times 10^7$ K and $\leq 1 \times 10^6$ K, for the combination of filters chosen. In order to reduce the noise influence, the preliminary results here shown do not consider pixels with temperature below 1×10^6 K (on the other hand Yohkoh/SXT has virtually no sensitivity to plasma below such a temperature), or with less than 10 photons. The technique of pixel summation could also show how the DEMT distribution depends on the pixel size and on the unavoidable inclusion of plasma components at different temperatures in the same pixel.

We will take into account these improvements in an on-going work (Peres *et al.* 1997).

By the same token, we plan to include information from other instruments, most notably Yohkoh/BCS, and to fold the synthesized spectra through the spectral response of other non-solar instruments, so as to allow the comparison with a broader set of data. Future work will also include a more extensive analysis and possibly spectra of structures or regions imaged on the Sun's disk.

Yohkoh data, beyond giving highly detailed images and broadband photometry, provide the additional advantage of monitoring the Sun with good time coverage over several years with adequate time resolutions, thus show-

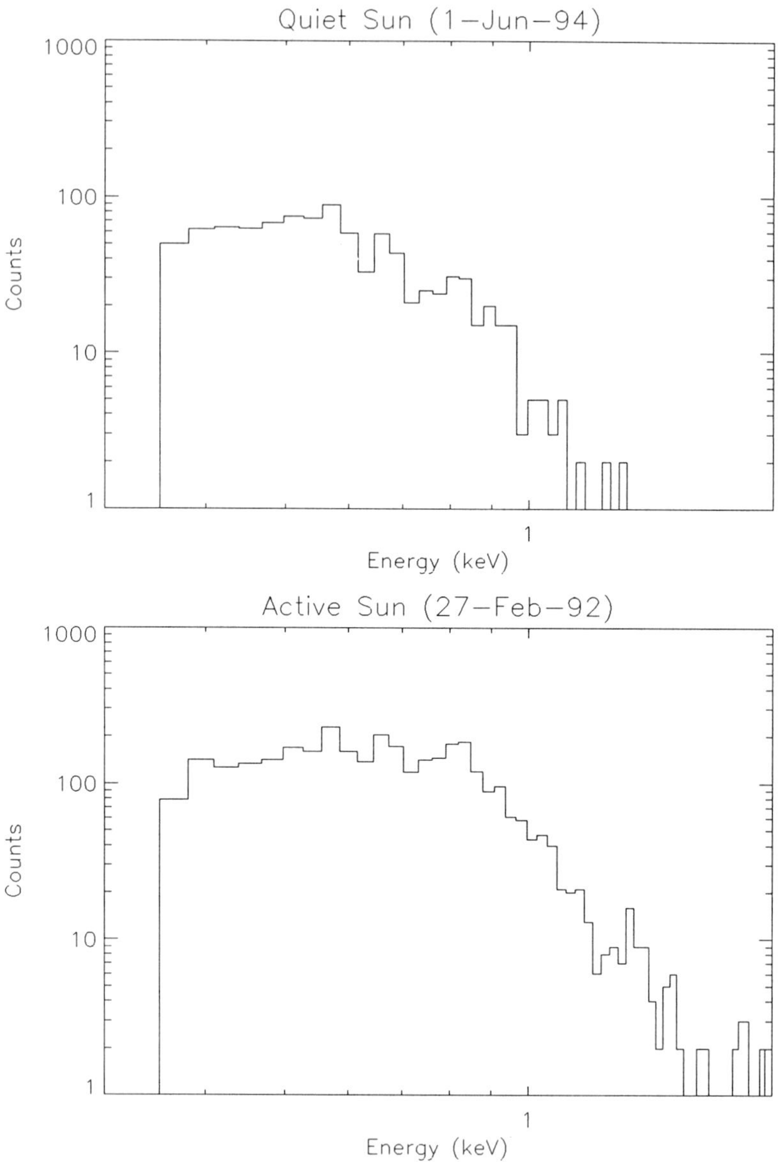

Figure 4. X-ray spectra of the Sun as would be observed with the ASCA/SIS, derived from the two DEMT's shown in Fig.2 and folded through the instrument spectral response. The 965 photon counts of the quiet-Sun-like star would be collected over 1.5×10^5 s. The 3364 photon counts of the active-Sun-like star would be collected in 10^4 s.

ing many aspects of solar activity and different phases of the solar cycle. It is worth comparing all these different views of the same star, the Sun, with the large diversity found on stars, each star, instead, being observed only occasionally and for short periods, and to evaluate the extent to which such a diversity, found even among very similar stars, may be ascribed to intrinsic variability of individual stars.

This project opens the possibility to use the wealth of the solar X-ray data to provide diagnostics as well as phenomenological models for many events and spectral features observed on stars. It is a direct way to bridge the gap between the solar and the stellar coronal physics and to understand better how the physical conditions on the stars influence the phenomena we see on the Sun, i.e. to investigate the solar-stellar connection.

Acknowledgements

We acknowledge partial support from the Agenzia Spaziale Italiana and from the Italian Ministero dell'Università e della Ricerca Scientifica e Tecnologica. The first analysis and the selection of the data here reported was performed at the Istitute of Space and Astronautical Science (Japan). We are indebted with S. Tsuneta, T. Watanabe and L. Acton for useful discussions.

References

Linsky, J. and Serio, S. eds. 1993, *Physics of Solar and Stellar Coronae*, Kluwer Academic Publishers

Maggio, A., Reale, F., Peres, G., Ciaravella, A. 1994, *Comp. Phys. Comm.*, **81**, 105

Noyes, R. W. 1985, *Solar Physics*, **100**, 385.

Noyes, R. W. 1986, *Solar and Stellar Magnetic Activity*, in Physics of the Sun, vol. III, P. A. Sturrock (ed.), p. 125, D. Reidel

Peres, G., Orlando, S., Reale, R., Rosner, R., Hudson, H. et al. 1997, *Ap. J.*, in preparation

Tsuneta, et al. 1991, *Solar Phys.*, **136**, 37.

Vaiana, et al. 1981, *ApJ*, **245**, 163.

Watanabe, et al. 1995, *Solar Phys.*, **157**, 69.

Wilson, P. R., 1994, *Solar and Stellar Activity Cycles*, Cambridge University Press

CORONAL HEATING DUE TO TWISTING FACULAE OF 1 KM RADII

T. HIRAYAMA
Meisei University, Tokyo, Japan

Abstract. We propose a scenario for coronal heating as a result of body Joule heating of a twisting magnetic tube of ≈10 km radius due to anomalous resistivity, whose effective collision frequency is 10^{-2} times the electron plasma frequency. The kilo-Gauss faculae of 0.2 arcsec diameter in the photosphere are expected to split further by "collisions" with the non-magnetic eddies of various size owing to the large Reynolds number. The smallest size of a kilo-Gauss fluxtube which can be twisted lies just between frozen and non-frozen states, and it is ≈1 km. When expanded into the corona as a ≈10 km tube, the twisting inevitably makes a charge separation. The latter in turn produces a large electric field parallel to the magnetic field, which makes runaway electrons, and an anomalous resistivity is expected from instability between the runaway and bulk electrons. The anomalous resistivity will stop the runaway electrons and the process will be repeated. From these, we can obtain the correct order of magnitude of the coronal heating rate $\approx 10^{-4}$erg $cm^{-3}s^{-1}$. Facular body Joule heating as a result of small sizes is also proposed.

1. Small Scale Magnetic Flux of 1 km Size and Twisting

Observations show many instances of splitting and merging of 0.2" – 0.5" sized faculae (e.g. Berger, 1996). We propose that further splitting may be occurring beyond the present observational limit. Outside of these faculae and still in the intergranular lanes, plasmas must be in a turbulent state due to a high Reynolds number of $R \approx 10^9$, so that there should be various sizes of non-magnetic elements from 100 km down to a size of 1–10 cm ($R \approx O(1)$) as would be the case in a stationary turbulent spectrum.

T. Watanabe et al. (eds.), Observational Plasma Astrophysics: Five Years of Yohkoh and Beyond, 39–44.

We define the splitting of the magnetic elements as a process of the horizontal penetration of a non-magnetic gas blob of smaller size into a magnetic flux tube of larger size. The vertical extension of the flux thus splitting is of cource limited by the vertical extension r_z of the penetrating blob. For the splitting to be realized, it is necessary that the original kinetic energy of the blob, $(\rho v_y^2/2)\times$volume, moving in the horizontal y-direction, should be larger than the work done by the larger flux tube against the blob through the magnetic tension, $\frac{B_z}{4\pi}\frac{\partial B_y}{\partial z}\cdot\frac{r_y}{2}\times$ area $\times$ length(Δy). Since we are considering velocities much smaller than the Alfvén velocity $v_A \approx 8\,\mathrm{km\,s^{-1}}(B_z \approx 1500\,\mathrm{G}$ and $\rho = 3\times10^{-7}\mathrm{g\,cm^{-3}})$, the gas and magnetic pressures should always be balanced. Taking the ratio of the two, we obtain

$$\alpha \equiv \frac{\text{work by tension}}{\text{original kinetic energy}}$$

$$= \frac{\frac{B_z}{4\pi}\frac{\partial B_y}{\partial z}\frac{r_y}{2}\cdot\frac{\Delta y}{r_y}}{\rho v_y^2/2} = (\frac{v_A}{v_y})^2(\frac{r_y}{r_z})^2\frac{\Delta y}{r_z} \leq (\frac{v_A}{v_y})^2(\frac{r_y}{r_z})^3. \tag{1}$$

The last inequality shows that the maximum penetration at $\Delta y = r_y$, though $\alpha < 1$ can hold if Δy is sufficiently small. We used $\partial B_y/\partial z \equiv \Delta B_y/r_z \approx \Delta B_z r_y/r_z^2 \approx B_z r_y^2/r_z^3$ (the second equality is from div $\boldsymbol{B}$=0), and a prolate (spindle) shape $(r_x \approx r_y)$ was assumed. Keeping the volume, the penetrating blob will become more elongated vertically due to the magnetic tension. A maximum allowable r_z may be on the order of the pressure scale height $H \approx 150$ km, otherwise the blob will lose its identity. For $v_y = 1\,\mathrm{km\,s^{-1}}$ and $r_y = 30\,\mathrm{km}$, $\alpha < 1$ is satisfied.

Hence although the details should be worked out, there is a good possibility that the magnetic flux of 0.2" size will split further down to much smaller scales. When split, the magnetic flux will generally rotate due to not the viscous, but rather to the inertia force of the non-magnetic blob rotating with velocity $v_\theta \approx v_y$.

The twisting of the magnetic field inside the split flux tube is only effective until the magnetic Reynolds number reaches $R_M \approx O(1)$, because for tubes of $R_M < 1$ only the gas inside can be twisted (slipping field); we define R_M in magnetic flux tube (subscript m),

$$R_M \equiv \left|\frac{(\mathrm{rot}\boldsymbol{v}\times\boldsymbol{B})_\theta/c}{(\mathrm{rot}\ \eta\boldsymbol{j})_\theta}\right| = \frac{r_m v_m}{\eta_m} \quad . \tag{2}$$

We expect that the smaller the radius, the smaller the twisting velocity in the magnetic tubes as in gas without magnetic fields:

$$r_m/r_f = (v_m/v_f)^n \tag{3}$$

Here, for simplicity, we take $n = 3$ (Kolmogorov spectrum) and $R_M = 2$. With facular values of $r_f = 75$ km, $v_f = 1.6\text{km s}^{-1}$and $\eta_m(\text{cm}^2\text{s}^{-1}) \approx 10^9 (\equiv \eta c^2$, η=electrical resistivity in Gaussian unit) for T=6400 K (Hirayama, 1992), we obtain

$$r_m = 0.6\,\text{km and } v_m = 0.3\,\text{km s}^{-1} \quad . \tag{4}$$

This is for a flux tube whose magnetic field can still be twisted $[\partial B_\theta/\partial t = -c(\text{rot } \boldsymbol{E})_\theta \neq 0$, see (7)]. Similar values are obtained for $2 < n < 5$. Because even at these values we still have $R \equiv r_m v_m/\nu_{kin} = 9.3 \times 10^5 (\nu_{kin}$ = kinematic viscosity), the gas inside the 0.6km fluxtube is further rotated and cascaded to $R \approx O$ (1). Note that $B_z = 1500$ G is retained for these small tubes due to pressure balance while they are splitting (and merging). Since we are assuming a nearly stationary spectrum, sizes of flux tubes between 75 km and 0.6 km are not only co-existent, but have naturally finite depth extensions.

2. Coronal Heating

Now we proceed to study the effect of the twisting flux tube of 1 km size in the photosphere on coronal heating. With $r_m = 0.6$ km, $B_m = 1500$ G, and $B_c = 10$ G, the radius of the expanded elementary flux tube in the corona becomes $r_c = \sqrt{1500/10} \cdot 0.6$ km=7.3 km (subscript c for the corona). Energy flux density at the coronal base is $\rho_c v_c^2 (v_A)_c = 2.3 \times 10^5$ergs $\text{cm}^{-2}\text{s}^{-1}$, using $v_c = 25\,\text{km s}^{-1}$. Dividing this by the scale height 10^5km, the coronal heating rate becomes $\epsilon_H = 2.3 \times 10^{-4}\text{erg cm}^{-3}\text{s}^{-1}$. This is sufficient to heat the quiet corona of 10^6K, if damped rapidly as shown below.

First we derive the charge separation between ions n_i (protons) and electrons ($n_e = 10^8\text{cm}^{-3}$), and then calculate the electric field E_z parallel to the main magnetic field component B_z along the coronal tube of radius of 7.3 km or so.

When a magnetic field is being twisted in a cylindrical tube, there appears a (rather constant) charge separtaion, $n_i \neq n_e$ (Alfvén, 1950). Writing down Gauss' law explicitly,

$$4\pi e(n_i - n_e) = \text{div } \boldsymbol{E} = \frac{1}{r}\frac{\partial}{\partial r}(rE_r) + \frac{1}{r}\frac{\partial}{\partial \theta}E_\theta + \frac{\partial}{\partial z}E_z \quad . \tag{5}$$

If the last two terms are negligible, and if $\frac{1}{r}\frac{\partial}{\partial r}(rE_r) \neq 0$, then $n_i \neq n_e$. This is essentially what is realized. In fact if $\boldsymbol{E} = -\, \boldsymbol{v} \times \boldsymbol{B}/c$ as expected in the corona and if $|\boldsymbol{B}| \approx B_z$ ($B_\theta = v_\theta B_z/v_A \ll B_z$; $B_r \approx 0.1 B_z$ for example), then $|\boldsymbol{E}| \approx |E_r| = |v_\theta B_z/c|$ even if $v_z (\gg v_r)$ were assumed to be comparable to v_θ. We obtain

$$\sigma_n \equiv \frac{n_i - n_e}{n_e} = -\frac{\partial(r v_\theta B_z)/r\partial r}{4\pi e c n_e} \approx -\frac{B_z v_\theta}{2\pi e c n_e r_c}$$

$$= \quad -3.8 \times 10^{-9} \frac{B_z}{10\,\mathrm{G}} \frac{v_\theta}{25\,\mathrm{km\ s^{-1}}} \frac{7.3\,\mathrm{km}}{r_c} \frac{n_e}{10^8}. \tag{6}$$

Here we assumed $\partial B_z/\partial r \approx 0$ (low plasma β) and $v_\theta \propto r$ (other forms lead to similar results), and $e = 4.8 \times 10^{-10}$and $c = 3.0 \times 10^{10}$cm s^{-1}. Also $\partial E_\theta/r\partial\theta \approx B_z \partial v_r/cr\partial\theta$ is negligible with the above ($\partial/r\partial\theta \approx 0$ may well be the case). $\partial E_z/\partial z \approx (\partial/c\partial z)(B_r v_\theta - B_\theta v_r)$ is negligible because B_z does not appear and $\partial/\partial z$ in a slender tube gives much smaller values. For reference we also record the generalized Ohm's law (≈equation of electron motion):

$$\boldsymbol{E} = -\frac{\boldsymbol{v}\times\boldsymbol{B}}{c} + \eta\boldsymbol{j} - \frac{\nabla P_e}{en_e} + \frac{4\pi}{w_{pe}^2}\frac{\partial\boldsymbol{j}}{\partial t} - \frac{m_e\boldsymbol{g}}{e} - \frac{m_e}{e}(\boldsymbol{v}\cdot\nabla)\boldsymbol{v}. \tag{7}$$

Here since we are considering that the σ_n-value is constant or slowly varying ($\Delta t^{-1} \ll v_A/r_c \approx 3 \times 10^2 \mathrm{s}^{-1}$) as compared to, e.g. the plasma frequency $w_{pe} = 5.6 \times 10^8 \sqrt{n_8}\mathrm{s}^{-1} (n_8 \equiv n_e/10^8)$, it is easy to see $\boldsymbol{E} \approx -\boldsymbol{v}\times\boldsymbol{B}/c$ (frozen), $\boldsymbol{v}$ being the average plasma velocity. The terms after $\eta\boldsymbol{j}$ are much smaller than $v_\theta B_z/c$, and the last non-linear term is also negligible. Even if the resistivity η becomes anomalous (see below), $\eta^*|\boldsymbol{j}| \approx \eta^* j_z$ is by a factor $60(\approx R_M)$ smaller than $v_\theta B_z/c$. The second equality in (6) can also be obtained by expanding $\mathrm{div}(-\boldsymbol{v}\times\boldsymbol{B}) \equiv \boldsymbol{v}\cdot\mathrm{rot}\ \boldsymbol{B} - \boldsymbol{B}\cdot\mathrm{rot}\ \boldsymbol{v} \approx -\boldsymbol{B}\cdot\mathrm{rot}\ \boldsymbol{v}(\mathrm{rot}\ \boldsymbol{B} \propto \boldsymbol{j}\perp\boldsymbol{v})$, being the same as Alfvén (1950) and Spicer (1992). The latter author tried to use this for the coronal heating as in this paper (see also Tsuneta, 1995).

An important point in the coronal application is that inside any point in the coronal tube where $\boldsymbol{B}\cdot\mathrm{rot}\ \boldsymbol{v} \approx B_z v_\theta/r_c$ is not zero, **the charge separation $\sigma_n \equiv (n_i - n_e)/n_e$ cannot be dispersed, nor cancelled, but maintains the value of (6) all the time**. An average value of σ_n is also kept when waves having with changing n_e are produced as we introduce below. Another noticeable point is that even though $\Delta n \equiv n_i - n_e$ is only 0.38 cm^{-3}, this should be considered very large! Namely if we assume $\boldsymbol{v} = 0 = \partial/\partial t$, and $\eta\boldsymbol{j} = 0$ in (7) and with a similar equation for ions of mass $m_i = 1.7 \times 10^{-24}$g, one gets an upward directed electric field of $E_z = m_i g/2e = 1.4 \times 10^{-8}$V cm^{-1} (e.g. Hundhausen, 1972). Here the gravitational acceleration $g = 2.7 \times 10^4 R_\odot^2/(R_\odot + z)^2$cm s^{-2}, $R_\odot$ is the solar radius, and z is the height above the photosphere. Without this electric field, electron densities would be uniform over enormous height lengths of $kT/m_e g = 80 R_\odot$. Taking the divergence ($g = g(z)$), we obtain $\Delta n = 2 \times 10^{-13}$cm^{-3}($\ll 0.38$ cm^{-3}), which must be present and necessary in the corona.

The net charge will certainly produce an electric field, $E_z(\sigma_n)$, parallel to the magnetic field B_z. This can be calculated again using Gauss' law as a very small perturbation: in fact through $E_z(\sigma_n,\ z) \equiv -d\psi/dz$ and $\psi(z) = e\sigma_n \int_{-H/2}^{H/2} \int_o^{r_c} dr_1 dz_1 2\pi r_1/[(z-z_1)^2+r_1^2]^{1/2}$ on the z-axis($r = 0$). Since $\sigma_n \approx$const is always present, we neglected the retarded time. Using $r_c \ll H$, we find

$$E_z(\sigma_n,\ |z| = H/4) \approx 1 \times 10^{-5}\text{V cm}^{-1} \times \frac{B_c}{10\,\text{G}} \cdot \frac{v_\theta}{25\,\text{km s}^{-1}} \cdot \frac{r_c/H}{10^{-4}}. \quad (8)$$

This is on the order of the Dreicer field, $E_D = 1.2 \times 10^{-5} n_8 T_6^{-1}\text{V cm}^{-1}$ for $\ln \Lambda = 20$. H is the scale height $\approx 10^5$km ($|E_r| \approx |v_\theta B_z/c| = 0.25\text{V cm}^{-1} \gg E_z(\sigma_n)$). As a typical value of z, measured this time in the middle, in E_z, we took $|z| = H/4[E_z(z=0) = 0]$. We expect the following process, whose details will be given in a full paper together with a discussion of (8). Even before E_z reaches the Dreicer field, some of electrons are known to begin to runaway (e.g. Wiley and Hinton, 1980), and this occurs from all places where $\sigma_n \neq 0$ is kept constant. When seen at a fixed location, runaway electrons, having velocities several times with the electron thermal speed of 5.5×10^3km s^{-1}, will become a significant fraction of plasma density of $n_e = 10^8\text{cm}^{-3}$. Then it will be expected that the beam instability sets in even for a beam density of, e.g., $10^{-2} n_e$ (e.g. Gary, 1993). The effective collison frequency will become of the order of magnitute $\nu^* \approx 0.02 w_{pe}(\text{s}^{-1})$ or so. The electrical resistivity is $\eta^*(\text{s}) = 4\pi\nu^*/w_{pe}^2$. the resulting Joule heating is

$$\epsilon_H = \eta^* |j|^2 = 2.3 \times 10^{-4} \frac{\sqrt{n_8} \cdot (v_\theta/25\,\text{km s}^{-1})^2}{(r_c/7.3\,\text{km})^2} \cdot \frac{\nu^*/w_{pe}}{0.02} \text{erg cm}^{-3}\text{s}^{-1} \quad (9)$$

where we have used $|j| \approx j_z = cB_\theta/2\pi r_c$, and $B_\theta = v_\theta B_z/v_A \equiv v_\theta \sqrt{4\pi m_i n_e}$. This is the same as above if $\nu^*/w_{pe} = 0.02$ is adopted. The corona may be heated in this way.

The new point is that $r_c \approx O(10\ \text{km})$ is adopted from the foregoing section, and electron runaway is invoked. Then the heating rate becomes just enough for coronal heating at $T = 10^6$K. Note that once the effective collision frequency becomes much larger than the classical collision frequency, ν_{cl}, there will be no runaway electrons anymore (Dreicer field $\propto \nu_{cl} \rightarrow \nu^*$), and the process may well become intermittent (Davidson and Krall, 1977).

In an active region, the heating rate becomes larger: $\epsilon_H \propto B_c$ because of $\pi r_c^2 B_c = \pi r_m^2 B_m$, where neither r_m nor B_m changes from the quiet to active regions [we neglected weaker dependency on B_c due to $n^{1/2}$ in (9)]. Also these arguments on heating hold for both closed and open magnetic fields.

3. Photospheric Facular Heating

If the radius of the elementary flux tube is in fact $r_m \approx 1\,\mathrm{km}$, we can expect a large Joule heating because the azimuthal current J_θ is large due to a steep field gradient $\partial B_z/\partial r \propto J_\theta$ almost everywhere inside the tube (body heating). Using the vertical field distribution in the form $B_z = B_{zo}\exp[-(r/r_m)^2]$ (Hirayama, 1992), integrating $\int_0^\infty \eta J_\theta^2 2\pi r dr/\pi r_m^2$, and multiplying by Δz, we obtain the body Joule heating to be

$$\begin{aligned} E_J &= \frac{\sqrt{2\pi}}{4}\eta_m \frac{B_{zo}^2}{8\pi}\frac{\Delta z}{\pi r_m^2} \\ &= 7.4\times 10^{10}\left(\frac{B_{zo}}{1500\,\mathrm{G}}\right)^2\left(\frac{0.6\,\mathrm{km}}{r_m}\right)^2\frac{\Delta z}{150\,\mathrm{km}}\mathrm{erg\ cm^{-2}s^{-1}}. \end{aligned} \quad (10)$$

Here $\eta_m = 1\times 10^9\mathrm{cm^2s^{-1}}$ is the electrical resistivity at $\tau_5 = 1$, and Δz is the photospheric pressure scale height. On the other hand observations (e.g. Hirayama, 1992) indicate that the excess flux of a 0.2"facula is

$$E^{obs} \approx 1.3\times 10^{10}\mathrm{erg\ cm^{-2}s^{-1}}. \quad (11)$$

Hence the fractional area of the km-size flux in 0.2"-facula should not be too small ($f \approx 1.3/7.4$ from above). In fact Mehltretter's observation (1974) indicates that a magnetic flux of $4.4\times 10^{17}\mathrm{G\ cm^2}$ is contained in a 0.5" diameter facula: $4.4\times 10^{17} = f\times 1500\,\mathrm{G}\times \pi r_f^2$ is satisfied with $f = 1.3/7.4$ and $2r_f = 0.64$", showing good agreement (perhaps too good).

In concluding, we assert that the excess energy flux of the photospheric faculae comes from the body Joule heating of the 1 km size flux tubes of 1500 G which are rather densely packed in a 0.2"facula. The similar chromospheric heating will be treated in a full paper.

References

Alfvén, H., 1950, in *Cosmical Electrodynamics*, Caredon Press, Oxford, pp.6.
Berger, T.E., 1996, Dissertation, Stanford University.
Davidon, R.C. and Krall, N.A., 1977 in *Nuclear Fusion* **17**, 1313.
Gary, S.P., 1993 in *Theory of Space Plasma Microinstabilities*, Cambridge Univ. Press.
Hirayama, T., 1992, *Solar Phys.* **137**, 33.
Hundhausen, A.J., 1972, in *Coronal Expansion and Solar Wind*, Springer-Verlag, Berlin, p.16.
Mehltretter, J.P., 1974, *Solar Phys.* **38**, 43.
Spicer, D.S., 1991, in P. Ulmschneider, E.R. Priest, and R. Rosner(eds.) *Mechanisms of Chromospheric and Coronal Heating*, Springer-Verlag, Berlin, p.547.
Tsuneta, S., 1995, *Publ. Astron. Soc. Japan* **47**, 691.
Wiley, J.C. and Hinton, F.L., 1980, *Phys. Fluids* **23**, 921.

COMPARISON BETWEEN STATISTICAL FEATURES OF X-RAY FLUCTUATIONS FROM THE SOLAR CORONA AND ACCRETION DISKS

SATORU UENO
Kwasan Observatory, Kyoto University
Ohmine-tyo, Kita-kazan, Yamashina-ku, Kyoto city, Japan
ueno@kwasan.kyoto-u.ac.jp

1. Introduction

Since the 1970's, solar physicists have been performing statistical studies of temporal variations of flux in the solar atmosphere in various wavelength ranges, including radio, H-alpha, soft X-ray, hard X-ray, and gamma-ray, mainly in order to explain coronal heating and periodicities of flares (e.g., Drake 1971; Datlowe, Elcan, & Hudson 1974; Lin et al. 1984; Dennis 1985; Rieger et al. 1984; Kile & Cliver 1991; Bai & Sturrock 1993). Also, following the launch of various X-ray satellites, astrophysicists have tried to observe and statiscally analyze the violent temporal variations of high energy flux from black hole candidates, Active Galactic Nuclei, etc. (e.g., Makishima 1988; Pounds & McHardy 1988; Negoro et al. 1995).

So far, these solar and astrophysical studies have been done almost independently of each other. However, when the accretion disk plasmas are optically thin and have non-negligible magnetic pressures, the circumstances are likely to be very similar to solar coronal plasmas. Therefore, it is also possible to imagine some similarity between the mechanisms of flares in the solar corona and X-ray shots in accretion disks.

I will discuss the following two subjects:

1. Global mechanism of solar flare occurrence distribution

   I focus on the mechanism which controls the spatial and temporal distributions of flares on the whole solar surface, rather than the mechanism which just controls the occurrence of individual flares. Such a grobal mechanism may be reflected in various statistical fea-

T. Watanabe et al. (eds.), Observational Plasma Astrophysics: Five Years of Yohkoh and Beyond, 45–50.

tures. These features can help determine the source of coronal heating.

2. Relationship between the solar corona and accretion disks

   I will discuss the possibility that flare-like phenomena occur on the accretion disks which have very low gas density.

2. Data and Light Curves

I first compare light curves from the solar corona and from accretion disks. For the solar corona, I use the soft X-ray fluctuation from the entire solar surface, as detected by the GOES-6 satellite over the 32 month period from September 1991 (during solar maximum) to April 1994 (solar minimum). The X-ray energy band of GOES is about 2 - 10 Kev. The light curve shows large fluctuations, from large flares to small flares. And generally, the time profile of an individual flare shows very rapid growth and a gradual exponential decay. For comparison, consider the hard X-ray fluctuation data from Cyg X-1 (a famous black hole candidate) obtained by the GINGA satellite. The energy band of GINGA is 1.2 - 58.4 Kev, and the hard X-ray contribution is large. Because none of these data overlap very well with the GOES energy band,I use this GINGA data reluctantly. The time variation of these data show rather low contrast and symmetrical profiles.

3. Power Spectral Densities

Next, we consider the Power Spectral Densities (PSD) calculated from the above data with a modified temporal Fourier transform (UeNo et al. 1997). The PSD for solar flares can be roughly separated into two parts. One is associated with long term periodicities ($< 10^{-4.5}$Hz), and the other is associated with noise properties ($10^{-4.5} < f < 10^{-3.0}$Hz). The latter part is composed of three regions with different power-law indices, a flat part, a $1/f$-like part, and a steep part, running from low frequency to high frequency. The two break frequencies which separating the PSD into three parts, and the power-law index of the intermediate ($1/f$-like) part are three important parameters which characterize the PSD. The total PSD is formed by a superposition of the individual PSDs of many flares with various time scales. The lowest break frequency corresponds to the time scale of the longest flare, and the highest break frequency corresponds to that of the shortest flare among all the observed flares. I studied the time variation of these parameters over a 32 month period, sampling the data every two months. While all the parameters fluctuate, they fluctuate around a certain average value. The most important fact is that all of the parameters are completely independent of solar activity which is due to the solar cycle.

This suggests that the global mechanism for solar flare generation is nearly completely independent of solar activity level. In contrast, the PSD of X-ray fluctuations (X-ray shots) from Cyg X-1 was calculated by Negoro (1992). He also found a three-part PSD, including a flat part, a $1/f$-like part, and a steep part. Moreover these inclinations are very similar to those of solar flare PSDs. This may suggest that the fundamental generation mechanisms for solar flares and for X-ray shots are almost the same.

4. Peak Interval Distributions

Another quantity of interest is the "peak interval distribution," i.e., the interval of time between two consecutive flares (Negoro et al. 1997). Plotting log peak interval and log occurrence frequency is expected to result in a linear distribution if the data are completely random. But I find that the solar flare distribution can not be fitted with one straight line. The flare occurrence tends to deviate toward high frequency in the short-interval time region. So, we can better fit this distribution with two different straight lines. This feature may suggest that solar flares can be separated into two different groups: flares which occur more intermittently and flares which occur more concentratedly. But the interpretation of what these two goups correspond to does not yet exist. On the other hand, the peak interval distribution of Cyg X-1 generally follows a straight line. But in the short-interval time region, however, the occurrence frequency of large X-ray shots is reduced. This suggests the existence of energy-accumulation structures, such as magnetic fields. That is, a second large shot does not occur soon after a previous large shot.

5. Peak Intensity Distributions

Generally, the peak intensity distribution of solar flares (Negoro et al. 1997) obeys a power-law. Moreover, the power-law index is nearly constant independent of solar activity. Again, this suggests that the generation mechanism for solar flares does not vary with solar activity. The value of the power-law index indicates the existence of other coronal heating mechanisms besides flares. Generally, the peak intensity distribution of Cyg X-1 also obeys a power-law. The power-law index of Cyg X-1 is, however, rather different from that of solar flares.

6. Models

A few models have been suggested in order to explain the above statistical features. At present, the Self-Organized Critical state model (SOC model) is one of the likeliest candidates.

Lu & Hamilton (1991, 1993) simulated solar flares on 3-dimensional lattices. They added random magnetic fields at random points, and assumed an instability resulting in reconnection occured when the local gradient of the magnetic field exceeded a certain critical magnetic shearing level. They further assumed that the magnetic shearing was dispersed to neighboring lattices, and that the released magnetic energy changed into X-ray emission. This simulation successfully reproduces flare light curves with various high-contrast flare intensities, and the peak intensity distribution and its power-law index. Power spectral densities, however, have not yet been calculated by anybody. Furthermore, this simple SOC model cannot express the two-component feature of the peak interval distribution.

For accretion disks, Takeuchi et al. (1995) also performed simulations on 3-dimensional lattice structures. They also added random masses at random points. And when the accumulated mass exceeded a critical mass, an instability appeared dispersing mass to inner cells. It is possible that the critical mass is determined by magnetic pressure, and that the instability is something like magnetic reconnection. In addition, in the case of accretion disks, steady mass dispersion toward the disk center is assumed. This simulation can also reproduce the time profile of X-ray shots with low-contrast intensities when the steady accreting mass value is set to be large. Interestingly, when this steady accreting mass is set to be small, the reproduced light curve is similar to that of solar flares (high contrast). In fact, in the solar corona, mass accretion of plasma hardly occurs. When the number of 3-dimensional cells is increased, we can find the $1/f$-like features better reproduced in the power spectrum diagram. The reproduced peak interval distribution roughly follows a straight line, and the reduction of large X-ray shots at the short-interval regime is successfully reproduced. Moreover, the power-law index in the peak intensity distribution is also very well reproduced.

Thus, as stated I introduced above, this SOC model can explain many statistical features of X-ray flautuations in both the solar corona and accretion disks.

7. Summary: Problems and Required Studies

As I introduced above, there are many interesting relationships between flares in the solar corona and X-ray shots in accretion disks. There are especially close similarities in power spectral densities and model reproductions. Of course, however, there are many differences and still unexplained features. Here, I present some problems to be further investigated in considering the relationship between the solar corona and accretion disks.

1. *Difference of X-ray intensities*

   X-ray emission from accretion disks is about 10^4 times larger than that from the solar corona.

2. *Difference of light curve time profiles*

   While solar soft X-ray profiles have rapid growth and gradual decay and high contrast, hard X-ray profiles from Cyg X-1 have symmetrical profiles and low contrast.

3. *Difference of power-law index of peak intensity distributions*

   The power-law index of solar flares is smaller than that of Cyg X-1.

   Maybe these differences ($1 \sim 3$) are due to the differences of the observed objects. In the case of solar flares, we observe magnetic energy release and thermal emission from heated plasma. But, in the case of X-ray shots, we may be observing accreting perturbed waves made by magnetic energy release and reflected shock waves. These waves emitt strong X-rays (Manmoto et al. 1996). Regarding the power-law index of peak intensity distributions, the simulations of Takeuchi et al.(1995) indicate that the index becomes small when the steady accreting mass value is small. Perhaps, this accreting mass ratio is one of the reasons for these differences.

   However, we must consider the more basic difference of the energy bands between YOHKOH and GINGA. According to YOHKOH data, there are some differences between soft X-ray and hard X-ray time profiles, too. It is possible that these difference influence statistical features. So, we have to analyze both soft X-ray data and hard X-ray data of YOHKOH.

4. *Interpretation of peak interval distribution of solar flare*

   Statistical features of peak interval distributions presents one of the biggest differences between the two objects. Distributions of solar flares are fitted by two different random distributions. But, actually, solar flares do not occur randomly. They tend to occur at locations where other flares appeared. We have to explain this statistical feature quantitatively.

5. *Universality of statistical features of solar flares for different spatial sizes*

   Especially in light of solar flare models, the dependence of statistical features on spatial scales is very important. For example, if the SOC model is correct, these features should be independent of spatial scale. So, we have to investigate these features for restricted fields, e.g., a single steady active region with as long a duration as possible. YOHKOH data is also usefull for this purpose.

6. *Reproduction of time profiles of individual solar flares at various energy bands by MHD simulations*

   Various workers are attempting to reproduce individual solar flares with MHD simulations. Most of these studies are restricted to morphorogy, temperature, and density. For the purpose of explaining differences between time profiles of solar flares and X-ray shots, we have to reproduce each time profile theoretically in various filter bands. For accretion disks, Manmoto et al. (1996) successfully reproduced the time profile of individual X-ray shots theoretically.

7. *Model construction to explain statistical features*

   The SOC model can reproduce many of the above statistical features. Nonetheless, especially regarding solar flares, there are some observational facts which cannot be explained by those simulations. If the SOC model cannot be modified to explain these features, it may be necessary to find a new model.

Yohkoh data will be of use in addressing several of the questions raised here.

The statistical features of solar flares are summarized and will be published in the near future in UeNo et al.(1997), and Negoro et al.(1997).

I am grateful to H. Negoro, S. Mineshige, K. Shibata, and H. Hudson for useful comments and advice for analysis.

References

Bai,T., Sturrock, P.A. 1991, *ApJ*, **409, 476**

Datlowe, D. W., Elcan, M. J., & Hudson, H. S. 1974, *Sol.Phys.*, **39, 155**

Dennis, B. R. 1985, *Sol.Phys.*, **100, 465**

Drake, J. F. 1971, *Sol.Phys.*, **16, 152**

Kile, J. N. & Cliver, E. W. 1991, *ApJ*, **370, 442**

Lin, R. P., Schwartz, R. A., Kane, S. R., Pelling, R. M., & Hurley, K. C. 1984, *ApJ*, **283, 421**

Lu, E. T., and Hamilton, R. J. 1991, *ApJ*, **380, L89**

Lu, E. T., Hamilton, R. J., McTiernan, J. M., & Bromund, K. R. 1993, *ApJ*, **412, 841**

Makishima, K. 1988, in *Physics of Neutron Stars and Black Holes* ed. Y. Tanaka (Tokyo, Universal Academy Press), p.175

Manmoto, T., Takeuchi, M., Mineshige, S., Matsumoto, R.,Negoro, H. 1996 *ApJ*, **464, L135**

Negoro, H. 1992, *Master thesis, Osaka University*

Negoro, H., Kitamoto, S., Takeuchi, M., and Mineshige, S. 1995, *ApJ*, **452, L49**

Negoro, H., UeNo, S., Mineshige, S., Shibata, K.,and Hudson, H. S. 1997, in preparation.

Pounds, K. A., and McHardy, I. M. 1988, in *Physics of Neutron Stars and Black Holes*, ed. Y. Tanaka (Universal Academy Press, Tokyo), p.285

Rieger, E., Share, G. H., Forrest, D. J., Kanbach, G., Reppin, C.,Chupp, E. L. 1984, *Nature*, **312, 623**

Takeuchi, M., Mineshige, S., Negoro, H. 1995, *PASJ*, **47, 617**

UeNo, S., Mineshige, S., Negoro, H., Shibata, K., Hudson, H.S., 1997, *ApJ*, **484, 920**

FROM THE SUN TO THE GALAXY CLUSTERS

K. MAKISHIMA
Department of Physics, University of Tokyo
7-3-1 Hongo, Bunkyo-ku, Tokyo, Japan 113

1. *Yohkoh* and *ASCA* – an Introduction

The *Yohkoh* Hard X-ray Telescope (HXT; Kosugi et al. 1991) forms images utilizing the Fourier-synthesis X-ray imaging technique (Makishima et al. 1978). It has unprecedented angular resolution of $\sim 5''$ over a wide energy range of 15–95 keV. With HXT, clear evidence of particle acceleration due to magnetic reconnection was obtained (Sakao et al. 1992; Sakao 1994; Masuda 1994; Masuda et al. 1994), and the flare morphology was mathematically quantified (Inda-Koide 1994; Inda-Koide et al. 1995).

It was only 1.5 years after the *Yohkoh* launch when the *ASCA observatory* (Tanaka et al. 1994) was put into orbit. Since then, *ASCA* has been expanding our view of the high-energy universe. For *ASCA*, we have developed the Gas Imaging Spectrometer (GIS; Ohashi et al. 1996; Makishima et al. 1996). Together with the X-ray CCD cameras (called SIS; Burke et al. 1994), the GIS is coupled to the novel X-ray Telescope (XRT; Serlemitsos et al. 1995), thus ensuring wide energy coverage of 0.6–10 keV.

The relation of the GIS to the *ASCA* SIS resembles that of the HXT to the *Yohkoh* SXT, because the GIS is more weighted towards higher energies than the SIS, having a time resolution up to 61 μsec. Thus, the *Yohkoh* HXT and the *ASCA* GIS may both be considered instruments for *hard X-ray imaging spectroscopy of dynamical phenomena in hot plasmas and energetic particles*, even though their observational targets are different.

With the advent of *Yohkoh*, it has become clear that magnetohydrodynamic (MHD) processes are nearly always responsible for plasma heating and particle acceleration, at least in the Sun. Are the same mechanisms operating on larger scales in the universe? This is the simple-minded question from which the present talk stems. Topics related to active galactic nuclei are excluded, since they are covered by T. Takahashi in this same volume.

T. Watanabe et al. (eds.), Observational Plasma Astrophysics: Five Years of Yohkoh and Beyond, 51–60.

2. High Energy Phenomena in Galaxies

2.1. STELLAR ACTIVITIES

Significant progress has been achieved with *ASCA* on understanding stellar X-ray emission. Here we quote one particular highlight, i.e. protostars, from a review by Koyama et al. (1997).

Observations with *ASCA* have established that protostars can often be luminous (up to $\sim 10^4$ times the Sun) hard X-ray sources. Since these objects are often embedded deep inside dark molecular clouds, the hard X-ray imaging capability of *ASCA* is essential. Violent X-ray flaring activities seen in these objects are closely connected with the presence of bipolar molecular flows, although the X-rays themselves are thought to be produced rather close to the stellar surface. These results jointly point to the possibility that MHD activity is much more enhanced in protostars than in main-sequence stars. This is presumably due to the combined effects of differential rotation and radial mass in-flow of the gaseous envelope, which feeds mass to the evolving central star.

2.2. SNR AND CRAB-LIKE PULSARS

Puzzlingly, extensive *ASCA* observations of Crab-like SNRs have led to almost null discoveries of new pulsars (Aoki 1995; Saito 1997). However, non-thermal X-ray pulses were detected with *ASCA* from the 237 ms γ-ray pulsar known as Geminga, and from the millisecond pulsar PSR B1821-24 for the first time (Saito 1997; Saito et al. 1997).

More assuringly, featureless non-thermal X-ray components have been detected from a number of SNRs with center-filled morphologies, even though they lack evidence of central pulsars in any wavelength. The spectrum of one such example, Kes 73, is shown in Fig.1 (left). The non-thermal component of this SNR becomes brighter towards the SNR center, providing evidence for a synchrotron nebula around a central unseen pulsar. As in the Crab Nebula (Kennel & Coroniti 1984; Pelling et al. 1987), these unseen pulsars are though to provide interstellar space with a large amount of magnetic flux, via pulsar winds. This process may contribute significantly to the creation of interstellar magnetic fields.

Spatially distinct thermal and non-thermal X-ray components were detected from SN1006 (Fig.1 right; Ozaki et al. 1994; Koyama et al. 1995). Its rim regions emit bright non-thermal X-rays, while its central region is filled with fainter thermal X-ray emission. Since this remnant lacks a central pulsar nebula, the non-thermal X-rays are presumably emitted by shock-accelerated particles via synchrotron process. This suggests a delicate competition between the acceleration and the heating.

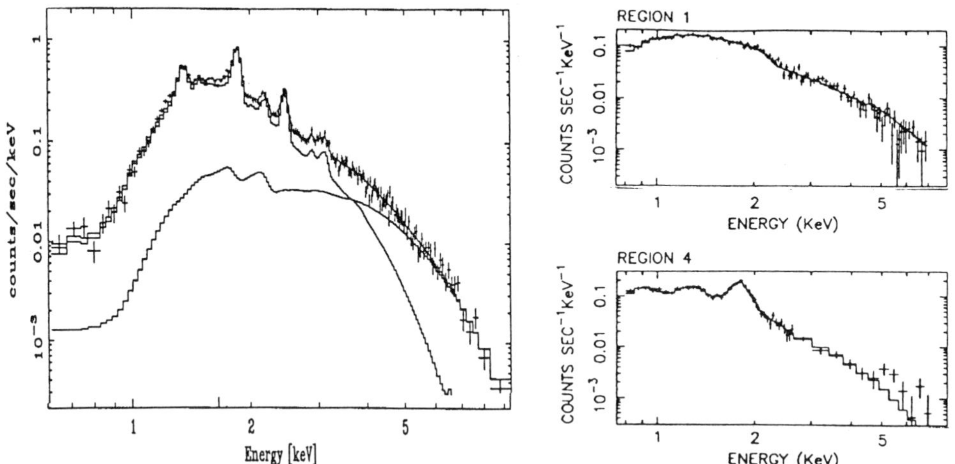

Figure 1. [Left] *ASCA* SIS spectrum of SNR Kes 73, exhibiting both a line-dominated thermal component of $kT \sim 0.86$ keV and a power-law component with photon index ~ 1.8 (Aoki 1995). [Right] *ASCA* GIS spectra of SNR SN1006 (Ozaki et al. 1994), from its bright rim region (top) and faint central region (bottom).

2.3. GALACTIC-RIDGE X-RAY EMISSION

An apparently diffuse X-ray emission, called Galactic-ridge X-ray emission (GRXE), is distributed along the Milky Way up to longitudes of about $\pm 60°$ (Worall et al. 1982). Its thermal nature was established (Koyama et al. 1986; Koyama 1989) with the *Tenma* GSPC experiment (Koyama et al. 1984; Tanaka et al. 1984), from which the *ASCA* GIS developed. Since then, GRXE has been studied extensively with *Ginga* (Yamauchi & Koyama 1993, 1995; Yamasaki 1996; Yamasaki et al. 1997) and *ASCA* (Yamauchi et al. 1995; Kaneda 1997; Kaneda et al. 1997).

GRXE in fact comprises two distinct thermal components, with characteristics as summarized in Table 1 (Kaneda 1997). Both components exhibit fairly smooth longitudinal brightness distributions, with at most 15% intrinsic fluctuations on angular scales of $10' \sim 60'$. However the soft-component brightness is more variable on larger scales ($\sim 10°$), up to a factor of 2. The two components otherwise exhibit very constant spectral properties along the Galactic longitude. Both clearly deviate from ionization equilibrium, as indicated by centroid energies of Fe-K, Si-K, Mg-K and Fe-L lines in comparison with the calculation by Masai (1984).

The cool component of GRXE was successfully explained as an assembly of SNRs of age $10^{4\sim5}$ years (Kaneda 1997). However, the same idea cannot explain the hot component, since SNRs as hot as 5–10 keV are too limited in number to reproduce the observed smooth brightness distribution. In addition, the deviation from ionization equilibrium rules out explanation

TABLE 1. Two thermal components of Galactic ridge X-ray emission (Kaneda 1997).

component	kT (keV)	H_G* (deg)	$n_e t$† (cm^{-3} s)	metallicity (solar)	N_H (10^{22} cm^{-2})	L_x‡ (erg s^{-1})
hotter	$4 \sim 10$	~ 0.5	$10^{10.8\pm0.3}$	$0.4 \sim 1$	~ 4	2×10^{38}
cooler	$0.6 \sim 1.0$	~ 2	$10^{9\sim10}$	$0.3 \sim 1.3$	~ 0.6	4×10^{39}

* : Typical scale height perpendicular to the Galactic plane.
† : Ionization parameter (Masai 1984), with n_e the electron density and t the time.
‡ : Spatially integrated 2–10 keV luminosity, corrected for absorption.

of the hot component as an assembly of stellar coronal sources.

The remaining possibility is that the hot component of GRXE actually comes from a very hot diffuse interstellar plasma. Then the plasma should have a mean density of $\sim 3 \times 10^{-3}\eta^{-1/2}$ cm^{-3}, and hence a pressure of $\sim 4 \times 10^{-11}\eta^{-1/2}$ erg cm^{-3}, where η is volume filling factor. This would be by far the highest energy density known in interstellar space. Obviously, such a plasma cannot be confined by either gravity or ordinary interstellar pressure. Nevertheless, the very low scale height and the relatively large $n_e t$ parameter (Table 1) of the hot component, together with energetics arguments, all require that the plasma be somehow confined to the Galactic plane (Kaneda 1997; Kaneda et al. 1997).

We have developed a model wherein the GRXE plasma is confined and heated by magnetic tubes of $\eta \sim 0.1$, locally amplified up to ~ 50 μG (Makishima 1994, 1995a, 1997ab) from an average interstellar value of ~ 6 μG (Ohno & Shibata 1993), like in the solar photosphere. Energetically, the bulk rotational energy of the Galaxy seems sufficient to amplify the interstellar magnetic fields up to this strength (Makishima 1997ab). A numerical simulation by Tanuma et al. (1997) shows that magnetic reconnection can heat the cool-component plasma up to a temperature of 5–10 keV.

The above MHD view is supported by further evidence. Without magnetic fields, the two plasma phases would be completely mixed in only $\sim 10^5$ years. Moreover, if the plasma were heated via kinetic process, unseen ions would carry away a dominant fraction of the input energy, because the electron-ion energy exchange is rather inefficient. Since this makes the energetics even more difficult, the electromagnetic processes that give similar energies to ions and electrons are favored. Then, the non-thermal hard tail seen in the GRXE spectrum (Yamasaki 1996; Yamasaki et al. 1997; Kaneda 1997), connecting to the diffuse Galactic γ-rays (Fig.2), can be accounted for naturally in terms of magnetic heating and acceleration. Thus, the GRXE phenomenon may be a Milky-Way version of the solar coronae.

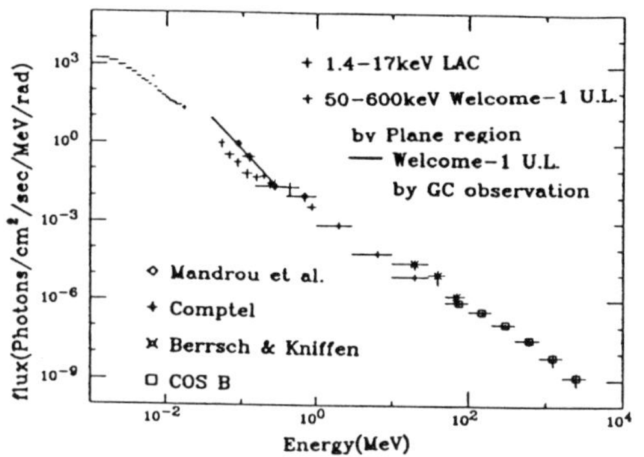

Figure 2. The wide-band spectrum of diffuse X-ray and γ-ray emission along the Galactic ridge, as taken from Yamasaki (1996). The region below 0.01 MeV represents the thermal GRXE.

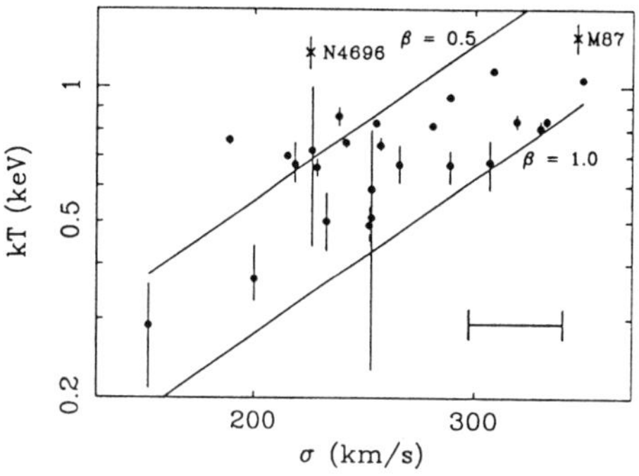

Figure 3. The ISM temperature of elliptical galaxies plotted against the stellar velocity dispersion (Matsushita 1997). The two lines indicate $\beta = 1$ and $\beta = 0.5$ in reference to equation 1.

3. Galaxies and Clusters of Galaxies

3.1. HOT INTERSTELLAR MEDIUM IN ELLIPTICAL GALAXIES

X-ray emitting plasmas are ubiquitously found in elliptical galaxies and galaxy clusters, in the form of interstellar medium (ISM) of $kT \sim 1$ keV and intra-cluster medium (ICM) of $kT = 2 - 10$ keV, respectively. Both ISM and ICM are thought to be hydrostatically confined by gravity. The ISM makes up a few percent of the stellar mass, while the ICM is often more massive than the total stellar mass of the cluster. The ICM contains a large amount of metal-rich gas ejected from member galaxies, exhibiting a metallicity of ~ 0.3 solar.

In an elliptical galaxy, mass is continuously supplied to the ISM through stellar mass loss; the gas shed from stars acquires a high temperature corresponding to the random stellar velocity. The supply of mass and heat is thought to be balanced by the radiative cooling of the ISM. In that case, the specific energy of the ISM would not exceed that of stars; in other words, the ratio

$$\beta \equiv \mu m_{\rm p} \sigma^2 / kT \tag{1}$$

would be larger than unity, where kT is the ISM temperature, σ is the stellar velocity dispersion, and $\mu \sim 0.6$ is the mean particle weight in units of the proton mass $m_{\rm p}$.

With *ASCA*, the ISM temperature and metallicity have been measured accurately in many elliptical galaxies (Awaki et al. 1994; Loewenstein et al. 1994; Makishima & Matsushita 1997; Matsushita 1997; Matsumoto et al. 1997). As shown in Fig. 3, kT generally correlates with σ, as expected.

Moreover, the ratio β in fact takes values closer to 0.5 than 1.0, implying that the plasma has about twice the specific energy of stars. Usually, this is taken as evidence of ISM heating by type I supernovae. However, in order to reconcile the very low metallicity of the ISM (0.5–1 solar; Awaki et al. 1994; Loewenstein et al. 1994) with the large amount of metals contained in the surrounding ICM, the supernova products must quickly be removed, without depositing much mass or energy in the ISM of the host galaxy (Matsushita 1997). If so, the ISM may be heated mainly via MHD process, in such a way that random stellar motion amplifies magnetic fields, which in turn heat the ISM via reconnection. Thus, the ISM may in some sense be an elliptical-galaxy version of the solar corona.

3.2. PHENOMENA SEEN AROUND CD GALAXIES

At the center of a galaxy cluster, often a central dominant (cD) elliptical galaxy is seen. A series of interesting phenomena have been noticed in the ICM near cD galaxies. These include:

1. Significant central enhancement in the soft X-ray surface brightness.
2. Decrease in the average ICM temperature towards the cluster center.
3. X-ray emission lasting longer than the nominal cooling time.
4. Filamentary X-ray morphology of the ICM, in some cases.
5. Sometimes excess X-ray absorption indicative of cool matter.
6. Emission of a huge number of H_α photons from filamentary sources.
7. Often enhanced magnetic field intensity, up to $\sim 30\mu$G.
8. Evidence of local particle acceleration, as suggested by radio halos.
9. Spatial co-existence of distinct cooler and hotter plasma phases.
10. Often enhanced ICM metallicity, up to one solar.
11. Detection of central excess brightness (item 1) in hard X-rays.
12. Presence of a dark gravitational potential of the cD galaxy itself.

Many of these facts had been known previously (e.g. Sarazin 1988), and interpreted as a result of a radiative cooling instability, called cooling flow, occurring in the densest ICM region (Fabian 1994). However this idea can no longer explain the last four items, which are new *ASCA* discoveries mainly referring to the Centaurus, Fornax, and Virgo clusters (Fukazawa et al. 1994; Mushotzky et al. 1995; Ohashi et al. 1995; Ikebe 1995; Ikebe et al. 1996, 1997ab; Makishima 1994, 1995ab, 1996, 1997ab; Matsumoto et al. 1996; Xu et al. 1997). Clearly, a paradigm shift is required.

As an alternative, we have developed a "cD corona" model (Makishima 1994, 1995ab, 1996, 1997ab) incorporating three ingredients. First, we assume that the excess gravitational potential of the cD galaxy (item 12) attracts an excess amount of hot ICM, which produces the nearly color-independent central X-ray excess (items 11,1). Second, we assume that the central potential dimple also confines the metal-enriched cooler ISM of the cD galaxy (item 10), producing the two-phase nature (items 9). Finally, we assume that the cooler ISM phase is confined inside closed magnetic loops (items 4,7) anchored onto the cD galaxy, while open-field regions ("cD coronal holes") are filled with the hotter ICM. Due to the magnetic insulation, the two plasma phases can stably co-exist (item 9), wherein an increase in the relative ISM contribution causes the central temperature drop (item 2). Orbital motions of non-cD galaxies are transferred to magnetic energy, which in turn is liberated to replenish the radiative cooling loss (item 3); Makishima (1997a) showed that such an MHD process can provide a sufficient energy input to sustain the cool-component luminosity. Items 5,6, and 8 may represent microflares (though on time scales of $10^{7\sim8}$ years!) and subsequent post-flare cooling loops. This novel view can be called a cluster version of the solar corona.

3.3. GALAXY METAMORPHOSIS VIA MHD INTERACTIONS

When a galaxy moves through intra-cluster space with a velocity of $v \sim 500$ km s^{-1}, its translational momentum may be transferred to the ICM on a time scale $\tau \sim M/(\pi r^2 \mu m_{\rm p} n v)$. Here, $n \sim 10^{-3}$ cm^{-3} is a typical ICM density, $M \sim 1 \times 10^{11}$ $M_\odot$ is a galaxy mass, and r is a radius of the galaxy's "magnetosphere". Taking $r \sim 30$ kpc leads to $\tau \sim 5$ Gyr. Therefore, a significant fraction of the galaxy's momentum can actually be transferred to the ICM over a Hubble time. Similarly, spiral galaxies may lose a considerable fraction of their angular momenta by launching helical Alfvén waves into the ICM. Binary galaxies may also merge more easily if the ICM is present, by dumping orbital angular momenta onto the ICM.

While these MHD interactions provide the ICM with sufficient energy input, the galaxies must receive considerable reactions from the ICM. In particular, the loss of angular momentum and the enhanced galaxy merging may drive the metamorphosis of spiral galaxies into elliptical galaxies. Actually, it has long been known that spiral galaxies are sparse in X-ray luminous clusters (Fig.4), particularly in the cluster center regions (Whitmore & Gilmore 1993). Furthermore, *HST* observations of distant rich clusters reveal that they contain numerous faint spirals with disturbed shapes (e.g Dressler et al. 1994), in contrast to nearby rich clusters which contain lots of ellipticals and dwarf spheroidals. Therefore the current dominance of elliptical galaxies in a gas-rich environment seems to have developed much later than the initial cluster formation. The overall scenario is most naturally explained in terms of the spiral-to-elliptical metamorphosis. For this purpose, MHD interactions are obviously much more efficient than gravitational close-encounter effects.

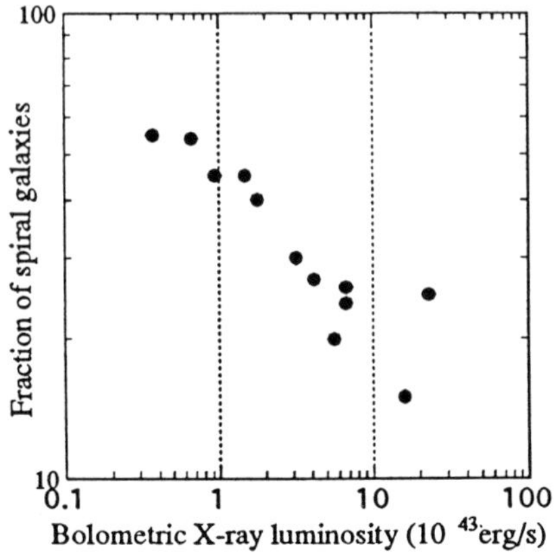

Figure 4. The spiral fraction of clusters, corrected for possible selection bias (Andreon 1993) and plotted against the X-ray luminosity.

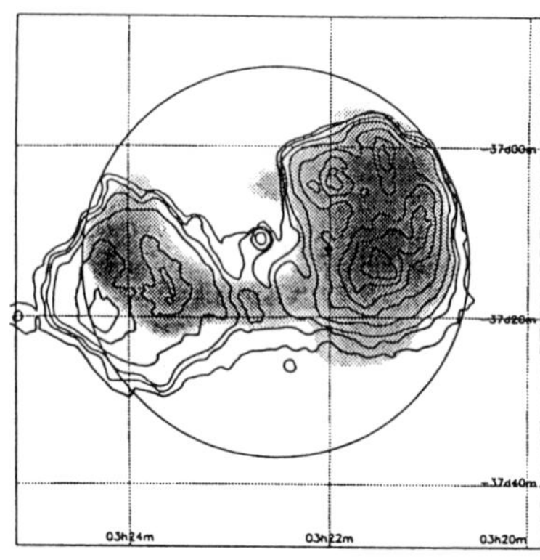

Figure 5. An overlay of the X-ray (gray scale) and 1.4 GHz radio brightness distributions, in lobes of the radio galaxy Fornax A (Kaneda et al. 1995).

Hickson's compact groups (HCGs; Hickson et al. 1989) provide another example of possible galaxy-ICM interaction. Although HCGs exhibit an amazingly narrow concentration in gravitating mass at $\sim 2 \times 10^{13}\ M_{\odot}$, their X-ray luminosities scatter very much, with a clear tendency of elliptical-rich HCGs to be X-ray luminous, hence gas-rich (Mulchaey et al. 1996; Fukazawa et al. 1996). However this effect cannot be explained by gas ejection from ellipticals, since the intra-group gas is not particularly metal rich (Fukazawa et al. 1996). Instead, we presume that some HCGs were gas-rich from the beginning while others were not, since HCGs are located near the threshold between "closed" and "open" systems (Fukazawa et al. 1996). Subsequently, the member galaxies in a gas-rich HCG presumably made a gradual metamorphosis into ellipticals.

3.4. MEASURING INTER-GALACTIC MAGNETIC FIELDS

In order to further develop the MHD view, accurate estimates of the magnetic field intensity in galaxies and clusters are indispensable. Through Faraday-rotation measurements, magnetic fields of $\sim 1\mu$G have been found to be ubiquitous in clusters (Kronberg 1994). However, this method tends to underestimate the magnetic field, because fields with opposite polarities cancel each other. A more reliable measurement may be available if there are adequate population of relativistic electrons, which are expected to emit both synchrotron radio emission and inverse-Compton (IC) X-ray emission. The IC X-rays arise when electrons scatter off cosmic microwave photons. The ratio of the surface brightness in these two emission channels gives magnetic field strengths (Harris & Grindlay 1979).

Although this simple method was proposed more than two decades ago, diffuse IC X-rays were too faint to be detected by previous experiments. At last, Feigelson et al. (1995) and Kaneda et al. (1995) succeeded in detecting extended IC X-ray emission from lobes of the radio galaxy Fornax A (NGC 1316), using *ROSAT* and *ASCA*, respectively (Fig.5). They measured the lobe magnetic fields to be 2–4 μG. In addition, Tashiro et al. (1996) report on the detection of IC X-rays from lobes of another radio galaxy, in which the deduced magnetic field is again $\sim 3\ \mu$G.

Now that the IC technique has been established, we plan to search for extended IC emission from nearby clusters (Rephaeli et al. 1994) to estimate the magnetic field intensity there. However, in order to avoid the bright ICM emission, we must use a much harder energy range than is available with *ASCA*. The hard X-ray detector (HXD; Kamae et al. 1996) experiment, which we are currently developing for ASTRO-E (to be launched in the year 2000), will be very powerful for this purpose.

References

Andreon, S. (1993) *Astr. Astrophys*, **276**, L17–L20

Aoki, T. (1995) PhD thesis, Department of Physics, Gakushuin University

Awaki, H. *et al.* (1994) *Publ. Astr. Soc. Japan*, **46**, L65–L70

Burke, B. E. *et al.* (1994) *IEEE Trans. Nuc. Sci.*, **41**, 375–385

Dressler *et al.* (1994) *Astrophys. J. Letters*, **435**, L23–L26

Fabian, A. C. (1994) *Ann. Rev. Astr. Astrophys.*, **32**, 277–318

Feigelson, E. D. *et al.* (1995) *Astrophys. J. Letters*, **449**, L149–L152

Fukazawa, Y. *et al.* (1994) *Publ. Astr. Soc. Japan*, **46**, L55–L58

Fukazawa, Y. *et al.* (1996) *Publ. Astr. Soc. Japan*, **48**, 395–407

Harris, D. E. and Grindlay, J. E. (1979) *Mon. Not. Roy. Astr. Soc.*, **188**, 25–37.

Hickson, P., Kindl, E. and Auman, J. R. (1989) *Astrophys. J. Suppl.*, **70**, 687–698

Ikebe, Y. (1995) PhD thesis, Department of Physics, University of Tokyo

Ikebe, Y. *et al.* (1996) *Nature*, **379**, 427–429

Ikebe, Y. *et al.* (1997a) *Astrophys. J.* , in press

Ikebe, Y. *et al.* (1997b) *Astrophys. J.* , submitted

Inda-Koide, M. (1994) PhD thesis, Department of Astronomy, University of Tokyo

Inda-Koide, M. *et al.* (1995) *Publ. Astr. Soc. Japan*, **47**, 661–676

Kamae, T. *et al.* (1996) *SPIE*, **2806**, 314–328

Kaneda, H. (1997) PhD thesis, Department of Physics, University of Tokyo

Kaneda, H. *et al.* (1995) *Astrophys. J. Letters*, **453**, L13–L16

Kaneda, H. *et al.* (1997) *Astrophys. J.* , in press (November 1 issue)

Kennel, C. F. and Coroniti, V. F. (1984) *Astrophys. J.*, **283**, 710–730

Kosugi, T. *et al.* (1991) *Solar Physics*, **136**, 17–36

Koyama, K. (1989) *Publ. Astr. Soc. Japan*, **41**, 665–678

Koyama, K. *et al.* (1984) *Publ. Astr. Soc. Japan*, **36**, 659–666

Koyama, K. *et al.* (1986) *Publ. Astr. Soc. Japan*, **38**, 121–131

Koyama, K. *et al.* (1997) *Imaging and Spectroscopy of Cosmic Hot Plasmas*, eds. F. Makino and K. Mitsuda (Universal Academy Press: Tokyo), p.509–514

Koyama, K. *et al.* (1995) *Nature*, **378**, 255–258

Kronberg, P. P. (1994) *Rep. Prog. Phys.*, p.325–382

Loewenstein, M. *et al.* (1994) *Astrophys. J. Letters*, **436**, L75–L78

Makishima, K. (1994) *New Horizon of X-Ray Astronomy*, eds. F. Makino and T. Ohash

(Universal Academy Press: Tokyo), p.171–180
Makishima, K. (1995a) *Elemetnray Processes in Dense Plasmas*, eds. S. Ichimaru and S. Ogata (Addison-Wesley), p.47–60
Makishima, K. (1995b) *Dark Matter* (AIP Proceedings No.336), ed. S. S. Holt, C. Bennet (AIP, New York), p.172–181
Makishima, K. (1996) *UV and X-ray Spectroscopy of Astrophysical and Laboratory Plasmas*, eds, K. Yamashita and T. Watanabe (Universal Academy Press: Tokyo), p.167–173
Makishima, K. (1997a) *Imaging and Spectroscopy of Cosmic Hot Plasmas*, eds. F. Makino and K. Mitsuda (Universal Academy Press: Tokyo), p.137–144
Makishima, K. (1997b) *Plasma Physics and Controlled Fusion*, in press
Makishima, K. and Matsushita, K. (1997) *Proc 2nd Stromlo Symposiumon the Nature of Elliptical Galaxies*, ed. G. da Costa and M. Arnaboldi, in press
Makishima, K. *et al.* (1978) *New Instrumentation for Space Astronomy*, eds. K. van der Hucht and G. S. Vaiana (Pergamon Press: Oxford), p.277–289
Makishima, K. *et al.* (1996) *Publ. Astr. Soc. Japan*, **48**, 171–189
Masai, K. (1984) *Astrophysics and Space Science*, **98**, 376–381
Masuda, S. (1994) PhD thesis, Department of Astronomy, University of Tokyo
Masuda, S. *et al.* (1994) *Nature*, **371**, 495–497
Matsumoto, H. *et al.* (1996) *Publ. Astr. Soc. Japan*, **48**, 201–210
Matsumoto, H. *et al.* (1997) *Astrophys. J.*, in press
Matsushita, K. (1997) PhD thesis, Department of Astronomy, University of Tokyo
Mulchaey, J. *et al.* (1996) *Astrophys. J.*, **456**, 80–97
Mushotzky, R. F. (1994) *New Horizon of X-ray Astronomy*, eds. F. Makino and T. Ohashi (Universal Academy Press: Tokyo) p.243–252
Ohashi, T. *et al.* (1994) *New Horizon of X-ray Astronomy*, eds. F. Makino and T. Ohashi (Universal Academy Press: Tokyo) p.273–278
Ohashi, T. *et al.* (1996) *Publ. Astr. Soc. Japan*, **48**, 157–170
Ohno, H. and Shibata, S. (1993) *Mon. Not. Roy. Astr. Soc.*, **262**, 953–962
Ozaki, M. *et al.* (1994) *New Horizon of X-Ray Astronomy*, eds. F. Makino and T. Ohash (Universal Academy Press: Tokyo) p.481–482
Pelling, R. E. *et al.* (1987) *Astrophys. J.*, **391**, 416–425
Rephaeli, Y., Ulmer, M. and Gruber, D. (1994) *Astrophys. J.*, **429**, 554–556
Saito, Y. (1997) PhD thesis, Department of Physics, University of Tokyo
Saito, Y. *et al.* (1997) *Astrophys. J. Letters*, **477**, L37–L40
Sakao, T. (1994) PhD thesis, Department of Astronomy, University of Tokyo
Sakao, T. *et al.* (1992) *Publ. Astr. Soc. Japan*, **44**, L83–L87
Sarazin C.L. (1988) X-Ray Emission from Clusters of galaxies (Cambridge Univ. Press)
Serlemitsos, P. *et al.* (1995) *Publ. Astr. Soc. Japan*, **47**, 105–114
Tanaka, Y., Inoue, H. and Hold, S. S. (1994) *Publ. Astr. Soc. Japan*, **46**, L37–L41
Tanaka, Y. *et al.* (1984) *Publ. Astr. Soc. Japan*, **36**, 641–658
Tanuma, S. *et al.* (1997) *Astrophys. J*, submitted
Tashiro, M. (1996) *Imaging and Spectroscopy of Cosmic Hot Plasmas*, eds. F. Makino and K. Mitsuda (Universal Academy Press: Tokyo) p.83–86
Whitmore, B. and Gilmore, D. M. (1993) *Astrophys. J.* , **407**, 489–509
Worall, D. M. *et al.* (1982) *Astrophys. J.*, **255**, 111–121
Xu, H. *et al.* (1997) *Publ. Astr. Soc. Japan*, **49**, 9–16
Yamasaki, N. (1996) PhD thesis, Department of Physics, University of Tokyo
Yamasaki, N. *et al.* (1997) *Astrophys. J.*, **481**, in press
Yamauchi, S. and Koyama, K. (1993) *Astrophys. J.*, **404**, 620–624
Yamauchi, S. and Koyama, K. (1995) *Publ. Astr. Soc. Japan*, **47**, 439–443
Yamauchi, S. *et al.* (1995) *Publ. Astr. Soc. Japan*, **48**, L15–L20

HIGH ENERGY EMISSION FROM BLAZARS

– ASCA Observation of Blazars and Multiband Analysis –

TADAYUKI TAKAHASHI AND HIDETOSHI KUBO
Institute of Space and Astronautical Science
Sagamihara, Kanagawa, Japan

GREG MADEJSKI
Laboratory for High Energy Astrophysics
Code 666, NASA GSFC, MD 20771, U.S.A

MAKOTO TASHIRO
Department of Physics, University of Tokyo
Bunkyo-ku, Tokyo, Japan

AND

FUMIYOSHI MAKINO
Institute of Space and Astronautical Science
Sagamihara, Kanagawa, Japan

1. Introduction

Blazars, a subclass of AGN, include BL Lac objects and OVV quasars. These objects are characterized by non-thermal continuum spectra and strong time variability, with pronounced occasional flaring in most observable wavelengths. One of their defining characteristics is high polarization in the radio and optical spectral regimes (cf. Angel & Stockman 1980). In the radio band, VLBI studies of these objects show compact radio structures on milli-arcsecond angular scale. These jet-like structures often show superluminal motion, where individual components are observed to move away from the radio cores with the apparent speed v/c of a few (see, e.g., Vermeulen & Cohen 1994).

The EGRET instrument onboard the *Compton* Gamma-Ray Observatory (*CGRO*) has so far detected emission in the GeV range from 60 blazars (see, e.g., Fichtel et al. 1994). Emission in the TeV range has also been detected from nearby BL Lac objects Mkn 421 and Mkn 501 with Whipple Observatory, which uses Cerenkov detectors (see Punch et al. 1992; Quinn

T. Watanabe et al. (eds.), Observational Plasma Astrophysics: Five Years of Yohkoh and Beyond, 61–70.

et al. 1996). Rapid γ–ray time variability commonly observed in these objects (see, e.g., Kniffen et al. 1993) implies very compact emitting regions. Since these sources show large luminosities in the spectral range where the opacity to pair production via $\gamma\gamma \rightarrow e^+e^-$ is large, it is generally accepted that the γ–ray emission is anisotropic. Both the VLBI studies and consideration of opacity to pair creation strongly suggest that the entire electromagnetic emission is anisotropic, and most likely relativistically beamed in one direction, forming a jet. An observer aligned closely to the jet axis sees emission from the radiating matter intensified, and time scale of variability shortened by the relativistic motion toward the observer (see, e.g., Blandford & Konigl 1979; Lind & Blandford 1985). In order to estimate the physical conditions in these objects, a "beaming" (Doppler) factor is defined to be $\delta = \frac{1}{\Gamma_j(1-\beta\cos\theta)}$, where Γ_j is the Lorentz factor of the emitting matter, $\beta = v/c$, and θ is the angle of motion with respect to the line of sight; apparent variability time scale and luminosity amplification depend on various powers of δ (see, e.g., Sikora et al. 1997).

Multi-frequency studies are very important tools to identify major emission processes in blazars. Recent studies show that the overall spectra have at least two pronounced components: the low energy peak (LE) and the high energy peak (HE) (see, e.g., von Montigny et al. 1995). The peak frequency of the LE component is distributed over 4 decades, $10^{13} - 10^{17}$ Hz. For the blazars hosted in quasars, and for BL Lac objects discovered via radio-selection techniques, the LE component peaks in the infrared, while in the BL Lacs found using X–rays, it peaks in the ultraviolet or even in soft X–rays (see, e.g., Sambruna, Maraschi, & Urry 1996). The HE component, on the other hand, peaks in the γ–ray band, in the MeV - to - GeV range, but in the case of BL Lacs, it sometimes extends to the TeV range. The local power-law shape and the smooth connection of the entire radio - to - UV/soft X–ray spectrum, as well as the relatively high level of polarization observed from radio to the UV implies that the emission from the LE component is most likely produced via the synchrotron process of relativistic particles radiating in a magnetic field. The HE component is generally thought to involve Comptonization of lower energy photons by the same particles that radiate the LE component. On the basis of non-contemporaneous data for a small number of sources, Maraschi, Ghisellini, & Celotti (1994) suggested that the ratio of the power in the HE to the LE components appears to be systematically larger for QHB than for XBLs.

It is very intriguing to study inter-relations between processes that are responsible for the LE and HE components. Since the launch of *ASCA*, we have carried out extensive multiband campaigns of blazars. In this paper, we study the relation between the LE component and the HE component, in terms of the peak frequency and the ratio of the luminosity of these

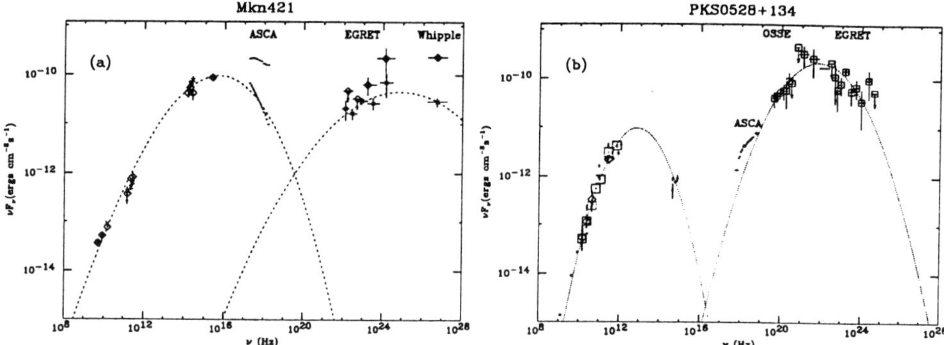

Figure 1. The multiband spectrum obtained from the simultaneous observations of (a) Mkn 421 and (b) Quasar-Hosted Blazar PKS 0528+134. The solid line shows the results of fit to a third-order polynomial function for low and high energy components.

components (L_{HE}/L_{LE}).

2. ASCA Observations

The X-ray satellite *ASCA* observed 18 blazars, of which 10 were also observed contemporaneously with EGRET as parts of multiband campaigns; such simultaneous observing campaigns are essential for these highly variable objects. Multiband analysis was made based on above data and published results to study the radiation process of blazars. Detailed analysis procedures and results are described in Kubo (1997) and Kubo et al. (1997).

Two examples of blazar spectra plotted in $\nu F(\nu)$ representation (giving the emitted power per decade of energy) are shown in Figures 1. Fig. 1(a) shows the spectrum of Mkn 421 (Macomb et al. 1996), a fairly typical XBL; its X-ray spectrum is soft (steep) and on the extrapolation of the ultraviolet, suggesting that the synchrotron emission extends into the X-ray range (see, e.g., Makino, Fink, & Clavel 1992; Edelson et al. 1995; Tashiro 1992). Fig. 1(b) shows PKS 0528+134 (Collmar et al. 1997), a strong lined QHB; in contrast to Mkn 421, the X-ray spectrum of PKS 0528+134 is harder than the UV spectrum, implying that the X-rays are already the onset of the second, HE component. Fig. 2 shows the distribution of photon indices of the blazars obtained with *ASCA* . This figure clearly shows that the X-ray spectra of XBLs are the softest, with power law index $\alpha \sim 1-1.5$, and they form the highest observable energy tail of the LE peak. X-ray spectra of the QHBs are the hardest ($\alpha \sim 0.7$) and are consistent with the lowest observable energy end ("onset") of the HE peak. For RBLs, the X-ray spectra are intermediate; in one case, 0716+714, the spectrum shows

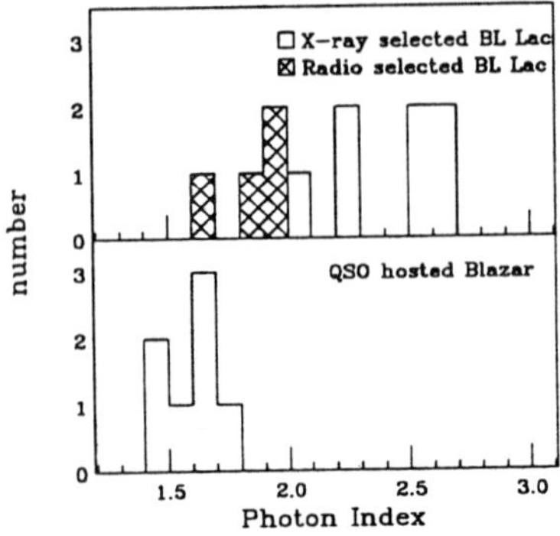

Figure 2. Distribution of averaged power law indices observed by *ASCA* . In all cases, the spectra were assumed to be simple absorbed power laws.

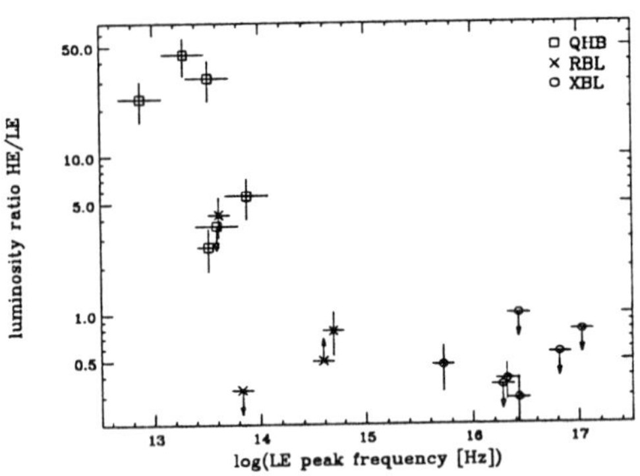

Figure 3. Distribution of the ratio L_{HE}/L_{LE} as a function of the peak frequency of the LE component. Arrows indicate the upper or lower limits measured with γ–ray observations.

hardening with an increasing energy. It is worthwhile to note that while the photon index for XBLs is often variable on a short time scale (see, e.g., Takahashi et al. 1996a), for QHBs the index remains almost constant when the flux changes.

In order to determine the peak frequency and the luminosity of both the LE and HE components, we performed a third-order polynomial fit to the spectra in $\nu F\nu$ as shown in Figs. 1(a) and 1(b). As Fig. 3 demonstrates – where half of the data are from simultaneous contemporaneous observations – the peak frequency of the LE component of the QHBs is lower and the $L(HE)/L(LE)$ ratio is greater than that measured in XBLs. For RBLs, these properties are, again, intermediate.

3. Multiband Analysis

Although it is almost certain that the process responsible for the HE component is the Comptonization of lower energy photons, the origin of these "seed photons" is still unresolved. The leading models propose either the synchrotron photons produced internally in the jet (the Synchrotron-Self Compton, or SSC model: Rees 1967; Konigl 1981; Marscher & Gear 1985; Ghisellini & Maraschi 1989), or photons external to the jet (the External Radiation Compton model or ERC model), either from the accretion disk (Dermer, Schlickeiser, & Mastichiadis 1992), or the central source radiation scattered and reprocessed by the emission line clouds and/or the intercloud medium (Blandford & Levinson 1995; Sikora, Begelman, & Rees 1994)

In the following analysis, we assume that *both* SSC and ERC processes may take place in blazars. We then estimate the contribution of the SSC emission in the HE component. In order to calculate the predicted luminosity due to the SSC emission, we assumed a simple homogeneous model, in which photons are produced in a region of radius R and with a constant magnetic field B; we estimate R from the shortest observed variability time scale Δt, assuming that $R = c\delta\Delta t/(1+z)$. We considered the radiation by a single population of relativistic electrons, with a broken power law in energy distribution, and a break point at γ_b, where γ is the Lorentz factor of the electrons (similar to e.g. Sambruna, Maraschi, & Urry 1996). We also assume that the radiation spectrum of the LE component peaks at a frequency corresponding to that radiated by the electrons with γ_b.

Peak frequencies of the synchrotron component (ν_{sync}) is then given as:

$$\nu_{sync} = 1.2 \times 10^6 \gamma_b^2 B\delta \quad \text{Hz}. \tag{1}$$

Therefore the wide distribution of ν_{LE} observed in blazars could be attributed to the distribution of either γ_b or B or δ. If the electron energy is still in the Thomson regime ($\gamma_{el} \times h\nu_{sync} << m_e c^2$), the expected peak of the SSC component is :

$$\nu_{SSC} = \frac{4}{3}\gamma_b^2 \nu_{sync} \tag{2}$$

The ratio of the luminosity of the SSC component L_{SSC} to the synchrotron luminosity L_{sync} is:

$$\frac{L_{SSC}}{L_{sync}} = \frac{u_{sync}}{u_B}, \tag{3}$$

where the $u_{sync} = \frac{L_{sync}}{4\pi R^2 c\delta^4}$ is the energy density of the synchrotron photons, and $u_B = \frac{B^2}{8\pi}$ is the energy density of magnetic field.

To check the validity of the assumption that the observed HE component is solely due to the SSC emission, we calculated the beaming factor (δ), which is obtained from the equation:

$$L_{ssc} = 1.6 \times 10^{12} \left(\frac{L_{synch}^2}{cR^2 \nu_{synch}^4 \delta^2} \right) \nu_{ssc}^2 \mathrm{erg/s} \tag{4}$$

The peak frequency and the luminosity of both the LE and HE components are calculated from a third-order polynomial fit to the spectra in $\nu F\nu$. Since we only have upper limits for γ-ray emission for 6 out of the 18 blazars, we use 12 blazars only for the multiband analysis. The upper limit of the size R can be estimated from the observed time variability (Δt) given by $R < c\Delta t\delta/(1+z)$. We also calculate values with a fixed value of $R = 0.01$ pc.

With the assumption that the SSC process dominates the HE peak – or ν_{HE} and L_{HE} are equal to ν_{ssc} and L_{ssc} – the values of δ calculated for 7 blazars are in the range of $\sim 5-10$. These values are consistent with the superluminal expansion results (cf. Vermeulen & Cohen 1994),and the limits obtained from arguments involving the γ–ray opacity (cf. Dondi & Ghisellini 1996). However the analysis for 5 QHBs and RBLs shows the values of δ to be much too large, sometimes exceeding $\delta \sim 100$. This suggests that an additional emission mechanism – such as the ERC process – contributes significantly in the γ–ray regime.

However, the fact that the QHBs have a hard spectrum of $\alpha \sim 0.7$ in X–ray band implies that the main X–ray emission of the QHBs is due to a separate emission process, different from the synchrotron process responsible for the IR - to UV spectrum. Previous work (see, e.g., Hoyle, Burbidge, & Sargent 1966) suggests that the SSC process must be important, and we thus adopt it as the dominant mechanism of the production of X–rays. In this case, we expect that the peak luminosity of SSC radiation is higher than the X–ray luminosity observed with *ASCA* , and lower than the extrapolated luminosity from the *ASCA* spectra by assuming beaming factor $5 < \delta < 20$ (Kubo et al. 1997).

Figure 4 shows the L_{SSC} obtained from the multiband analysis. L_{SSC} for 5 QHBs are estimated from the *ASCA* spectrum. Once we obtain L_{SSC}, we can calculate the strength of the magnetic field B as follows:

$$B = 0.27 \left(\frac{R}{10^{-2}\mathrm{pc}} \right)^{-1} \left(\frac{\delta}{10} \right)^{-2} \sqrt{\left(\frac{L_{sync}}{10^{46}\mathrm{erg/s}} \right) \left(\frac{L_{sync}}{L_{ssc}} \right)} \quad \mathrm{gauss} \tag{5}$$

γ_b is then calculated using Eq.1,

From our analysis, B is inferred to be $0.1 \sim 1$ gauss (see Fig. 4(b)) for δ=10, an intermediate value. With these values of B, we estimate γ_b is $10^3 \sim 10^4$ for QHBs, and 10^5 for XBLs. This difference of γ_b between

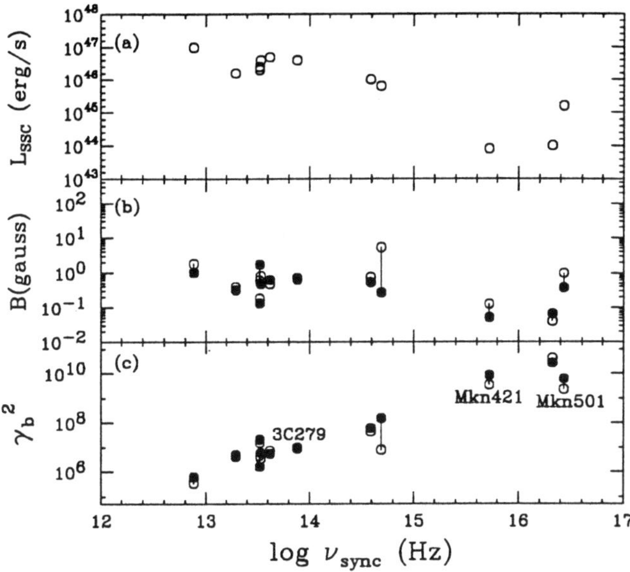

Figure 4. The distribution of parameters derived from the multi-band analysis. (a) γ–ray luminosity (L_{SSC}) , (b) magnetic field (B), and (c) the electron Lorentz factor γ_e as a function of the peak frequency of the LE component. For XBL and RBL, we use the L_{HE} as L_{SSC}. L_{SSC} for QHB are estimated by extrapolating the *ASCA* spectrum. In the calculation, we use a beaming factor of δ=10. The size R is estimated from the observed time variability (open symbols). We also plot values calculated with R = 0.01 pc (filled symbols).

the two sub-classes of blazars suggests that the relativistic electrons are accelerated to higher energies in blazars with higher LE peak frequencies (e.g. HBLs) than in those with lower LE peak frequencies (e.g. QHBs). Alternatively, since B and γ_{el} depend on the value of δ, the results might be affected by changing δ. But the effect of varying δ is at most a factor of 4 in B and a factor of 1.4 in γ_b for δ ~5–20. which is small compared with the difference of γ_b. Although the larger value of δ would be a way to obtain a larger γ_b, the larger δ also leads to ae higher ν_{sync} (Eq. 1), which is not consistent with the lower ν_{sync} for the higher γ_b objects in our results.

For QHBs, where dense external radiation fields exist, the ERC emission most likely contributes equivalently with or even more significantly than the SSC emission in the γ–ray band (see. e.g., Sikora et al. 1997). Therefore, our analysis indicates that the difference of γ_{el} is most likely due to the large photon density in QHBs as compared with that of XBLs. It should be noted that the TeV photons have been observed only from XBLs which turned out to have higher γ_{el}.

4. Dynamics: the case of BL Lac object Mkn 421

Among the GeV-emitting blazars, the BL Lac object Mkn 421 is unique as the first – and so far, the brightest – member where the γ–ray emission extends up to the TeV energies at a level allowing detailed spectral and variability studies in the broadest range of wavebands. Mkn 421 is one of the best known and studied BL Lacs; it shows optical polarization, a flat radio spectrum, and significant time variability, all of which are characteristics of the blazar class (e.g. Tashiro 1992).

The previous multi-frequency observation of Mkn 421 from the radio to TeV γ–ray bands in 1994 resulted in the detection of a simultaneous keV/TeV flare by the *ASCA* and *Whipple* Observatory (Takahashi et al. 1994; Takahashi et al. 1996a; Kerrick et al. 1995). Notably, the GeV γ–ray flux observed by EGRET, as well as the radio and UV fluxes showed less variability than keV or TeV bands (Macomb et al. 1995). The *ASCA* observation lagged the onset of the TeV flare by one day but recorded a high level of 2 – 10 keV X–ray flux peaking at 3.7 $\times 10^{-10}$ ergs cm^{-2} s^{-1}, a 10-fold increase over the quiescent level (Takahashi et al. 1996a). The 1995 multiwavelength campaign revealed another coincident keV/TeV flare (Takahashi et al. 1996b; Buckley et al. 1996; Wagner et al. 1997). Importantly, both 1994 and 1995 *ASCA* data clearly show that in addition to the keV flux variability, the X–ray *spectrum* is variable, such that the soft X–ray photons lag the harder X–rays by about 1 hr (Figure 5).

The discovery of the soft X–ray lag in the X–ray variability of Mkn 421 allows further constraints on the parameters of the emitting region. If this is due to cooling of electrons by synchrotron radiation, we can calculate the magnetic field (see, e.g., Tashiro et al. 1995; Takahashi et al. 1996a). Since the lag is the difference of t_{sync} , the time when an electron loses a half of its energy, at different X–ray energies, we use the data shown in Fig. 5 to infer $t_{sync} \sim 6000$ s at 1 keV. With this value, we calculate the magnetic field B of $\sim 0.2\ (\delta_5)^{-1/3}$ Gauss (where $\delta_5 = \delta/5$). With this B, we calculate the γ_{el} of electrons responsible for keV emission to be $\sim 5 \times 10^5$ $E^{1/2}$, where E is the energy in units of keV (Takahashi et al. 1996a). It should be noted that the value derived here is gratifyingly consistent with the value obtained from the multiband analysis discussed above.

The energy conservation (or, equivalently, the Klein-Nishina limit) requires that Lorentz factors γ of $\sim 10^6$ are the minimum necessary to produce TeV photons, in agreement with that derived from the delay between the hard X–rays and the soft X–rays. For the TeV radiation to be produced by Compton scattering, the energy of the "seed" low energy photons E_{seed} (as measured in the co-moving frame of the bulk plasma) must be around 0.1 - 1 eV. We note that the energy density of the radiation spectrum peaks

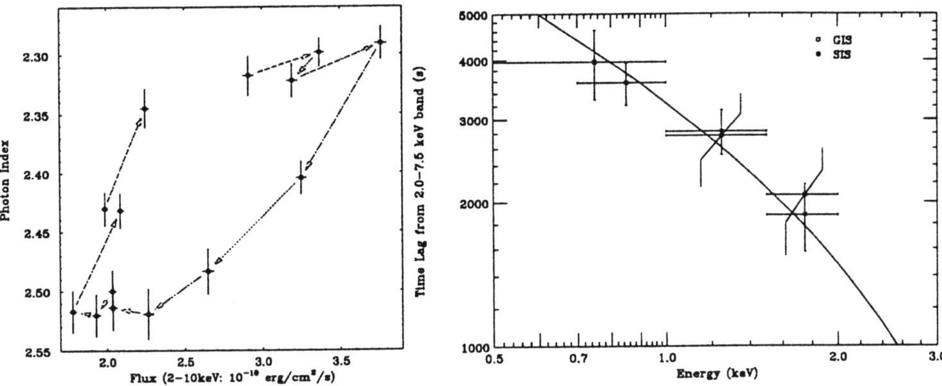

Figure 5. Right: The evolution of the X–ray spectrum of Mkn 421 as a function of the X–ray flux. The model is a power law with free absorption. Left: Time lag of photons from Mkn 421 of various X–ray energies against the 2 - 7.5 keV band photons , calculated using the discrete correlation function (DCF) with data obtained in 1994 observation, The solid line corresponds to a fit with $t_{sync}(E) - t_{sync}(5\text{keV}) = 1.2 \times 10^3\ B^{-3/2}\ \delta^{-1/2}$ $(E_{keV}^{-1/2} - (5)^{-1/2})$, which, for $\delta = 5$, yields $B \simeq 0.2$ Gauss, and $t_{sync}(1\text{keV}) \simeq 6000$ sec.

around 1 eV (Figure 2 of Macomb et al. 1995). However, $\gamma \sim 10^6$ put the kinematical limit at the upscattered photon energy at $E_c \sim 1$ TeV. These simple estimates can be refined only by introducing a detailed model of the jet, including its dynamics as well as its surroundings.

Very recently, the *Whipple* group reported an extraordinary outburst of γ–ray activity (Gaidos et al. 1996). In two hours of observations on 7 May 1996, a signal was detected at the 56σ level with a burst doubling time of one hour. The size of the emission region inferred from the observation is at most $\sim 10^{14}$ cm for the beaming factor of $\delta = 10$, suggesting a very compact region for the TeV γ–rays.

5. Summary

The non-thermal emission from blazars, observed from radio to GeV/TeV γ–rays, is thought to be the result of radiation of very energetic particles via both synchrotron and Compton processes. While the overall spectra of all blazars over this wide range of energies appear similar from one object to another, the X–ray regime is a very important band, as it is where the the emission due to both processes overlap: it is either the high energy end of the synchrotron emission or the lowest observable energy end of the Comptonized spectrum. The *ASCA* data – and, specifically, the spectral variability detected with its sensitive instruments, in the context of the multiband analysis – allows an investigation of the details of the emission and an estimate of the physical parameters, such as B and γ_{el}. However, the question of particle acceleration in jets still remains open, with the hope that the future missions – such as the upcoming X-ray satellite *ASTRO-E* – will further advance our knowledge of the details of physical processes in these extreme, enigmatic objects.

References

Angel, R., & Stockman, H. S. 1980, ARAA, 18, 321
Blandford, R. D., & Konigl, A. 1979, ApJ, 232, 34
Blandford, R. D., & Levinson, A. 1995, ApJ, 441, 79
Buckley, J. H., et al. 1996, ApJ, 472, L9
Collmar, W., et al. 1997, in preparation
Dermer, C. D., Schlickeiser, R., & Mastichiadis, A. 1992. A & A, 256. L27
Edelson, R. A. & Krolik, J. H. 1988, ApJ, 333, 646
Fichtel, C. E., et al. 1994, ApJS, 94, 551
Dondi, L., & Ghisellini, G. 1996, MNRAS, 273, 583
Edelson, R., et al. 1995, ApJ, 438, 120
Gaidos, J. A., et al. 1996, Nature, 383, 319
Hoyle, F., Burbidge, G., & Sargent, W. 1966, Nature, 209, 751
Kerrick, A. D., et al. 1995, ApJ, 438, L59
Kniffen, D., et al. 1993, 411, 133
Konigl, A. 1981, ApJ, 243, 700
Kubo, H., 1997, Ph.D.Thesis, University of Tokyo
Kubo, H. et al., 1997, in preparation
Lind, K. R., & Blandford, R. E. 1985, ApJ, 295, 358
Macomb, D., et al. 1996, ApJ, 449, L99 (erratum 459, L111 [1996])
Makino, F., Fink, H. H., & Clavel J. 1992, in Frontiers of X–ray Astronomy, eds. Y. Tanaka & K. Koyama (Tokyo: Universal Academy Press), 543
Marscher, A. & Gear, W. 1985, ApJ, 198, 114
Maraschi, L., Ghisellini, G., & Celotti, A. 1994, in IAU Symp. 159, p.233
Mattox, J. R., et al. 1993, ApJ, 410, 609
Punch, M., et al. 1992, Nature, 358, 477
Quinn, J., et al. 1996, ApJ, 456, L83
Rees, M. 1967, MNRAS, 137, 429
Sambruna, R. M., Maraschi, L., & Urry, C. M. 1996, 463. 444
Sikora, M., Madejski, G., Moderski, R., & Poutanen, J. 1997. ApJ, in press.
Sikora, M., Begelman, M. C., & Rees. M. J. 1994, ApJ, 421, 153
Takahara, F. et al. 1995, in Towards a Major Atmospheric Cerenkov Detector III, ed. T. Kifune (Tokyo: Universal Academy Press), 131
Takahashi, T., et al. 1996a, ApJ, 470, L89
Takahashi, T., et al. 1996b, Mem. Soc. Astron. Ital., 67, 533
Tashiro, M. Makishima, K., Ohashi, T., Inda-Koide, M. Yamashita. A., Mihara, T., & Kohmura, Y. 1995, PASJ, 47, 131
Tashiro, M. 1992, Ph.D. Thesis, University of Tokyo
Vermeulen, R. C., & Cohen, M. H. 1994, ApJ, 430, 467
von Montigny, C., et al. 1995, ApJ, 440, 525
Wagner, S. J., et al. 1997, ApJ, submitted

II. Origin of Cosmic Plasma Dynamics

MAGNETIC RECONNECTION AND DYNAMOS IN LABORATORY PLASMAS

P. K. BROWNING
Dept of Physics, UMIST,
P.O. Box 88, Manchester M60 1QD, U.K.

1. Introduction

Magnetic reconnection is a process of major importance in astrophysics. It is often invoked in the solar corona, but also in the Earth's magnetosphere and in many other contexts. One of the achievements of YOHKOH has been to provide observational evidence for magnetic reconnection in the solar corona (e.g. Tsuneta et al, 1992). As well as large scale reconnections which may reconfigure the global field, many theories of coronal heating propose that coronal plasma is heated by a superposition of small localised reconnection events (which may be identified with nanoflares; Parker, 1988). One way to quantify the heating by many such small scale events is to use the idea of relaxation theory, first developed by Taylor (1974) to explain the presence of reversed toroidal field in Reverse Field Pinch devices, whereby a plasma tends to minimise its magnetic energy whilst conserving its global magnetic helicity (Heyvaerts and Priest, 1984; Browning, 1988, 1991; Vekstein et al, 1993). This hypothesis has recently been strongly supported by evidence from numerical simulations (Kusano et al, 1994). Helicity conservation has also proved useful in modelling filament eruptions (Rust and Kumar, 1996) and other astrophysical phenomena such as galactic jets (Vekstein et al, 1994).

The magnetic fields in stars such as the Sun, as well as the magnetic field of the galaxy itself, are thought to be generated by dynamos. This is a process whereby plasma motions generate magnetic fields. The solar cycle, whose manifestation in the corona has now been well observed by YOHKOH, is a consequence of the non linear nature of the dynamo process. Dynamos have more recently been invoked in laboratory plasmas, to explain phenomena such as the generation and maintenance of reversed

T. Watanabe et al. (eds.), Observational Plasma Astrophysics: Five Years of Yohkoh and Beyond, 73–82.

toroidal field in a Reverse Field Pinch (RFP) device (Gimblett and Watkins, 1975). A dynamo process has often been postulated as the dynamic mechanism by which a relaxed state is attained, and results from MHD simulations support this (Ortolani and Schnack, 1993). Magnetic reconnection is intrinsically involved in the dynamo (Rusbridge, 1991).

Magnetic reconnection, relaxation and dynamos are thus inter-related phenomena. All naturally involve activity at small scales which are difficult to resolve with solar or other astrophysical observations; furthermore, it is unlikely that the solar dynamo, which occurs in the solar interior, could ever be directly observed. Therefore, results from laboratory plasmas, where direct measurement of relevant quantities is possible, should be of great interest to astrophysicists. Such measurements may both provide a test of established models and suggest novel theoretical ideas. This paper outlines some results from the SPHEX spheromak which elucidate the physics of the processes of relaxation, reconnection and the MHD dynamo.

2. SPHEX and Spheromaks

Conceptually, a spheromak is a magnetically confined plasma contained within a topologically spherical vessel. Indirect means must be used to create such a configuration, since no conductors may link the plasma to generate magnetic fields or currents (as in a conventional tokamak, for example). For a comprehensive review of spheromaks, see Jarboe (1994). The SPHEX device at UMIST (Rusbridge et al, 1997), shown in Figure 1, uses gun injection in a coaxial Marshall gun as a formation scheme. A ring of plasma, with an azimuthal magnetic field due to the gun current, is created between two concentric cylindrical electrodes and ejected by the Lorentz force towards the gun muzzle. A solenoid inside the inner electrode generates an approximately radial field at the mouth of the gun, which is dragged out by the emerging plasma ring, carrying with it the azimuthal (toroidal) field, and reconnects to form closed (poloidal) flux. The ring expands to fill a conducting shell, the flux conserver, where the plasma is confined by the magnetic fields originating from its own internal currents and eddy currents in the walls. The magnetic field has both toroidal and poloidal components, of similar magnitudes.

Typically, a gun current of 60 kA is supplied in SPHEX with a solenoid flux of 2.8 mWb, with a pulse length of about 1 ms. The flux conserver is made of copper and has a volume of $0.25\,\mathrm{m}^{-3}$. The plasma has a typical electron temperature of 30 eV and density of $8 \times 10^{19}\,\mathrm{m}^{-3}$. The magnetic fields are of the order 0.1 T and the plasma carries a net toroidal current of about 60 kA. The total poloidal flux is about a factor of 5 times the solenoid flux. This so-called flux amplification means that the poloidal field

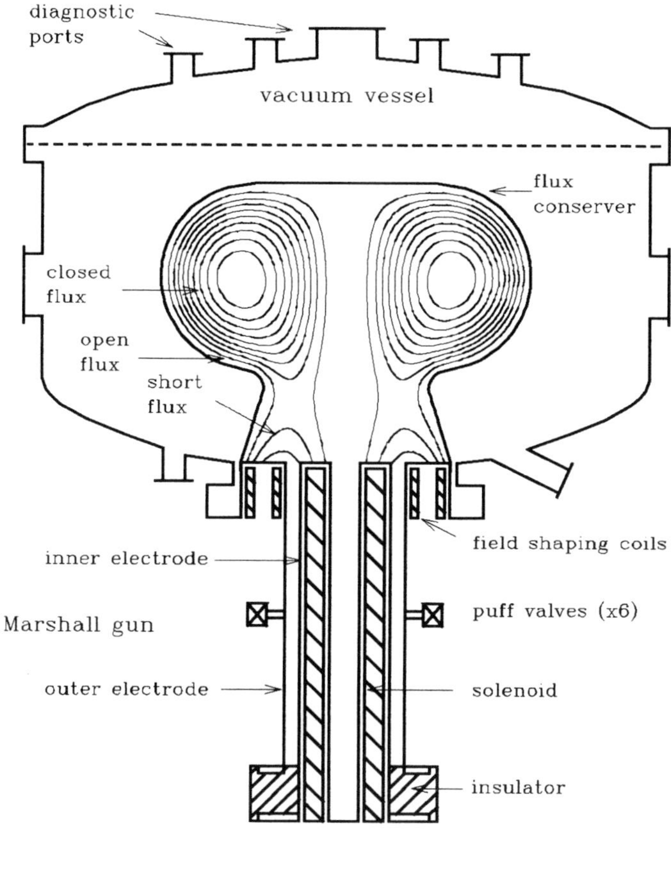

Figure 1. **A diagram of the SPHEX spheromak showing model magnetic flux surfaces from a force free equilibrium code.**

cannot originate solely due to two-dimensional reconnection of the emerging solenoid field lines. It is thought that a process of flux conversion must be involved, whereby the toroidal flux is asymmetrically distorted and then reconnects into poloidal flux. The structure (Fig 1) of the time-averaged magnetic field consists of open field lines which are rooted in the gun and closed poloidal flux surfaces. Gun injected spheromaks such as SPHEX can operate in a driven mode, whereby the plasma and its confining magnetic field are sustained so long as a voltage is applied at the gun electrodes. The pulse duration in SPHEX is several resistive decay times. A major question is to explain how a toroidal plasma current on the closed flux may be maintained by supplying a radial voltage on the open flux.

Early studies on SPHEX identified two regions of plasma, on the basis of measurements of the time-averaged electrostatic floating potential (Browning et al, 1992). These are: the central column, with high mean electric field, located along the geometric axis and directly connected electrically to the gun; surrounded by the toroidal annulus with weak mean electric fields. A clue towards understanding the sustainment of the annulus was supplied by measuring the net Poynting flux due to coherent fluctuations (at around 20 kHz, with toroidal mode number $n = 1$) in the electric and magnetic fields. It was shown that the $n = 1$ mode provides a net power flow from column to annulus, sufficient in magnitude to balance Ohmic dissipation in the annulus.

3. Relaxation and Reconnection in Spheromaks

Magnetic reconnection is essential to the formation of the spheromak configuration, both as the stretched poloidal field lines reconnect into a closed plasmoid and as the toroidal flux is asymmetrically distorted and reconnects. The former process is strongly reminiscent of processes believed to occur in the solar atmosphere, for example, during prominence eruptions (e.g. Priest and Forbes, 1990). Reconnection must also occur during sustainment, since the gun continually supplies new toroidal flux to the column which is somehow converted by reconnection to poloidal flux to replenish the annulus. This flux conversion is an essential feature of global astrophysical dynamos; however, in the spheromak we have an external source of toroidal flux whereas in a self sustaining dynamo, a process of conversion from poloidal to toroidal flux is also required.

The dynamical processes are extremely complex, but a simple explanation for the creation of the spheromak configuration is provided by relaxation theory (Taylor, 1974). It is hypothesised that a highly conducting plasma subject to small scale reconnections relaxes towards a state of lowest possible magnetic energy whilst conserving its global helicity. Helicity is defined as

$$K = \int_V \mathbf{A} \cdot \mathbf{B} dV \tag{1}$$

where $\mathbf{A}$ is the vector potential ($\nabla \times \mathbf{A} = \mathbf{B}$). This is a topological property of the magnetic field, representing the interlinkage of flux; for two interlinked flux tubes with fluxes φ_1 and φ_2, the helicity is

$$K = 2\varphi_1\varphi_2. \tag{2}$$

In a perfectly conducting plasma, the helicity of every flux tube is conserved, but magnetic reconnection can transfer helicity between flux tubes so that only global helicity survives as an invariant. The minimum energy or relaxed state is described by

$$\nabla \times \mathbf{B} = \mu \mathbf{B} \tag{3}$$

where $\mu = \mu_0 \mathbf{j} \cdot \mathbf{B}/B^2$ is spatially constant.

In an ideal spheromak, with no flux crossing the boundary, (3) is an eigenvalue problem with solutions only for discrete eigenvalues μ_i dependent on the container geometry. In a real or "flux cored" spheromak, in which some flux is line-tied to conducting boundaries, there are solutions for all values of μ but with resonances at the eigenvalues where the field magnitude becomes infinite (Taylor, 1986). The first eigenvalue of the SPHEX flux conserver is $\mu_1 \cong 11.1\,\mathrm{m}^{-1}$. A further consequence of flux linking the electrodes is that helicity is no longer simply conserved, but can be injected if a voltage is applied on the open flux. This leads to a helicity balance of the form

$$\frac{dK}{dt} = 2V_{gun}\psi_{sol} - \int_V 2\eta \mathbf{j} \cdot \mathbf{B} dV. \tag{4}$$

The first term on the right hand side of (4) represents helicity injection by the gun voltage V_{gun} on the open (solenoid) flux ψ_{sol}, while the second is Ohmic dissipation of helicity (which is negligible over relaxation time scales but not over the longer times of a sustained pulse). This shows how the spheromak may be maintained in a steady state ($dK/dt = 0$) by electrostatic helicity injection balancing Ohmic decay, giving a method of DC current drive which is a very desirable process for the fusion programme. Physically, (4) may be understood by noting from Faraday's Law that the gun voltage generates new toroidal flux V_{gun} each second, which interlinks poloidal flux ψ_{sol}; in accordance with (2), this supplies new helicity $K = 2V_{gun}\psi_{sol}$ each second. The effectiveness of this mechanism requires a relaxation process to redistribute the current which is driven directly on the open flux to the annulus - equivalently, to convert the injected toroidal flux into poloidal flux. Hence, magnetic reconnection is required.

A consequence of helicity injection is that a fully relaxed state is never attained. Rather, there is a balance between relaxation (tending to flatten the μ profile), external driving (tending to increase μ on the open flux) and Ohmic dissipation (tending to lower μ). Thus, the equilibrium field is a nonlinear force-free state

$$\mathbf{j} \times \mathbf{B} = 0 \quad \Longrightarrow \quad \nabla \times \mathbf{B} = \mu(\mathbf{r})\mathbf{B} \tag{5}$$

with a spatially varying μ profile; in the driven phase, it is expected that μ should be high on the open flux and lower on the closed flux. SPHEX equilibria have been modelled by solving the force free Grad Shafranov equation with μ depending linearly on the flux function (Kitson and Browning, 1990). In a flux amplified spheromak, the μ profile spans the linear eigenvalue μ_1.

A major problem is to predict self consistently the μ profile that should arise when there is an external input of energy and helicity. One resolution of this is to develop a modification to relaxation theory, taking account of the extra constraints imposed by the driving (French et al, 1994). It is assumed that the spheromak field consists of open flux, on which the gun current flows and therefore the value of μ is constrained to be

$$\mu = \mu_{gun} = \mu_0 \frac{I_{gun}}{\psi_{gun}}, \tag{6}$$

and closed flux, where the current profile is determined by relaxation. The total helicity (on open and closed flux) is fixed. The minimum energy state is not (in general) a constant μ field: if it were, the value of μ would have to be μ_{gun}, but this is unphysical (since this exceeds the first eigenvalue μ_1) and in any case the helicity corresponding to this μ value would be unlikely to equal the actual helicity content given from integrating (4). Thus, a family of μ profiles is assumed on the closed flux, and the magnetic field with lowest total energy and fixed helicity is then determined by a numerical minimisation. The resulting μ profile is found to be non monotonic, with a deep minimum at the edge of the closed flux compensating for the imposed high value of μ on the open flux. This agrees very well with measured profiles obtained using a specially designed insertable probe (Martin et al, 1993).

4. The Relaxation Mechanism and the Dynamo

Relaxation theory does not say anything about the dynamic mechanism which leads to a relaxed state. Numerical MHD simulations of RFPs suggest that fluctuations in magnetic field and velocity can cause relaxation (Ortolani and Schnack, 1993). An apparent paradox (Cowling, 1934) arises on consideration of a closed field line in an axisymmetric plasma, such as the magnetic axis of a spheromak or the reversal surface of a RFP. For integrating the MHD Ohm's Law

$$\mathbf{E} + \mathbf{v} \times \mathbf{B} = \eta \mathbf{j} \tag{7}$$

around the closed circular field line of radius r ($\mathbf{B} \parallel d\mathbf{l}$; $\mathbf{B} = B\hat{\theta}$) leads to

$$2\pi r E_\theta = 2\pi r \eta j_\theta.$$

But the left hand side vanishes for steady fields by Faraday's law whereas the right hand side is non-zero. This paradox may be resolved by supposing that fluctuations in magnetic field and velocity give an effective mean electric field, with Ohm's law for the mean fields given by

$$\mathbf{E} + \langle \tilde{\mathbf{v}} \times \tilde{\mathbf{b}} \rangle = \eta \mathbf{j} \tag{8}$$

where $\sim$ denotes a fluctuating quantity and $\langle\rangle$ denote an average. The dynamo is

$$\mathbf{E}_d = \langle \tilde{\mathbf{v}} \times \tilde{\mathbf{b}} \rangle. \tag{9}$$

The MHD dynamo has been measured in SPHEX by first expressing (9) in terms of fluctuations in components of the electric and magnetic fields, then measuring these simultaneously with an insertable probe consisting of an array of (almost) co-spatial Langmuir probe tips and magnetic coils. The results show a strong "antidynamo" along the geometric axis, $E_{d\|} = -425 \pm 20\,\mathrm{V/m}$, which is consistent with a mean field Ohm's Law $E_{d\|} + E_{\|} = \eta j_{\|}$, given $E_{\|} = 480 \pm 30\,\mathrm{V/m}$ and $\eta j_{\|} = 55 \pm 40\,\mathrm{V/m}$. Likewise, the dynamo field at the magnetic axis of about $10\,\mathrm{V/m}$ is consistent with the estimated $\eta j_{\|} \cong 9\,\mathrm{V/m}$, since the mean electric field necessarily vanishes there. It is further shown that the dynamo in the column is mainly associated with the $n = 1$ mode, whereas that in the annulus is due to broadband turbulence. However, the MHD dynamo does not fully explain the current drive. For at the column-annulus boundary, it is negative. The interpretation is that an additional (presumably non-MHD) process, whose nature is as yet uncertain, drives current at the edge of the annulus. The MHD dynamo then redistributes this current throughout the annulus, in particular supporting current at the magnetic axis. Further measurements of the MHD dynamo and a related process arising from the Hall effect, involving fluctuations in electron pressure, have been carried out in RFPs (Ji et al, 1995).

A better understanding of the process by which the open flux and closed flux regions are coupled is gained by investigating the structure of the central column (Duck et al, 1997). The coherent $n = 1$ perturbations are assumed to be rotating spatial structure. By mapping a time sequence of measurements at a point to the phase of the mode at a reference location, a spatial map in three dimensions (r, θ, z) is obtained, where θ is identified with the phase of the mode. This has been done for the floating electrostatic potential ϕ (Figure 2). This shows a strong helical structure winding around the geometric axis. The helical column has a width of about 9 cm and its axis is displaced from the geometric axis by about 7 cm (at the equatorial plane); the flux within the column is equal to the open flux from the gun. The whole helical structure is rotating toroidally at 20 kHz. We identify the helical column as arising from a current driven kink instability of the open flux tube, along which most of the gun current flows, which saturates at finite amplitude. There is a non-linear coupling, as yet not well understood, between column and annulus, whereby the rotating column drives current in the annulus and the annulus causes the column kink to saturate.

One possibility for the process by which current is driven at the annulus column boundary is kinetic ion drive (Rusbridge et al, 1997). It is noticed

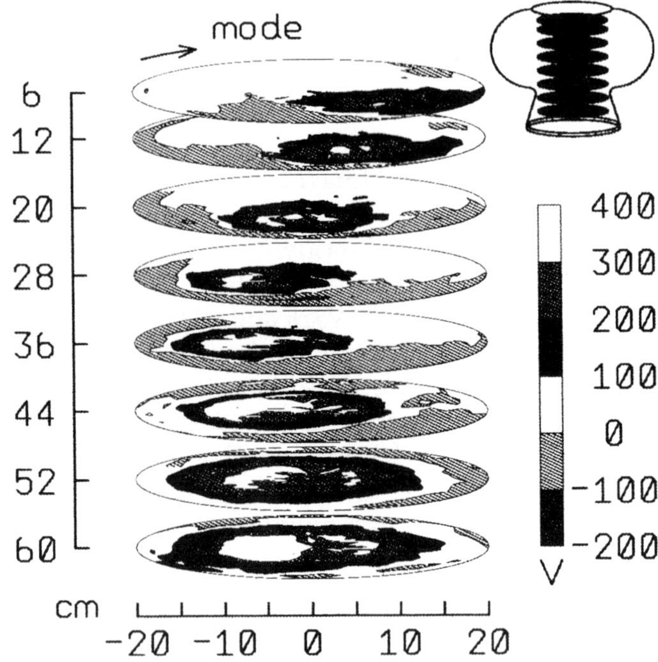

Figure 2. The helical structure of the central column, shown through contour maps of the floating electrostatic potential at successive slices along the axis.

that there is a population of fast ions with energies of several hundred eV, which appear to originate in the central column in the driven phase (Gibson et al, 1995). An observed large loss of gun current to the flux conserver wall, which requires current to cross the magnetic flux surfaces, may be due to the large Larmor radius fast ions leaving the helical field lines. It is suggested that current drive in the annulus arises as rotating ions from the column cross into the annulus.

5. Summary and Discussion

We have seen that magnetic reconnection is crucial to the formation and sustainment of the magnetic field configuration in a spheromak, such as the SPHEX device at UMIST. The field resulting from multiple reconnections can be modelled using relaxation theory. The spheromak can be sustained in a steady state by applying a voltage across the open flux which intersects the gun electrodes, a process known as helicity injection. A toroidal plasma current on the closed flux is indirectly maintained in this way. The resulting magnetic equilibrium is determined by a balance between the external driving, which inputs energy to the system, and small scale activity which dissipates magnetic energy and moves towards a minimum energy state. The relaxation process, by which current is transferred from open to closed flux, is not fully understood, but a key element is the MHD dynamo, which was measured directly for the first time in SPHEX. It has been shown that the open flux tube in the driven phase is a rotating helix, resulting from a saturated current-driven instability. The rotation of this structure within the annulus is an important part of the relaxation process.

The physics involved has many parallels in solar physics. In formation, an emerging plasmoid drags out the surrounding open magnetic field lines which reconnect behind it, a process very reminiscent of an erupting prominence or coronal mass ejection. Helical filaments are observed in the solar atmosphere, which are suggested to be non linearly kinked flux tubes (e.g. Rust, 1994). The existence of a stable helical flux tube in SPHEX lends support to this idea. Likewise, compact flares have been modelled as arising from the non linear kink instability of a flux tube.

Relaxation theory may be used to explain heating of the corona, by postulating that the heating is due to dissipation of free magnetic energy, supplied by photospheric footpoint motions, by small scale reconnections. The reconnections conserve magnetic helicity and thus the dissipation process is actually a relaxation to a minimum energy (constant μ) field (Heyvaerts and Priest, 1984; Browning et al, 1986; Browning, 1988; Vekstein et al, 1993; Kusano et al, 1994; Lothian and Browning, 1995). The results from SPHEX throw some light on the dynamics whereby this relaxation and energy dissipation may be achieved. An important element is the identification of the dynamo effect arising both from coherent MHD fluctuations and from turbulence. An interesting feature in the spheromak is that current is driven indirectly on certain field lines (the closed flux) when different field lines are driven. One may speculate that a similar process may arise in the corona, where currents can be driven by motions of footpoints of the closed magnetic field lines in loops, so that currents may be indirectly generated on adjacent open field. The fast ions seen in SPHEX, which may originate in magnetic reconnection, may also have counterparts in the corona.

It is evident that the physics of plasmas in the solar corona and in laboratory devices such as the spheromak has many similarities, and that a comparison of the two may further our understanding of many fundamental physical processes.

A. Acknowledgements

The results described are attributable to past and present members of SPHEX team, in particular A. al-Karkhy, G. Cunningham, R. Duck, P.

French, S. Gee, K. Gibson, J. Hugill, R. Martin, M. Rusbridge, M. Willett, S. Woodruff. The author is very grateful to the Organising Committee of the YOHKOH Anniversary Symposium for providing financial assistance which allowed this work to be presented at the meeting.

References

1. Boozer, A. H.; *J. Plas. Phys.* **35**, 133 (1986)
2. Browning, P. K., Sakurai, T. and Priest, E. R.; *Astron Astrophys.* (1986)
3. Browning, P. K.; *Plas. Phys. Cont. Fus.* (1988)
4. Browning, P. K.; *Plas. Phys. Cont. Fus.* **33**, 539 (1991)
5. Browning, P. K., Cunningham, G., Gee, S. J., Gibson, K. J., al-Karkhy, A., Kitson, D. A., Martin, R. and Rusbridge, M. G: *Phys. Rev. Lett.* **68**, 1718 (1992)
6. Duck, R. C., Browning, P. K., Cunningham, G. C., Gee, S. J., al-Karkhy, A., Martin, R. and Rusbridge, M. G.; *Plas. Phys. Cont. Fus.*, in press, (1997)
7. Cowling, T. G.; *Mon. Not. Roy Astr. Soc.* **94**, 39 (1934)
8. French, P., Browning, P. K. and Jarboe, T. R.; *Proc 20th Eur. Conf. Cont. Fus. Plas. Phys.*, European Physical Society, Lisbon(1993)
9. Gibson, K. G., Gee, S. J. and Rusbridge, M. G.; *Plas. Phys. Cont. Fus.* **37**, 32 (1995)
10. Gimblett, C. G. and Watkins, M. L. *Proc 7th Eur. Conf. Cont. Fus. Plas. Phys.*, European Physical Society, Lausanne, Vol.1, p. 103 (1975)
11. Jarboe, T. R.; *Plas. Phys. Cont. Fus.* **36**, 945 (1994)
12. Ji, H., Yagi, Y., Hattori, K., Almagri, A. F., Prager, S. C., Hirano,Y., Sarff, J. S., Shimada, T., Maejima, Y. and Hayase, K.; *Phys. Rev. Lett.* **75**, 1086 (1995)
13. al-Karkhy, A., Browning, P. K., Cunningham, G. C., Gee, S. J. and Rusbridge, M. G.; *Phys. Rev. Lett.*
14. Kitson, D. A. and Browning, P. K.; *Plas. Phys. Cont. Fus.* **32**, 1265 (1990a)
15. Kitson, D. A. and Browning, P. K.; *Proc 17th Eur. Conf. Cont. Fus. Plas. Phys.*, European Physical Society, Amsterdam, (1990b)
16. Kusano, K., Suzuki, Y., Kubo, H., Miyoshi, T. and Mishikawa, K.; *Astrophys. J.* **433**, 361 (1994)
17. Lothian, R. M. and Browning, P. K.; *Sol. Phys.* **161**, 289 (1995)
18. Martin, R., Gee, S. J., Browning, P. K. and Rusbridge, M. G.; *Plas. Phys. Cont. Fus.* **35**, 269 (1993)
19. Ortolani, S. and Schnack, D. D.; "Magnetohydrodynamics of plasma relaxation", World Scientific (1993)
20. Parker, E. N.; *Astrophys. J.* **330**, 474 (1988)
21. Priest, E. R. and Forbes, T. G.; *Sol. Phys.* **126**, 319 (1990)
22. Rusbridge, M. G.; *Plas. Phys. Cont. Fus.* **33**, 1381 (1991)
23. Rusbridge, M. G, Gee, S. J., Browning, P. K., Cunningham, G. C., Duck, R. C., al-Karkhy, A., Martin, R. and Bradley, J. W.; *Plas. Phys. Cont. Fus.*, in press, (1997)
24. Rust, D. M. and Kumar, A. J.; *Astrophys. J.* **464**, L199 (1996)
25. Rust, D. M.; *Geophys. Res. Lett.* **21**, 241 (1994)
26. Taylor, J. B.; *Phys. Rev. Lett.* **33**, 1139 (1974)
27. Taylor, J. B.; *Rev. Mod. Phys.* **58**, 741 (1986)
28. Tsuneta, S., Takahashi, T., Acton, L. W., Bruner, M. E., Harvey, K. and Ogawara, Y.; *Pub. Astr. Soc. Jap.* **44**, L211 (1992)
29. Vekstein, G., Priest, E. R. and Steele, C. D. C.; *Astrophys. J.* **417**, 781 (1993)
30. Vekstein, G., Priest, E. R. and Steele, C. D. C.; *Astrophys. J. Sup.* **92**, 111 (1994)

A MECHANISM DRIVING SOLAR FLARES

ATTILA GRANDPIERRE
Konkoly Observatory, Budapest, Hungary

1. Extended Abstract

Solar flares are thought to be the result of magnetic reconnection. Nevertheless, a clear understanding of the generation of the conditions necessary for reconnection is practically missing, indicating a basic deficiency of the flare models and theories. Here I classify the reconnection configurations according to their possible topologies. Considering the physical mechanisms, which would be able to generate these different topologies, I present a criticism of the existing reconnection models. I show that theoretical and observational facts suggest that the magnetic topology is a simple one, in which the reconnection occurs in the loop at the loop-top from where an elongated feature outgrows towards the corona. I point out that a whole series of problems and results have arisen recently, which strengthens this picture, clearing up the most fundamental physics of the flare phenomena. The theoretical direction towards the convective flare theory - in which mass upflows drive loop expansion and generate high energy particle beams that shoot into the loop-tops from below - is able to solve all these problems, completing the present day reconnection theories at their most crucial point. show, that energetic particle beams injected upwards from below the photosphere by a shock generated by high-speed convective cells can provide a mass and energy source to supply the flares on the proper scales.

According to widespread views, the magnetic field is in itself enough to explain the generation of the flares. Nevertheless, observations strongly suggest the role of reconnection only in the later phases of the flare (see e.g. Kopp, Pneuman, 1976; Forbes, 1992). Therefore, to understand the generation of the flare, we have to clarify how the pre-flare conditions lead to flare onset and the generation of conditions for magnetic reconnection. Hudson and Ryan (1995) in their review state clearly that "the key question

T. Watanabe et al. (eds.), Observational Plasma Astrophysics: Five Years of Yohkoh and Beyond, 83–84.

remains open, namely: Does the reconnection that we are now beginning to see quite clearly really provide the flare energy from a coronal reservoir?" Tsuneta (1997) pointed out that the mechanism for the formation of the neutral sheet (X-point) has not yet been understood (see also Tsuneta, 1996) and this is currently the most important question (Tsuneta, personal communication, 1996) . I will show here that if we do not understand how the neutral sheet configuration is produced, then we do not understand the most basic aspect of the flare phenomenon. I discuss the key elements of the flare generation and attempt to present a picture, which is in more satisfactory agreement with the basic physics.

The results show that the flare reconnection topology is generated as an elongated structure developing from the loop-top. The mechanism generating the elongated vertical structure is the key element of the flare phenomena. It is shown that energetic particle beams injected from below the photosphere are able to drive the production of these elongated vertical structures. The energetic particle beams are a natural product of the sudden destruction of high-velocity convective cells which are produced by the fundamental thermonuclear instabilities in the solar core (Grandpierre, 1990, 1991, 1996). The observational possibilities of the upward injected particle beams from below the photosphere are discussed (Grandpierre, 1997).

Acknowledgement. This research is partly supported by the Hungarian Scientific Research Foundation under No. T 014 224.

References

Forbes, T. G. 1992, in Eruptive Solar Flares, eds. Z. Svestka, B. V. Jackson, M. E. Machado, Lecture Notes in Physics, Springer-Verlag, 399, 79
Grandpierre, A. 1990, Sol. Phys. 128, 3
Grandpierre, A. 1991, Mem. Soc. Astron. Ital. 62, 401
Grandpierre A. 1996 A&A 308, 199
Grandpierre, A. 1997, Astrophvs. J. (submitted)
Hudson, H. and Ryan, J. 1995, Annu. Rev. Astron. Astrophys. 33, 239
Kopp, R. A. and Pneuman, G. W. 1976, Sol. Phys. 50, 85
Tsuneta. S. et al. 1992, PASJ 44, L63
Tsuneta, S. 1993, in The Magnetic and Velocity Fields of Solar Active Regions, ASP Conference Series 146, 1993, H. Zirin, G. Ai and H. Wang (eds.), 239
Tsuneta, S. 1995, PASJ 47, 691
Tsuneta, S. 1996 June, electronic mail
Tsuneta, S. 1996, in Yohkoh Conference on Observations of Magnetic Reconnection in the Solar Atmosphere, eds. B. Bentley and J. Mariska, PASP Conference series, in press
Tsuneta, S. 1997, ApJ, April 1

ON THE ORIGIN OF HELICITY IN ACTIVE REGION MAGNETIC FIELDS

ALEXEI A. PEVTSOV AND RICHARD C. CANFIELD
Department of Physics, Montana State University
Bozeman, MT 59717-3840, U.S.A.

Extended Abstract

The magnetic helicity $\mathcal{H}$ of flux tubes can be separated into twist $\mathcal{T}$ and writhe $\mathcal{W}$ (Moffatt & Ricca 1992). If magnetic flux rises in Ω-shaped loops through the convection zone (Babcock 1961), it acquires writhe through the Coriolis force on internal longitudinal flows (Fan, Fisher, & DeLuca 1993). The tilt θ of active regions with respect to the equator is an observable manifestation of such writhe, at photospheric levels. If $\mathcal{H}$ is conserved, as we expect (Taylor 1974), rising flux should acquire equal and opposite values of $\mathcal{T}$ and $\mathcal{W}$. We expect $\mathcal{W} > 0$ ($\theta < 0$) in the Northern hemisphere, since the leading polarity of active regions tends to be closer to the equator. Hence, we expect $\mathcal{T} < 0$ there.

Using a dataset of Mees Solar Observatory vector magnetograms for about 100 active regions, we have measured the tilt θ, the separation of leading and following polarities L, and the best-fit force-free field parameter α_{best} (proportional to $\mathcal{T}/L$ in a simple model), at large scales. This dataset clearly shows two well-known phenomena, the latitude dependence of θ, known as Joy's law (Hale *et al.* 1919), and the hemispheric dependence of α_{best} (Pevtsov, Canfield & Metcalf 1995). Joy's law can be seen in Fig. 1. For regions that are within $\pm\, 6\sigma$ of Joy's law in our data, we find no relationship between θ/L and α_{best}. However, for regions that strongly disobey Joy's law, we find a clear relationship between θ/L and α_{best}, as shown in Fig 2. However, the sign is opposite the expectation of the previous paragraph. Both figures imply that the twist seen in active region magnetic fields is not a consequence of the rise of flux through the convection zone. We conclude that this twist is the signature of a deep-seated phenomenon, presumably that of the solar dynamo itself.

T. Watanabe et al. (eds.), Observational Plasma Astrophysics: Five Years of Yohkoh and Beyond, 85–86.

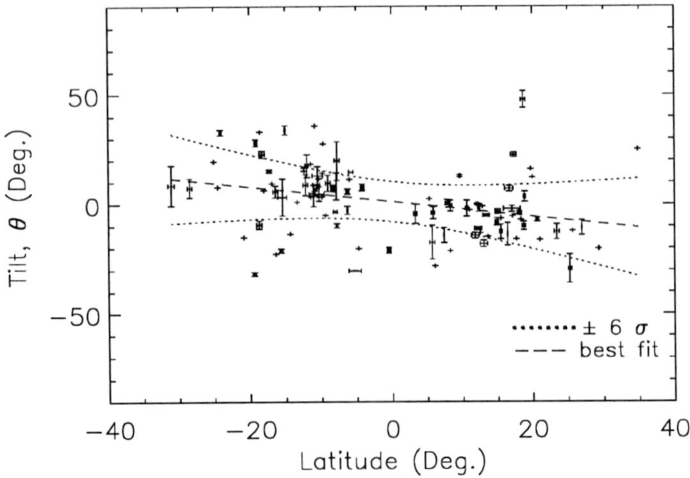

Figure 1. The dependence of the active region tilt on latitude. The dashed line is the best least squares fit to the full dataset (Joy's law). The dotted lines show the $\pm\ 6\sigma$ error band of the fit.

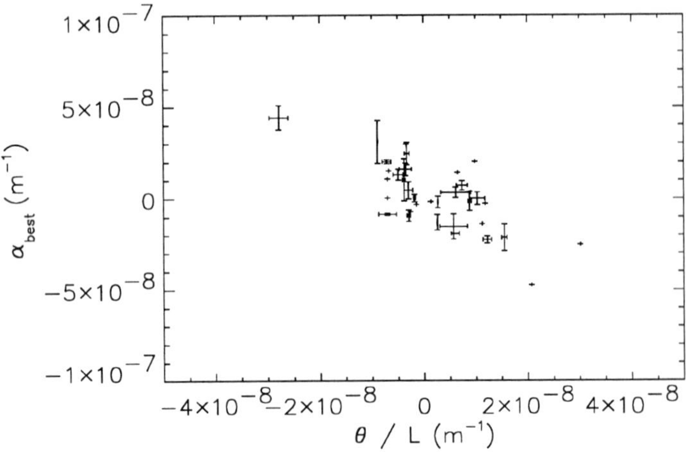

Figure 2. The dependence of α_{best} on the ratio of active region tilt to characteristic length, for the the data points that are beyond $\pm\ 6\sigma$ of the best fit in Figure 1..

References

Babcock, H. 1961, ApJ, 133, 572.
Fan, Y., Fisher, G. H., & DeLuca, E. E. 1993, ApJ, 405, 390.
Hale, G. E., Ellerman, F., Nicholson, S. B., & Joy, A. H. 1919, ApJ, 49, 153.
Moffatt, H.K., & Ricca, R.L., 1992, Proc. R. Soc. Lond. A, 439, 411.
Pevtsov, A. A., Canfield, R. C., and Metcalf, T. R. 1995, ApJ, 440, L109.
Taylor, J. B., 1974, Phys. Rev. Lett., 33, 1139.

SXR CORONAL POLAR JETS AND RECURRENT FLASHES

S. KOUTCHMY
Institut d'Astrophysique–CNRS
98 Bis Bd Arago, F–75014 Paris (France)

H. HARA, K. SHIBATA AND Y. SUEMATSU
National Astronomical Observatory
Mitaka, 181 Tokyo (Japan)

AND

K. REARDON
Osservatorio Astronomico di Capodimonte
via Moiariello 16, I–80131 Napoli (Italy)

Abstract. We give evidence for the existence of a new type of soft X-ray (SXR) brightening event that we call coronal polar jets and flashes. The phenomenon was observed on a long-exposure time series taken with the SXT of Yohkoh in a north polar coronal hole, near the activity minimum. Events last as little as 1.5 minutes and the detected SXR photon fluxes span a broad range, from small scale brightenings, at the minimum detectable level near 10^{24} erg, to more smeared multiple pixel brightenings still an order of magnitude fainter than the known small SXR bright points and/or transient brightenings that have been observed. The typical observed occurrence rate of flashes is one event per super-granule per 10 mn period. At least some of the coronal flashes are recurrent and some of them have associated SXR jets. It is not clear what, if any, is their optical counterpart. Polar jetlets are observed extending radially without showing a significant radial gradient up to the 50 Mm heights. Their occurence rate is typically one order of magnitude lower than flashes and their lifetime seems to be longer when they are broader. The average upward proper motion observed in polar jets is 200 to 400 km s^{-1}.

Keywords: Solar corona; MHD; SXR; jets; bright points; brightening events; reconnections; magnetic dissipation; nano-flare.

T. Watanabe et al. (eds.), Observational Plasma Astrophysics: Five Years of Yohkoh and Beyond, 87–94.

1. Introduction

To understand coronal heating, we need a mechanism that maintains the high temperature of the plasma. This mechanism must be related to the transfer of momentum that produces the solar wind. This is especially true in coronal holes (CH), where the energy transferred to the corona is converted primarily into the acceleration of the fast solar wind observed emanating from these regions of open field lines.

Small-scale impulsive mechanisms were proposed to provide the necessary energy input through magnetic-dissipation processes in explosive events (e.g. Dere et al. 1989). Frequent UV brightenings (Porter et al. 1984) and even more commonly, coronal bright points (Golub, 1980) and transient brightenings (Shimizu et al. 1994) are observed in coronal regions. Outside of CH, in more active regions of the Sun where loops and flares are recorded, SXR coronal jets have been discovered, see Shibata et al. (1992). Much further down the energy scale, nano-flares were proposed by Parker (1988) as a mechanism to input energy into the quiet corona without waves. They would have to be rather ubiquitous and are presumed to be produced by fast magnetic reconnections occurring in tangential discontinuities (TD) among the coronal field lines. These discontinuities are difficult to identify in the very inner corona, but are observed in the white-light intermediate corona at larger scale (Koutchmy, 1972) and in the solar wind (Neugebauer et al. 1984); they seem not to dissipate and produce heating. If the partition of energy in these small-scale events can be assumed to be similar to that of small flares, it is likely that these tiny reconnections would primarily dissipate their energy in the acceleration of high energy particles, and would generate only a small radiative signature amenable to observations. Also, fast, small-scale events may provide information on scales that are at the limits of the current observational capabilities.

A bright coronal background would obscure the signature of such small events because the simple statistical distribution of photon emission from the background would be greater than the signature from the small events. Deep SXR observations performed on Yohkoh near solar minimum provides a good opportunity to observe the tiniest events that can be found in the region of the darkest background of polar CH. The more extended events possibly connected to the well known from total eclipse observations W-L polar plumes are also better recorded at that time.

2. Observations

Two runs were made on Aug. 25 and Sept. 2, 1995 using the soft X-ray telescope (SXT) aboard Yohkoh (Tsuneta et al. 1991) in order to observe the signature of white-light (W-L) coronal spikes of the inner corona that were detected during solar total eclipses (Koutchmy and Stellmacher, 1978) and that could be similar to the recently discovered SXR jets (Shibata et al. 1994) associated with flares in X-ray bright points. Later, Shimojo et al. (1996), in their statistical analysis of jets found that 11% of jets occurred in XBP of CH. Another possibility was to look at the faintest transient brightenings (TB), similar to those seen by Shimizu et al. (1994).

Two sets of data were obtained in a limited region near the heliocentric north pole, in the 'partial-frame mode' with a frame size of $256x64$ pixels at the highest possible cadence. Data were obtained on Aug. 25, 1995 (92 images with 15 second exposures during 2 orbits) and on Sept. 2, 1995 (202 images of 30 s exposures during 5 orbits). For each run, several continuous time series were taken at approximately 35 s intervals with the AlMg or 'thick filter' that has a peak temperature sensitivity near 3 to 5 MK (Yoshida et al. 1995). Here we concentrate on a small part of these observations and looked at the smallest events above the noise level, see Koutchmy et al. (1997) (hereafter called KHSR 97).

Using visual correlations of intensities recorded over pixels (the pixel size was 2.45 arcsec), we identified many very weak events. The small motions of the satellite were accurately taken into account using the standard Yohkoh IDL package available at ISAS. The dark current was first removed using the standard Yohkoh procedure that uses a weighted average of the two nearest dark images before and after the time the data was taken. However, this failed to correct the dark current "spike pixels" that appear in the data. These spikes are produced by damaged pixels that do not respond linearly to dark current, and thus are very difficult to correct. Instead, we isolated these pixels on the CCD by finding the large deviations in the mean image and using a mask on each image to zero each of these "bad" pixels for the entire dataset. Further, to facilitate the search, we use a negative display and a high magnification to increase the visibility of the pattern. Surprisingly, we found many apparently real events that are seen on at least 3 consecutive frames, without appreciable motion; these events are eventually seen again in the time series, 20 min or more later.

For obvious reasons, on the disk, in active regions, or in quiet regions at solar maximum, the SXR background is considerably brighter on average and such long exposure times could not have been used. These events have presumably not been reported in the literature (see, e.g., Shimizu, 1995). As an example see Figure 1 in KHSR 97 where these events that we call flashes are reported for the first time. Frames were taken 32 s apart and each exposure lasted 30 s. Light curves are also given there.

3. Analysis and results

Different types of event are seen over a short time series: a/ faint bright points (BP), usually with a lifetime of many hours and quite extended, or faint transient brightenings (TB), also extended but with a duration of only a few min; b/ faint brightening with associated jet or jetlet (elongated structures) which sometimes is quite broad and can be shifted in position angle compared to the position of the associated foot point brightening; c/ flashes occuring over 1 or more pixels. These last are the smallest signal that we can reliably measure on this set of data; let us examine their properties.
First, note that a rather continuous distribution of SXR intensities is observed from BP towards flashes and that, approximately twenty min later, approximately 40% of the flashes could still be found. Further, we note that over the very dark background of the coronal hole, the noise is due to random events most likely of instrumental origin, such as the readout noise of the CCD. The photon shot noise of the true background photons is completely negligible and we described above how the "spike pixels" were eliminated.

From our preliminary calibration based on the comparison with the bright points, it is clear that flashes correspond to data numbers (DN) of a single or of several photo-electron events per pixel. Since the SXT uses the CCD in an energy-integrating mode (Yoshida et al. 1995), it is not possible to calculate the exact number of detected photons to which a given count of data numbers corresponds. In such cases, single pixel events are easier to identify: the probability of having random noise events overlapping it is limited, and the probability of having such events on two or three consecutive frames exactly on the same pixel is very low. We conclude that when a single pixel brightening occurs consecutively on three or more frames, it must be real. Additionally, the recurrence of each event seems to be a more fundamental property of flashes. In order to evaluate the level of the faintest meaningful events observed with SXT, we compared the pattern on a full-frame SXT image (taken from the synoptic program) with a full-frame taken independently in one of the coronal channels (Fe XII) of the EIT experiment of the SOHO mission (data kindly provided by the EIT team for evaluation purpose only). We used several frames taken on May 26, 1996 at almost the same time by both instruments, with standard exposures. After sharpening the images, the EIT frames taken using a coronal filter show more small events than the SXT full-disk images taken with short exposure times. Doing a precise spatial correlation on properly scaled negative copies we convincingly see that rather faint events taken with SXT are real, because they show up on the simultanous EIT image, especially in the dark regions of the CH. It is clear that using the SXT time series made of sub-field frames of longer exposure time in an inherently darker coronal region, as those data analysed in this study, we see fainter events.

On Figure 1 we show a mosaic of selected frames in order to show at its best a large event with a jet very close to the North pole. The brightening near its foot is clearly occurring just a little bit before as confirmed by the analysis of the light curve (not shown here). More evident on Fig. 1 is the shift in position angle between the foot point TB and the jet. The upward proper motion is evident in the jet by watching the movie made of consecutive frames. The velocity of the upper part of the jet can only be estimated. We looked at several well observed jets and jetlets of narrower section and found that the average value of the velocity is 200 to 400 km s^{-1}. Another interesting point is the systematic dispersion we observed in the positioning of the jet structure with respect to the foot point brightening and also, in the shape and direction; even curved jetlets were noticed.

Let us give a more quantitative evaluation on flashes. A 30 s integration time is used to determine the typical 'density' of flashes (or occurrence rate) over a polar region to be on the order of: 1 event per $arcmin^2$ of the solar surface per 5 min interval (including the faintest flashes). The average lifetime is 3 to 5 minutes although shorter events certainly exist. This number includes the very bright flashes, which may be nothing but the faintest TBs discussed by Shimizu, 1995. Going toward the limb and beyond, it is difficult to claim any relationship with respect to cosine θ. Flashes seem to evenly cover the solar surface. Concerning jets, their variability is large and it seems difficult to give any quantitative interpretation of flux variations.

4. Discussion

We believe that these flashes are a new class of SXR transient brightenings for the following reasons:

a– The flux is typically 2 orders of magnitude smaller than the faintest TB reported by Shimizu et al (1995) and their 'density' or occurrence rate at the solar surface is 2 to 3 orders of magnitude larger.

b– Flashes also occur above the limb, although it is difficult to make a clear distinction between off-the-disk flashes and very faint jets.

c– They are significantly smaller in size than previously described events. Indeed the smallest flashes could be smaller than the pixel size ($\sim$ 1700 km at disk center), as suggested by the dispersion in our measurements of the location of the smallest flashes.

d– Their frequency distribution does approximately fit with the extrapolated value obtained from the frequency distribution graphs of TB as given in Shimizu (1995) provided the saturation effect at small values of fluxes due to the overlapping effect from the background emission is avoided.

If we want to propose a 'universal' frequency distribution for flare-like events, such as a single value power-law distribution, it is clear that no saturation effect should occur at any energy level, especially in the range of small energies which is considered here.

However, the sizes of the flash extension is definitely smaller than for a typical bi-polar magnetic region responsible for a bright point and/or a TB. It is closer to the size of a magnetic element. Their number 'density' and behavior make these events comparable to the chromospheric spicules (Suematsu et al., 1996) and to impulsive events seen in Hα above polar CH (Koutchmy and Loucif, 1991). Flashes could be due to a process proposed for nano-flares (e.g. Einaudi and Velli, 1994) or, more likely, could result from reconnection phenomena occuring at lower heights (Yokoyama and Shibata, 1995). The numerical modelling although limited in resolution, already showed the most remarkable details of the observed coronal jet phenomenon, see Shibata (1995), including the ejection of plasmoids.

Finally, polar jets and/or jetlets are also observed on the EIT sub-field time series of polar regions; Gurman (1996) in a preliminary study called them "micro-jets". In our opinion, it is now urgent to accumulate more observations in possibly other spectral and temporal ranges in order to better define these events, especially their energy distribution. This would allow the measurement of the total energy released by these small events and the determination as to their possible importance in the heating of the polar corona. Simultaneous observations taken with both SXT of Yohkoh and EIT of SOHO satellite would be interesting in this respect. White-light eclipse observations of spikes (elongated coronal density structure) and plasmoids (small closed magnetic structure) could also be helpful, as well as deep coronal Hα and He lines observations. To demonstrate whether jet– TB and/or jetlet– flash is or not the SXR counterpart of Hα spike-prominences with bright-knots at the feet is a challenge which has to be resolved if we want to understand the origin of the polar plumes.

Acknowledgements. We thank our Yohkoh colleagues for their many contributions to building and operating the SXT. The Yohkoh mission is supported by both ISAS (Japan) and NASA (USA). This investigation benefited from resources provided by NAOJ which permitted the stay of S.K. at Mitaka. Research partly supported by the NATO CRG 940291 and CNES (France). We thank Profs. P. Lorrain, G. Einaudi, Drs. J-P. Delaboudiniere, K. Dere, J. Gurman, G. DeForest, H. Hudson and L. Golub for discussions and V. Demailly for preparing the manuscript.

References

Dere K.P., Bartoe J.-D.F., Brueckner G.E., 1989, Solar Phys. **123**, 41–68
Einaudi G., Velli M., 1994, Sp. Sc. Rev. **68**, 97–102
Golub L., Krieger A.S., Silk J.K., Timothy A.F., Vaiana G.S., 1974, ApJ. **189**, L93
Gurman J., 1996, private communication
Koutchmy S., 1971, A&A **13**, 79
Koutchmy S. and Stellmacher G., 1976, Solar Phys. **49**, 253
Koutchmy S. and Loucif M., 1990, in "Mechanisms of Chromospheric and Coronal Heating", Ulmschneider, Priest, Rosner Ed. S-V, 152
Koutchmy S., Hara H., Suematsu Y. and Reardon K., 1997, A&AL, in press
Neugebauer M., Clay D.R., Goldstein B.E., Tsurutani B.T., Zwickl R.D., 1984, J. Geophys. Res. **89**, 5395
Parker E.N., 1988, ApJ **330**, 474
Porter J.G., Toomre J., Gebbie K.B., 1984, ApJ **283**, 879–886
Shibata K. et al., 1992, PASJ **44**, L173–L179
Shibata K., 1995, Adv. Sp. Res. COSPAR, **17**, 4/5, 9
Shimizu T., 1995, PASJ **47**, 251–263
Shimizu T. et al., 1995, ApJ **422**, 906–911
Shimojo M. et al., 1996, PASJ **48**, 123
Suematsu Y., Wang H., Zirin H., 1995, ApJ **450**, 411–421
Tsuneta S. et al., 1991, Solar Phys. **136**, 37–67
Yokoyama T., Shibata K., 1995, Nature **375**, 6526, 42–44
Yoshida T., Tsuneta S., Golub L., Strong K., Ogawara Y., 1995, PASJ **47**, L15–L19

Figure 1. Mosaic of five selected frames taken on Sept. 2, 1995 over the north pole region. The exposure time is 30 s and the filter used is AlMg or 'thick' filter on SXT, which has a peak temperature sensitivity covering the range 3 to 5 MK. Note the rather broad jet near the pole slightly at the left of a brightening close to the limb.

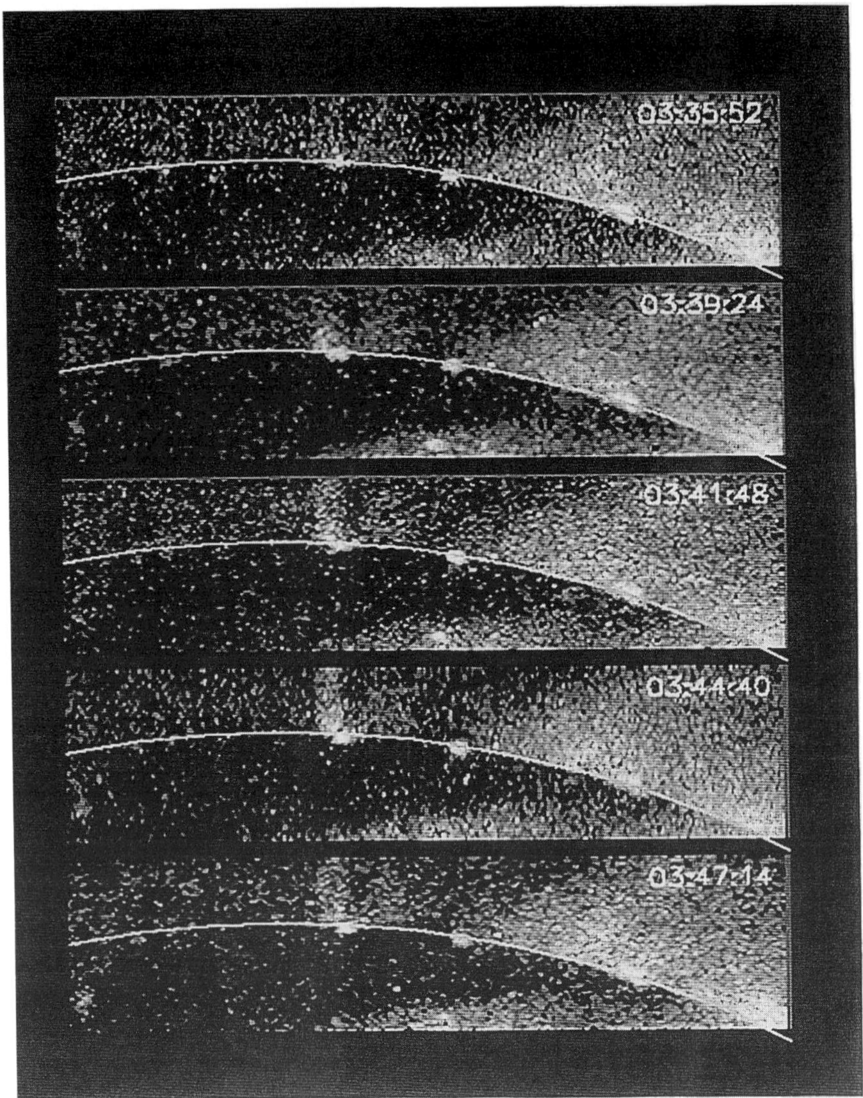
03:35:52
03:39:24
03:41:48
03:44:40
03:47:14

THE FILAMENT DISAPPEARANCE OF MAY 7, 1992

E. HIEI
Meisei University, Tokyo, Japan
Z. MOURADIAN AND I. SORU-ESCAUT
Meudon Observatory, Meudon, France
A. H. MCALLISTER
High Altitude Observatory, Boulder, USA
K. SHIBASAKI
Nobeyama Radio Observatory, Nobeyama, Japan
AND
M. OHYAMA
National Astronomical Observatory, Mitaka, Japan

The post-flare X-ray loops of the May 7, 1992 event are quite similar in shape to a shrimp's tail (thus it has been called the EBI [shrimp in Japanese] event). It was studied by Khan et al. (1993) from X-ray data obtained with SXT onboard the *Yohkoh* satellite. Recently Meudon observers recognised that they had Hα observations for the period of the EBI event. We report here our further research on the EBI event, based mainly on Hα- and radio-observations.

Fig. 1 shows Hα Lyot filtergrams. At 05:50 UT, when Hα observations began, the eastern end of the filament was already lifting. The ascending filament was located at around 55E-35E, and 20S. Another two small filaments, which were seen to the south-east of the ascending filament, were near 60E 28S, and 70E 32S, respectively. The EBI soft x-ray loops later appeared above these three filaments. If the filament had risen vertically, its trajectory, projected onto the plane perpendicular to the line of sight, would be southward because the filament was in the southern hemisphere. However, the ascending motion was in the north-eastern direction, and thus the upward motion was tilted northward of the vertical. The velocity perpendicular to the line-of-sight was slow at first, but after 06:20 the filament accelerated by about a factor of 3. The western end, however, appeared to remain anchored at the photosphere, probably because it was near an

T. Watanabe et al. (eds.), Observational Plasma Astrophysics: Five Years of Yohkoh and Beyond, 95–100.

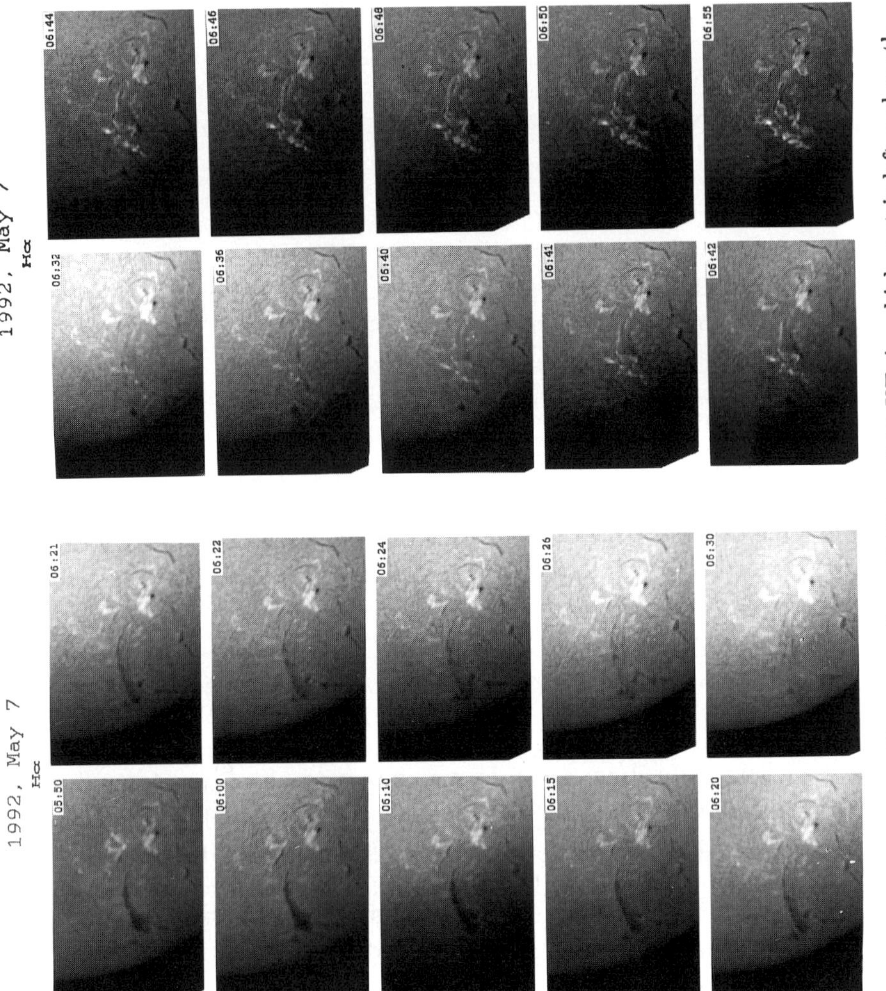

Figure 1. Hα Lyot filtergrams from 05:50 to 06:55 UT, in which east is left and north is up.

active region and the magnetic field was stronger than that at the eastern end. At 06:30 the main filament was gone, and the other two small filaments disappeared by 06:41.

Flaring bright points appeared on both sides of the disappearing filaments. The first bright point appeared at 06:30 and was seen for less than 10 min. The subsequent bright points continued to brighten and they eventualy formed two bright strands. The bright strands were dotted, and not the same as the usual smooth two-ribbon strands. The dotted bright regions corresponded to the foot points of the X-ray loops. Flaring bright strands after the disappearance of a filament are heated by conduction/penetration of high temperature plasma from X-ray loops, and not due to material falling from the lifted filament (Hyder 1967). The strands at the east side were brighter than that at the west side, probably due to the magnetic field at the east side being weaker than that on the west.

Fig. 2 shows a sequence of SXT images with the Hα filament location outlined by white crosses. The SXT image at 05:50 showed a twisted structure at the east end. Notice the small bright feature, which may be a plasmoid, which is seen above the main body. Such a bright feature was also observed in the events of January 24, 1992 (Hiei 1994) and of February 20, 1992 (Tsuneta et al. 1992). At 06:05-06:12 a bright straight structure appeared near the top of the X-ray arcade loop that Khan et al. have called the rail. The rail is just above the ascending filament. This rail structure may be related to the 'backbone' associated with the filament eruption of September 28, 1991 studied by McAllister et al. (1992). If the rail represents a part of the filament material heated to coronal temperatures, the upper edge of the filament should be affected and would be expected to fade. However, the edge of the filament was quite sharp and thus this was not the case. The rail seems to be an extention of a twisted loop. This whole structure rose, pulling upward the magnetic loops above the filament. This interaction probably contributed to the brightening of the twisted loop and the rail, but appears to have been high enough above the location of the cool filament material that it was not itself affected. In the last image at 6:30 the central portion of the rail and the distended arcade loops had vanished as the filament had risen above that level and broken loose. The extended magnetic loops around the erupting filament probably formed a vertical magnetic neutral sheet and an arcade structure would be reformed as a consequence of magnetic reconnection (Hiei et al. 1993). At 07:25, when SXT observations recommenced after the *Yohkoh* night, the bright flare loops of the Ebi structure had already appeared at the location above the three filaments.

Fig. 3 shows a plot of 2 GHz radio observations taken at the Toyokawa Observatory. This curve shows that the flux increased slowly and the event,

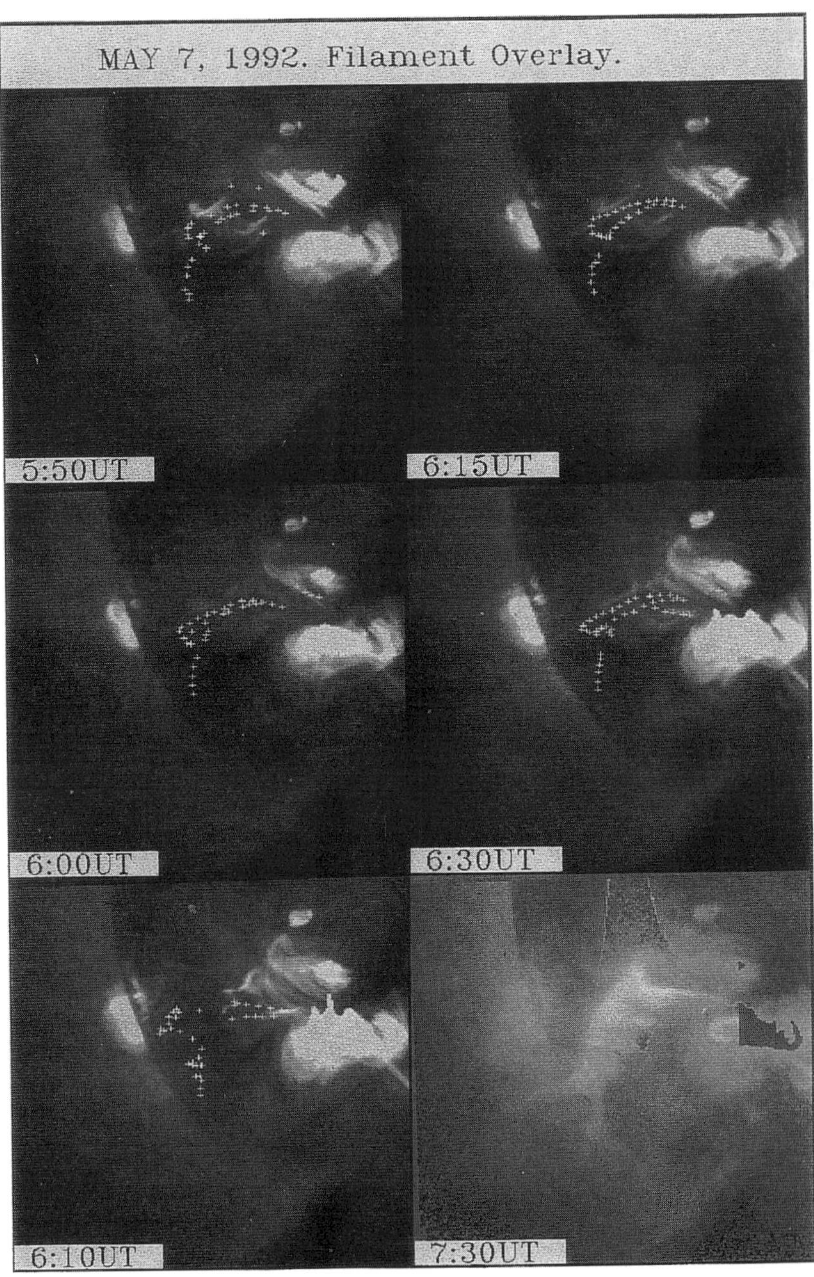

Figure 2. SXT images and Hα filament, which were plotted by white cross points on the SXT images.

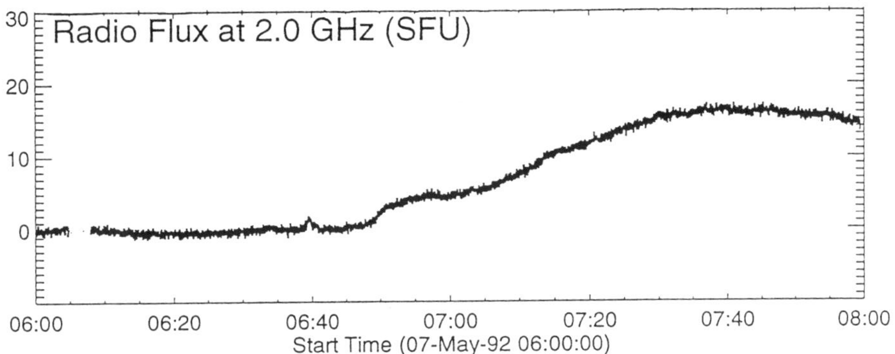

Figure 3. Radio plot at 2 GHz observed at the Toyokawa Observatory.

associated with this filament, was an LDE-type flare. The rise phase of the 2.0 GHz radio flux light curve was similar to the GOES flux except for a weak impulsive component at 06:40 UT. This impulse was not seen in 3.75 GHz and higher frequencies. A weak non-thermal process seems to be associated with initial Hα foot point brightening, as seen in Fig. 1. The gradual component can be interpreted as the combination of hot plasma, which was detected by GOES soft X-ray, and cooler and denser plasma, which is less effective at emitting soft X-ray flux. The contribution of the cooler and denser plasma became larger after 07:20 UT, which will be the result of the cooling of the hot plasma; the hot plasma cooled to become the EBI structure.

The combined Hα and SXT observations indicate the following results: i) An X-ray loop brightening occurred first (above the filament), and twisted X-ray loops appeared as the filament began to move. The filament disappeared completely, and then the main flare occured later (beneath the erupted filament). It is therefore inferred that the coronal magnetic configuration above the filament changed first, and the filament disappeared. Then an LDE-type flare occurred, ii) X-ray loops (Ebi structure) were seen above the position of the disappeared filament which would be interpreted as reformation of a magnetically closed loop structure as a consequence of magnetic reconnection, as observed in the helmet structure of the January 24, 1992 event (Hiei et al., 1993). iii) Hα bright strands at the chromosphere are heated by conduction from the hot X-ray loops, and not due to falling material. iv) The Ebi event occurred at the border of a magnetic peninsula in the synoptic chart. This magnetic peninsula with positive magnetic polarity continued for several months. The large helmet structure of the

January 24, 1992 event occurred at the same magnetic peninsula, and other disappearances of filaments with related SXT brightenings also occurred at the border of this magnetic peninsula. The repeated occurrence of eruptive events in this coronal configuration has been discussed by McAllister et al. (1996).

References

Hiei, E. (1994), in IAU Colloq. 144 *Solar Coronal Structures*, eds. V. Rusin, P. Heinzel, and J.-C. Vial, p163.

Hiei, E., Hundhausen, A.J., and Sime, D.G. (1993), *Geophys.Res.*, **L.20, No.24**, 2785.

Hyder, C. (1967), *Solar Phys.*, **2**, 49.

Khan, J.I., Uchida, Y., McAllister, A.H., and Watanabe, Ta. (1993), in *X-ray Solar Physics from Yohkoh*, eds. Y. Uchida, T. Watanabe, K. Shibata, and H.S. Hudson, Universal Academy Press, Inc., Tokyo, p201.

McAllister, A. H., Uchida, Y., Tsuneta, S., Strong, K.T., Acton, L.W., Hiei, E., Brunner, M.E., Watanabe, Ta., and Shibata, K. (1992), *Publ. Astron. Soc. Japan*, **44**, L205.

McAllister, A.H., Hundhausen, A.J., Burkpile, J.T., McIntosh, P., and Hiei, E.(1996) In *Magnetohydrodymanic Phenomena in the Solar Atmosphere, Prototypes of Stellar Magnetic Activity*, eds. Y. Uchida, H.S. Hudson and T. Kosugi, Kluwer, 123-124.

Tsuneta, S., Hara, H., Shimizu, T., Acton, L.W., Strong, K.T., Hudson, H.S., and Ogawara, Y. (1992), *Publ. Astron. Soc. Japan*, **44**, L63.

CORONAL AND INTERPLANETARY DISTURBANCES ASSOCIATED WITH AN ERUPTIVE PROMINENCE OF 28 AUGUST 1992

T. WATANABE AND M. YAMAMOTO
Ibaraki University, Mito 310, Japan
H. HUDSON
Solar Physics Research Corp., Tucson, U.S.A. and ISAS, Sagamihara 229, Japan
M. IRIE AND K. ICHIMOTO
National Astronomical Observatory, Mitaka 181, Japan
H. KUROKAWA
Hida Observatory, Kyoto University, Kamitakara 506-13, Japan
AND
H. YATAGAI
Fujitsu Co. Ltd, Tokyo, Japan

Abstract. We examine the dynamical characteristics of the eruption of a large (15 deg) north-south aligned quiescent prominence using soft X-ray (Yohkoh) and ground-based observations. The event occurred above the eastern solar limb near the equator on 28 August 1992. The eruption was preceded by a jet which was ejected from near the root of the southern end of the prominence. Changes of magnetic geometry at the southern end of the prominence appeared to trigger the eruption. The configuration of the eruption suggested helically twisted loops surrounding the prominence. This would be consistent with magnetic reconnection of a sheared arcade underneath the erupting prominence, forming a flux rope. Coronal dimming was observed, and this suggests the formation of the CME (and its interplanetary manifestation). An interplanetary disturbance was observed via interplanetary scintillation technique in the region to the east of the sun-Earth line on 29-30 August 1992.

T. Watanabe et al. (eds.), Observational Plasma Astrophysics: Five Years of Yohkoh and Beyond, 101–105.

1. Introduction

Many coronal disturbances associated with eruptions of quiescent prominences (filaments) have been observed by the Yohkoh SXT since the launch of the spacecraft in 1992 (e.g. Watanabe et al., 1992; McAllister et al., 1992; Hanaoka et al., 1994). A majority of these events showed bright X-ray arcades which were still being formed several hours after the eruption of the prominences. Since arcades are physically similar phenomena to H-alpha post- flare loops which are seen after solar flares, they are not very useful to study physical processes taking place before and during the prominence eruption. On the other hand, several, but not many, eruptive prominences appear to have been surrounded by complicated soft X-ray loop structures during their eruptions. These events are useful for studying temporal and spatial changes of magnetic structures which take place in eruptive-prominence events. In this article, we perform a case study on an eruptive prominence which was observed above the eastern solar limb on 28 August 1992 and for which the time coverage of the Yohkoh soft X-ray observations was remarkably good.

2. Observations and Data Analysis

2.1. SOFT X-RAY LOOPS SURROUNDING THE EUPTIVE

We use soft X-ray images (Yohkoh), H-alpha images (Boulder, Holloman, Norikura, and Hida), and 22 GHz radio images (Nobeyama) in the present analysis. An example of H-alpha images of the erupting prominence taken at Norikura (21:24 UT, 28 August 1992) is shown in Fig. 1. The latitudinal extent of the prominence was about 15 degrees. The eruption took place above the eastern solar limb, near the equator. According to H-alpha synoptic map from the Solar-Geophysical Data (NOAA/SESC), the magnetic neutral line underneath the prominence was aligned nearly in the north-south direction, so that we have a side view of the prominence. The ascending speeds of the prominence and the soft X-ray loops were 30-60 km/sec. Highly complicated soft X-ray loops, which were observed during the course of the prominence eruption (21:32:08 UT), are shown in the middle frame of Fig. 1. A composite sketch of the prominence and soft X-ray loops is also shown in this figure. The most interesting feature is the presence of kinked loops, which can be interpreted as the projected images of strands of a helical flux rope. Smoother loops passed along the axis of the larger-scale loops seen around the erupting prominence. Although it is difficult to know the true 3-D configuration of these loops, a combination of helically twisted loops on the outside and straighter loops passing through the axis of the helical loops is consistent with the projected view which we

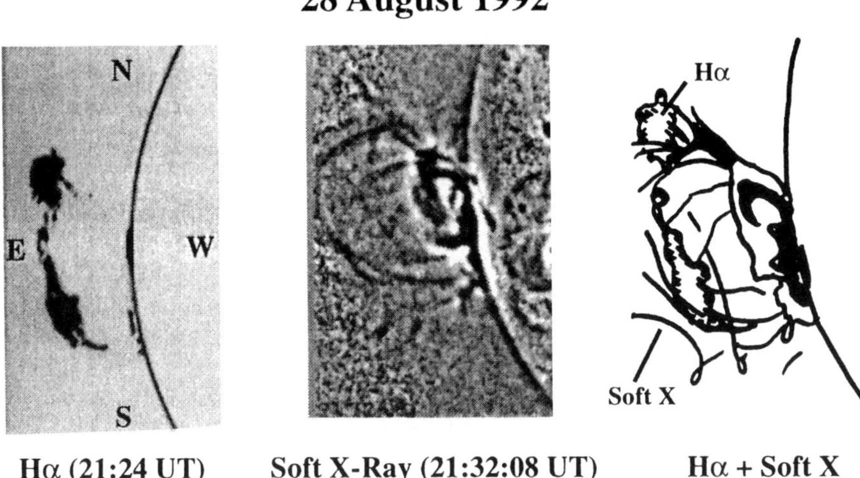

Figure 1. The eruptive prominence of 28 August 1992. The left panel is a negative representation of the H-alpha image taken at 21:24 UT (Norikura). The middle panel is an enhanced-contrast negative image of the soft X-ray loops surrounding the erupting prominence (21:32:08 UT). This image was taken by the soft X-ray telescope (SXT) of Yohkoh. The right panel shows a superposed sketch of the H-alpha and the soft X-ray images.

observed. This configuration is similar to that proposed by Van Ballegooijen and Martens (1989) to create twisted flux tubes above the magnetic neutral line. They showed that twisted flux tubes can be formed by the development of a sheared prominence-supporting field due to flux cancellation above the neutral line. They suggested also that prominence eruption can be caused by this process.

2.2. JET

Since the prominence did not show an appreciable ascending motion when H-alpha observations at Boulder and Holloman were performed at 14-17 UT on 28 August, the start time of the eruption must have been between 17 and 21 UT (the time of the first H-alpha observation in Japan on 28 August). The most remarkable soft X-ray feature observed in this time interval was the large jet shown in Fig. 2. This jet was ejected near the root of the southern end of the prominence at about 18:40 UT. The X-ray intensity of a bright region seen underneath the prominence (Fig. 1) began to increase after the appearance of the jet. We suggest that this jet, which was unusually large (see Shimojo et al., 1996), marked a precursor event of

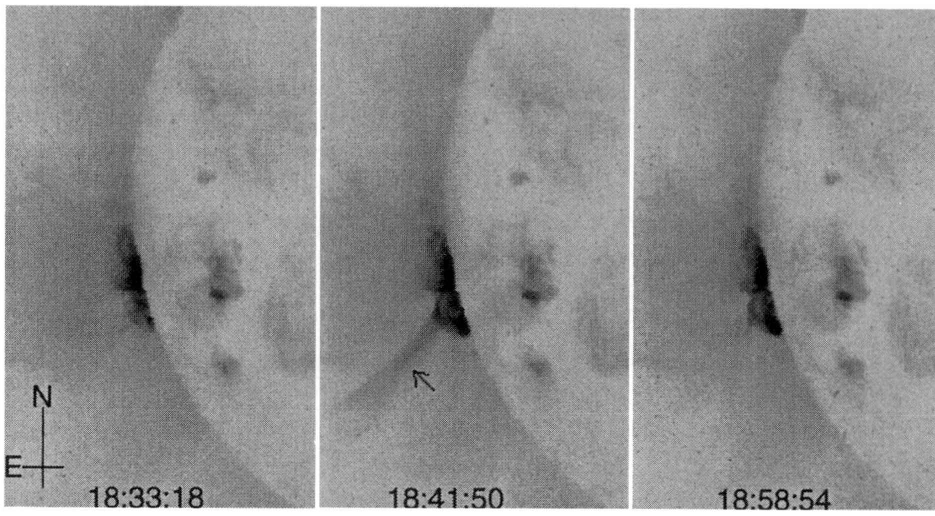

Figure 2. The soft X-ray jet which was ejected near the root of the southern end of the eruptive prominence of 28 August 1992. The left frame shows the coronal structures seen before the appearance of the jet (18:33:18 UT), and the central frame shows the jet at 18:41:50 UT (indicated by the arrow). The jet is not seen at 18:58:54 UT (right frame).

the prominence eruption. Since the prominence eruption began shortly after the appearance of the jet, the jet may reveal the start of the destabilization of the magnetic structures which supported the prominence.

2.3. CORONAL "DIMMING" AND CME

The soft X-ray intensity in the region above the arcade considerably decreased about three hours after the jet and approximately simultaneously with the flare brightening. The loops seen in Fig. 1 expanded to outside the field of view of the SXT. The dimming of the solar corona suggests the formation of a coronal mass ejection (Hudson and Webb, 1997). Although a CME cannot be directly seen in the SXT images, an interplanetary disturbance was detected by IPS (interplanetary scintillation) observations of several radio sources in the region to the east of the sun-Earth line on 29-30 August 1992. No geomagnetic activity relevant to the eruptive event was observed, but this would not be expected given the limb location of the solar event.

3. Concluding Remarks

This event is one of the best examples of prominence eruptions observed by Yohkoh. Since the eruption took place shortly (< 3 hours) after the appearance of the jet, we suppose that the trigger action required for the prominence eruption was the change in magnetic structure near the root of the southern end of the prominence. Perhaps emerging flux could have produced this magnetic restructuring, as suggested by Feynman and Martin (1995). The presence of helical loops surrounding the erupting prominence and an arcade following it can be explained by magnetic reconnection underneath the erupting prominence (Hirayama, 1974). The coronal dimming observed during the prominence eruption is an indication of the formation of a CME (and its interplanetary manifestation). Further detailed analysis is in progress.

Acknowledgment

The authors would thank Dr. K. Shibasaki (Nobeyama Radio Observatory, NAO) for his help in data processing of 22 GHz solar radio images. Thanks are also due to Dr. M. Kojima of STELAB, Nagoya University, who kindly provided us with his IPS data.

References

Feynman, J., and Martin, S. F.: 1995, *J. Geophys. Res.*, **100**, 3355.

Hanaoka, Y., Kurokawa, H., Enome, S., Nakajima, H., Shibasaki, K., Nishio, M., Takano, T., Torii, C., Sekiguchi, H., Kawashima, S., L. Bushimata, T., Shinohara, N., Irimajiri, Y., Koshiishi, H., M. Shimoi, Y., Nakai, Y., Funakoshi, Y., Kitai, R., Ishiura, K., and N. Kimura, G.: 1994, *Publ. Astron. Soc. Japan*, **46**, 205.

Hirayama, T.: 1974, *Solar Phys.*, **34**, 323.

Hudson, H. S., and Webb, D. F.: 1997, *in Proc. Chapman Conf. on Coronal Mass Ejections (in press).*

McAllister, A., Uchida, Y., Tsuneta, S., Strong, K., Acton, L. W., Hiei, E., Bruner, M. E., Watanabe, T., and Shibata, K.: 1992, *Publ. Astron. Soc. Japan*, **44**, L205.

Shimojo, M., Hashimoto, S., Shibata, K., Hirayama, T., Hudson, H., and Acton, L.: 1996, *Publ. Astron. Soc. Japan*, **48**, 123.

Van Ballegooijen, A. A., and Martens, P. C. H.: 1989, *Astrophys. J.*, **343**, 971.

Watanabe, T., Kozuka, Y., Ohyama, M., Kojima, M., Yamaguchi, K. , Watari, S., Tsuneta, S., Joselyn, J. A., Harvey, K., Acton, L. W., and Klimchuk, J.: 1992, *Publ. Astron. Soc. Japan*, **44**, . L199.

YOHKOH OBSERVATIONS OF SUPERHOT PLASMA IN SOLAR FLARES

NARIAKI NITTA
Lockheed Martin Solar and Astrophysics Laboratory
O/H1-12, B/252, 3251 Hanover Street, Palo Alto,
CA 94304, USA

Abstract.

We propose a new method to detect superhot ($\gtrsim$30 MK) plasma in solar flares. Our approach is to make use of *Yohkoh* soft X-ray images in three different filters and to calculate temperature (T_H) and emission measure (EM_H) of the hotter component assuming a two temperature model. The calculated distributions of T_H and EM_H are compared with temperatures obtained from the Fe XXV and Ca XIX diagnostics. In one flare, we find T_H of $\sim$30 MK in a loop structure away from the brighter loop.

1. Introduction

During cycle 21 (maximum around 1980), some solar flares were shown to contain superhot ($\gtrsim$30 MK) plasma, in addition to the ubiquitous 10–20 MK plasma. The superhot component was isolated using two independent techniques: hard X-ray continuum spectroscopy with high energy resolution (Lin *et al.*, 1981) and soft X-ray diagnostics on Fe XXVI lines (Tanaka, 1986). The latter has been extended into cycle 22 with the Bragg Crystal Spectrometer (BCS) aboard the *Yohkoh* satellite (Culhane *et al.*, 1994; Pike *et al.*, 1996), although no hard X-ray spectroscopy with resolution comparable to Lin *et al.* (1981) has been available.

In this paper, we propose a different approach for studying the superhot plasma in solar flares. We primarily make use of soft X-ray images in three filters obtained by the *Yohkoh* Soft X-ray Telescope (SXT). Our technique provides spatial information of the superhot plasma, making it easier to discuss the origin of the superhot plasma in terms of flare models.

T. Watanabe et al. (eds.), Observational Plasma Astrophysics: Five Years of Yohkoh and Beyond, 107–112.

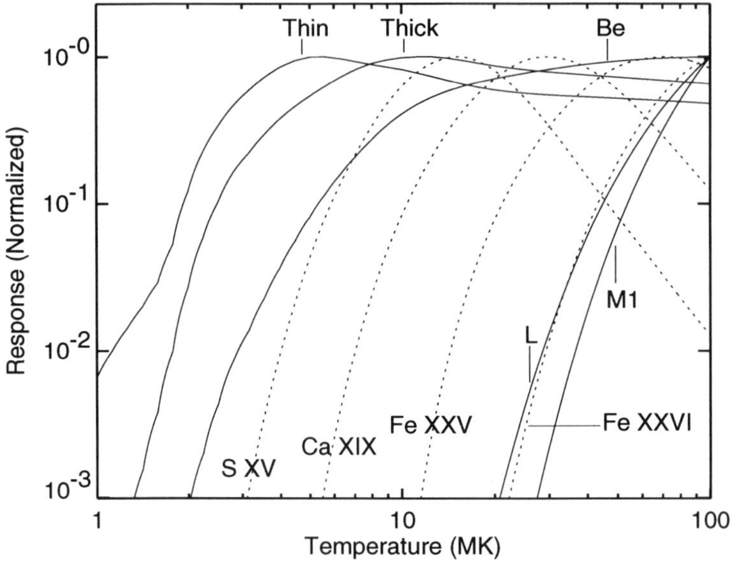

Figure 1. Response of SXT and HXT with temperature (solid lines). The curves for three filters (thin Al, thick Al and Be) of SXT and two channels (L and M1) of HXT are drawn. Dotted lines show contribution functions of the resonance lines of four ions that are observed with BCS. The curves for Be, L, M1 and Fe XXVI are normalized at 100 MK.

2. Responses of *Yohkoh* Instruments

Even within the *Yohkoh* measurements of a single flare, we usually find different maximum temperatures from different instruments, including the Hard X-ray Telescope (HXT), BCS and SXT. The typical values are summarized as: 10–15 MK from the BCS Ca XIX channel, 15–25 MK from the BCS Fe XXV channel, 30–50 MK from the ratio of the HXT L (14–23 keV) and M1 (23–33 keV) bands. The ratio of the thickest two filters of SXT almost always yields temperatures of 7–10 MK at loop footpoints and 10–20 MK at higher altitudes. Temperatures of $\gtrsim$20 MK are very unusual.

These values may be understood in terms of temperature sensitivities of the different measurements. Figure 1 shows normalized response curves for SXT and HXT together with the contribution functions for the resonance lines observed by BCS. We immediately notice that SXT is more sensitive to low temperatures than BCS and HXT but that its sensitivity does not drop significantly at superhot temperatures.

3. SXT Temperatures and Column Differential Measure (CDEM)

As described in Nitta and Yaji (1997), we analyze data of the M1 class flare that took place on 21 August 1992. The flare consisted of a bright compact loop on the eastern side and a separate diffuse loop structure on the western side. We first obtain T and EM maps using the ratio of Be and Thick Al filters. They persistently show that the western structure is hotter (15–20 MK) than the bright eastern loop (8–12 MK). The western structure is also aligned with strong hard X-ray emission.

Using the T and EM maps obtained above, we generate line profiles for the BCS Fe XXV and Ca XIX channels. We then fit these spectra in the same way we do the raw BCS data. It is found that the BCS temperatures calculated from the SXT T and EM maps are lower than the observed values. This result implies that the temperature from the SXT filter ratio is affected by cool material that exists along the line of sight, even if we use the ratio of the two thickest filters.

Another observational fact for the flare is the inequality $T_2 > T_1$, where T_2 and T_1 represent SXT observed temperatures from Be/Thick Al and Be/Thin Al, respectively. This would support the existence of plasma hotter than T_2, because the inequality requires plasma cooler than T_2 along the line of sight. This cooler plasma would reduce T_2 artificially, so plasma hotter than T_2 must exist. Here we refer to plasmas of different temperatures along the line of sight as column differential emission measure (CDEM).

In order to evaluate the effect of CDEM, we employ a two-temperature model with four parameters (T_L, EM_L) and (T_H, EM_H) for the cool and hot components. Three of them are uniquely determined when the remaining one is given, because we observe in three filters (Be, thick Al and thin Al). Here we calculate T_H, EM_H and EM_L for given values of T_L. The results for two values of T_L are given in the lower panels of Figure 2. In the uppermost panels of Figure 2, we show graphically on a (T_2, T_1) scatter plot how each pixel is matched by a particular value of T_H. As expected, the observed data are matched by T_H closer to T_2 for lower T_L.

Finally, we determine the value of T_L in such a way that it results in the set of T_H and EM_H that best explains the BCS Fe XXV and Ca XIX temperatures. We find that T_L is between 6.0 MK and 6.5 MK. Then the distribution of T_H for the western structure extends to ~30 MK, but not significantly beyond as in the case of T_L=7 MK. This is consistent with the temperatures from both HXT (M1/L) and Fe XXVI for the present flare. We also notice that the obtained EM_L is an order of magnitude larger than the preflare value even if the preflare temperature is close to T_L. Creation of such low-temperature plasma has been reported on the basis of radio observations (Chertok *et al.*, 1995).

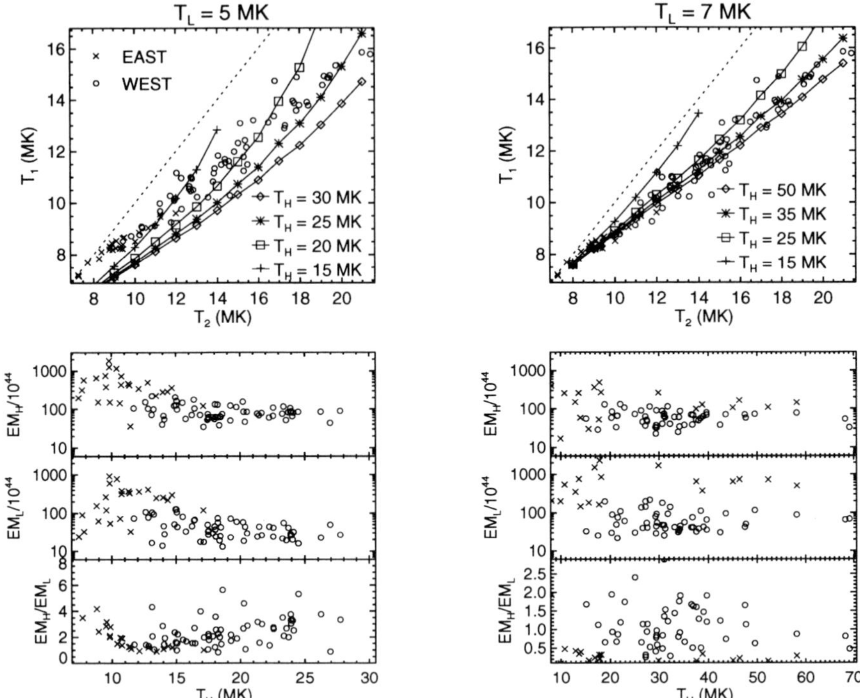

Figure 2. Uppermost panels compare the observed temperatures (T_2 and T_1) with those calculated from a two-temperature model of two values of T_L (5 MK and 7 MK) and four values of T_H (solid lines). Different symbols are used to indicate the eastern loop and the western loop structure. The lower panels plot (as scatter plots) the calculated parameters T_H, EM_H and EM_L, showing their distributions in the two areas.

4. SXT Temperatures of a Non-Flare Loop

Here we briefly discuss an example of non-flare loops accidentally observed in three filters. Outside flare periods, we normally have only two filters to obtain plasma parameters. We denote the temperature from the ratio of thick Al and thin Al as T_0. Figure 3 shows that there is a high chance of the existence of plasma hotter than T_0, which is compromised by cooler plasma along the line of sight. Such plasma is probably contained in the cool loops revealed in spectroscopy of coronal emission lines (Ichimoto *et al.*, 1995) and radio observations (Klimchuk *et al.*, 1995; Gary *et al.*, 1996).

5. Application to Other Flares

Although the 21 August 1992 flare (Nitta and Yaji, 1997) is a nice example of the superhot plasma as revealed by SXT data, there are other flares in which the superhot plasma is likely to be present outside a bright loop. One

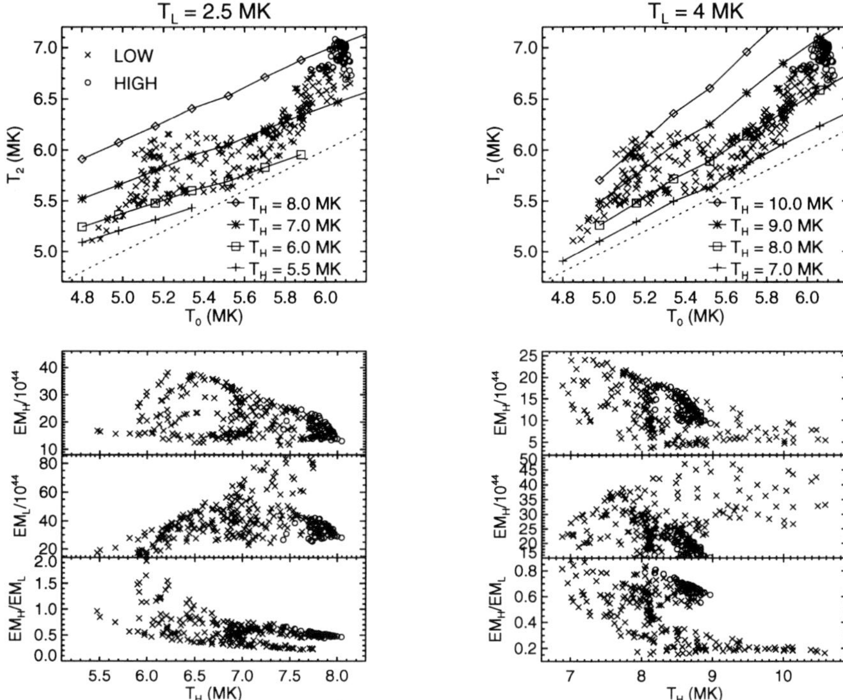

Figure 3. This figure is in a similar format to Figure 2, but deals with temperatures of a non-flare loop observed on the eastern limb on 29 January 1992. The x-axis in the uppermost panels is in terms of T(Thick Al/Thin Al), denoted as T_0, which is normally used for non-flare data from SXT, and the y-axis of the upper panels is in terms of T_2. The calculated parameters are given as scatter plots.

of them is the 13 January 1992 flare (Masuda *et al.*, 1994), where the hard X-ray source above the loop appears to be aligned with a high temperature region. In Figure 4, we show scatter plots of T_1 and T_2 for four intervals, separately for regions inside and above the bright loop. According to these plots, the region above the loop not only has higher SXT temperatures than the region inside but also shows larger deviations from T_1 and T_2, indicating even higher temperatures as illustrated in the uppermost panels of Figure 2. This is especially true for intervals (b) and (c). The same analysis as the 21 August 1992 is under way. We also plan to apply the same technique on about 50 out of 75 Fe XXVI flares as identified by Pike *et al.* (1996).

Acknowledgements

This work was supported by NASA grant NAS 8-40801.

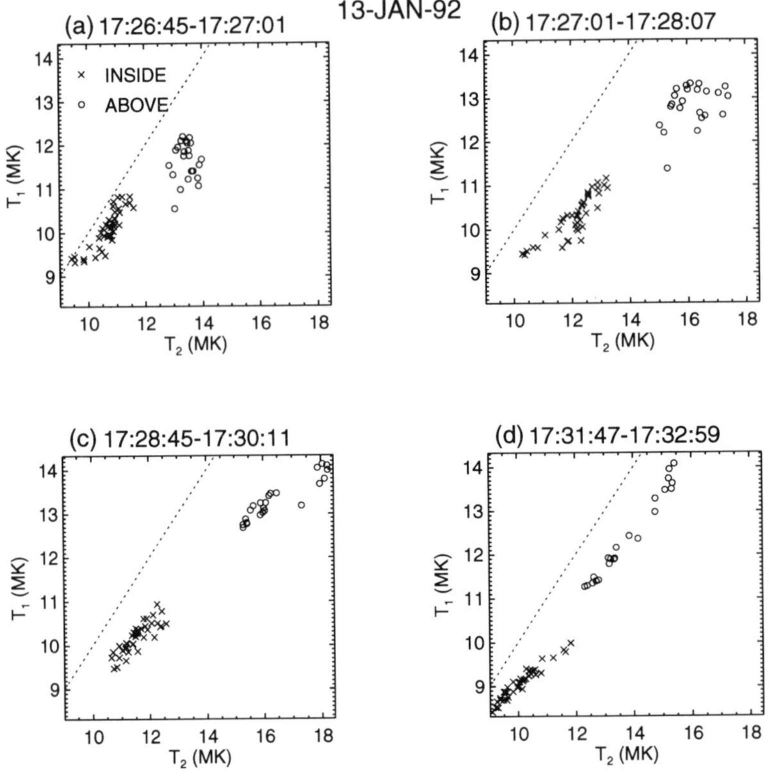

Figure 4. Scatter plots of T_1 and T_2 for four intervals during the flare of 13 January 1992.

References

Culhane, J. L. *et al.*: *Solar Phys.*, **153**, 307.

Chertok, I. M., Fomichev, V. V., Gorgusta, R. V., Hildebrandt, J., Krüger, A., Magun, A., and Zaitsev, V. V.: 1995, *Solar Phys.*, **160**, 181.

Gary, D. E., Wang, H., Nitta, N., and Kosugi, T.: 1996, *Astrophys. J.*, **464**, 965.

Ichimoto, K., Hara, H., Takeda, A., Kumagai, K., Sakurai, T., Shimizu, T., and Hudson, H. S.: 1995, *Astrophys. J.*, **445**, 978.

Klimchuk, J. A. and Gary, D. E.: 1995, *Astrophys. J.*, **448**, 925.

Lin, R. P., Schwartz, R. A., Pelling, R. M., and Hurley, K. C.: 1981, *Astrophys. J. (Letters)*, **251**, L109.

Masuda, S., Kosugi, T., Hara, H., Tsuneta, S.,and Ogawara, Y.: 1994, Nature, **371**, 495.

Nitta, N. and Yaji, K.: 1997, *Astrophys. J.*, in press.

Pike, C. D., Phillips, K. J. H., Lang, J., Sterling, A., Watanabe, T., Hiei, E., Culhane, J. L., Cornille, M., and Dubau, J.: 1996, *Astrophys. J.*, **464**, 487.

Tanaka, K.: 1986, *Publ. Astron. Soc. Japan*, **38**, 225.

NON-THERMAL VELOCITIES OBSERVED BY YOHKOH

L.K. HARRA-MURNION, J.I. KHAN AND S.A. MATTHEWS
Mullard Space Science Laboratory
Holmbury St. Mary, Surrey, RH5 6NT, UK

AND

D. ALEXANDER
Lockheed Palo Alto Research Laboratory
Dept. 91-30, Bldg. 252, 3251 Hanover St., Palo Alto, CA 94304, U.S.A

Extended Abstract

In this extended abstract we discuss the measurement and interpretation of the excess line broadening of soft X-ray lines above their thermal width, characterized by non-thermal velocity, and its importance in understanding the initial energy release process in solar flares. One important aspect which has not yet been addressed by *Yohkoh* is the relative timing of V_{nt}, with the hard X-ray bursts. The hard X-ray bursts are believed to be the response to the initial energy release and pre-*Yohkoh* observations suggested that V_{nt} and hard X-ray bursts peaked at the same time (e.g. Doschek et al. (1986)). One possible explanation was that the excess broadening was caused by mass motions due to chromospheric material being ablated into the corona. Here we investigate the relative timing of these two quantities using Yohkoh's Bragg Crystal Spectrometer (BCS) which has much higher sensitivity than similar spectrometers on previous missions.

The higher sensitivity of the BCS has provided us with the ability to observe V_{nt} earlier in the flare, providing important information regarding the energy release process of solar flares. We have analysed nine flares, all of which are C GOES class (to avoid the instrumental problem of line narrowing, Trow et al. (1994)) and quite simple in their morphology (one main flaring loop). Events were chosen with a significant count rate in the Hard X-ray Telescope (HXT) M1 channel, and with no measurable contamination from a secondary blue component in the soft X-ray line profiles.

T. Watanabe et al. (eds.), Observational Plasma Astrophysics: Five Years of Yohkoh and Beyond, 113–114.

We find in all cases that the V_{nt} peaks before the first significant hard X-ray burst. The delay between the peak of the V_{nt} and the hard X-ray (HXR) emission lasts for as long as one minute. Interestingly, irrespective of the number of HXR bursts which occur, there is in general only one significant peak of V_{nt}. This result differs from observations from previous missions. In addition to the Doschek et al. (1986) results noted above, Winglee et al. (1991) have studied five flares and found that the V_{nt} peaks at approximately the same time as the hard X-rays peak. However they do observe the V_{nt} peaking before the maximum in the blue shift velocity. It is possible that HXR emission is occuring earlier than the V_{nt} peak, but is as yet undetectable by the available information. However, what is apparent is that the V_{nt} is peaking well before the main HXR peak. The values of the V_{nt} peak between 100 km/sec $\rightarrow$ 240 km/sec.

Our observations in this paper suggest that the V_{nt} is not correlated strongly with the hard X-ray bursts. The broadening is caused by a process which occurs before the hard X-ray bursts. It has been suggested by Doschek et al. (1986) that non-thermal broadening is caused by chromospheric evaporation. From our observations it is not clear that the broadening is due to mass motions caused by chromospheric material being ablated by electrom beams. One alternative explanation is a form of pre-flare activity in the flaring loop. If the magnetic field slowly starts to re-arrange, building up to the impulsive phase, this could cause some motion in the plasma, which may be seen as broadening in the soft X-ray line profiles. Another possibility is that the broadening seen is due to the reconnection downflow as suggested by Tsuneta (1995). In this case Joule heating would be required to dominate over electron runaway for up to the order of minutes for this process to be successful.

References

Doschek, G. A. et al. 1986, The Solar Maximum Mission Workshop Proceedings, NASA Conference Publication 2439, chapter 4.

Trow, T.W., Bento, A.C., & Smith, A., 1994, Nucl Instr & Methods in Phys. Res. A, 348, 252.

Tsuneta, S., 1995, PASJ, 47, 691.

Winglee, R.M., Kiplinger, A.L., Zarro, D.M., Dulk, G.A. & Lemen, J.R., 1991, ApJ, 375, 366.

MAGNETOHYDRODYNAMICS OF ACCRETION DISKS

R. MATSUMOTO, T. MATSUZAKI
Department of Physics, Faculty of Science, Chiba University, 1-33 Yayoi-Cho, Inage-Ku, Chiba 263, Japan

T. TAJIMA
Institute for Fusion Studies, The University of Texas at Austin, Austin, TX78712, USA

AND

K. SHIBATA
National Astronomical Observatory, Mitaka, Tokyo 181, Japan

1. Introduction

Solar physics has close relationships to accretion disk physics. Galeev, Rosner & Vaiana (1979) proposed that accretion disks might have magnetically structured corona consisting of magnetic loops. Their model was based on Skylab soft X-ray observations of the solar corona. The soft X-ray telescope aboard the *Yohkoh* satellite revealed various dynamical activities in the solar corona which give clues to understanding the dynamics of accretion disk corona and X-ray time variabilities observed in black hole candidates (e.g., Miyamoto et al. 1994) and in star forming regions (e.g., Koyama et al. 1996). Recently, Ueno et al. (1997) showed that the power spectrum density of soft X-ray time variations of the solar corona have 1/f-like profiles similar to that of Cyg X-1. It is possible that the basic mechanism of X-ray fluctuations in accretion disks is similar to that of the solar corona.

It has been suggested that magnetic fields play essential roles in accretion disks as the origin of angular momentum transport (e.g., Shakura & Sunyaev 1973), formation of collimated outflows (e.g., Blandford & Payne 1982; Uchida & Shibata 1985) and X-ray flares (Hayashi et al. 1996). Figure 1 schematically shows various activities in accretion disks associated with magnetic fields.

T. Watanabe et al. (eds.), Observational Plasma Astrophysics: Five Years of Yohkoh and Beyond, 115–120.

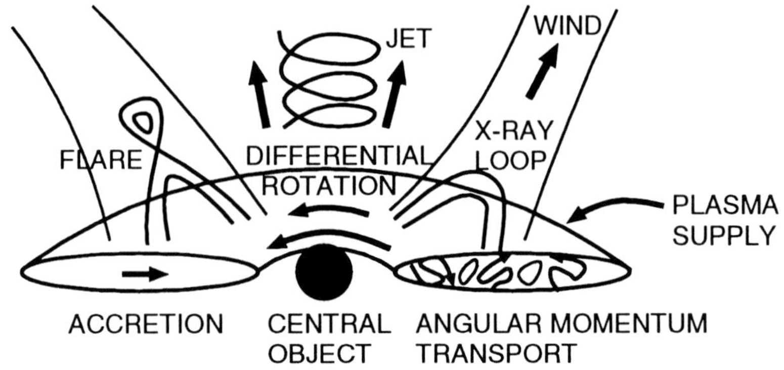

Figure 1. A schematic picture showing magnetic activities in accretion disks

In the following, we present some results of magnetohydrodynamic (MH D) simulations of accretion disks.

2. Angular Momentum Transport due to the Shear Flow Instability in Weakly Magnetized Accretion Disks

In conventional theory of accretion disks (e.g., Shakura & Sunyaev 1973), the α prescription of viscosity is adopted, in which the off-diagonal component of the stress tensor $\tau_{r\varphi}$ is assumed to be proportional to the pressure ($\tau_{r\varphi} = \alpha P$). By comparing with theory and observation of dwarf novae, it has been suggested that $\alpha \sim 0.02$ in the quiescent state (Cannizzo, Shafer, & Wheeler 1988). Since molecular viscosity cannot provide such a high rate of angular momentum transport, various models of anomalous angular momentum transport have been proposed. Balbus & Hawley (1991) pointed out the importance of the local magnetorotational (or Balbus & Hawley) instability on the generation of fluctuating magnetic fields which contribute to the angular momentum transport. When a differentially rotating plasma is threaded by weak poloidal magnetic fields, it becomes unstable against axisymmetric perturbations. Azimuthal magnetic fields are also subject to the non-axisymmetric magnetorotational instability (e.g., Balbus & Hawley 1992).

By adopting local shearing sheet approximation, the nonlinear growth and saturation of the Balbus & Hawley instability has been studied by three-dimensional local MHD simulations of Keplerian disks (Hawley, Gammie, & Balbus 1995; Matsumoto & Tajima 1995; Brandenburg et al. 1995; Stone et al. 1996; Matsuzaki et al. 1997). Through 3D MHD simulation including the effects of gravitational stratification of the disk, Stone et al.

(1996) showed that when the initial magnetic field is weak ($\beta = P_{gas}/P_{mag} \gg 1$) the disk evolves to a quasi-steady state with $\beta \sim 30$. In this state, the ratio of the spatial average of the off-diagonal component of magnetic stress to the pressure, $\alpha_B = -\langle B_r B_\varphi/(4\pi P)\rangle$ is $O(0.01)$. This value of α coincides with that in the quiescent state of dwarf novae.

3. Two Types of Magnetic Accretion Disks

X-ray observations of black hole candidates have shown that accretion disks have two spectral states; a soft state (or high state) and a hard state (or low state). The power spectral densities of time variations change between the high state and the low state; in the low state, low frequency time variations are much larger than in the high state (e.g., Miyamoto et al. 1994). One possible mechanism of large amplitude, sporadic time variabilities in the low-state is magnetic energy release in accretion disks.

The quasi-steady state of weakly magnetized (high-β) disks can explain various characteristics of high-state disks; e.g., angular momentum transport and small amplitude time variabilities. Here we present numerical results which indicate the existence of low-β states in accretion disks. It had been thought that low-β disks cannot exist because when magnetic pressure is comparable to the gas pressure, the buoyant escape of magnetic flux due to the Parker instability (Parker 1966) converts the disk into a high-β state. Shibata et al. (1990), however, showed through two-dimensional MHD simulation that in the presence of shear flow, once the disk becomes low-β, it can stay in a low-β state partly because the growth rate of the Parker instability decreases when $\beta \leq 1$ and partly because the shear flow amplifies magnetic fields.

When the effects of rotation are taken into account, the Balbus & Hawley instability generates fluctuating magnetic fields. Figure 2 shows the results of 3D MHD simulations of a local part of a Keplerian disk under shearing sheet approximation. The details of the numerical simulations are described in Matsuzaki et al. (1997). The curves in Figure 2 show the evolution of the mean square magnetic field strength $\langle B^2/(8\pi P_0)\rangle$ for $\beta_0 = 0.3, 1.0, 10.0, 30.0$, and 100.0 where β_0 is the initial β. Numerical results indicate that weakly magnetized ($\beta \gg 1$) disks approach a quasi-steady state with magnetic field strength $\langle B^2/(8\pi P_0)\rangle \sim 0.03$. This result is consistent with that reported by Stone et al. (1996).

We found that when the initial magnetic energy is comparable to the thermal energy ($\beta \sim 1$), the magnetic energy increases and the disk stays in a low-β state for time scales much longer than the rotation period. The model with $\beta_0 = 0.3$ also stays in a low-β state. The magnetic field amplification due to the coupling of the Balbus & Hawley instability and the Parker instability compensates the buoyant loss of magnetic flux. This field amplification is a kind of dynamical dynamo in which the exponential growth of radial magnetic fields due to the Balbus & Hawley instability efficiently re-generates toroidal magnetic fields. Numerical results confirmed the existence of two types of accretion disks; a high-β disk and a low-β disk.

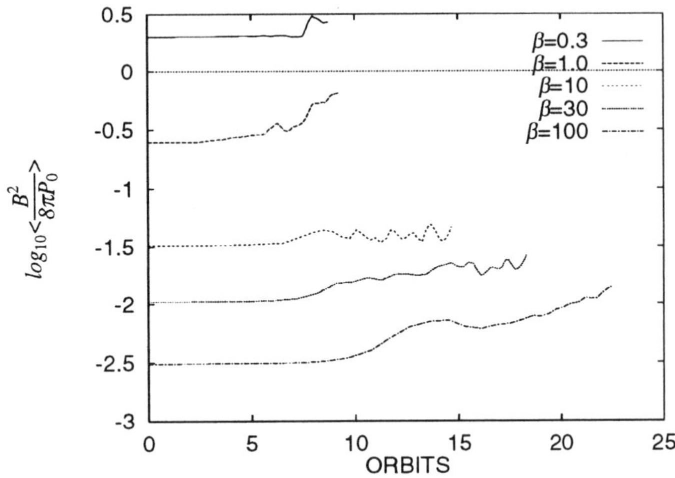

Figure 2. Time evolution of mean square magnetic field strength $\log_{10}\left\langle B^2/(8\pi P_0)\right\rangle$ for various β_0.

4. Global 3D MHD Simulations of Accretion Disks

When an accretion disk is threaded by a large scale poloidal magnetic field, a bipolar jet can be formed due to the accumulation and relaxation of magnetic twists generated by the rotation of the disk (Uchida & Shibata 1985; Shibata & Uchida 1986). In order to study the effects of non-axisymmetry on the structure and stability of the disk and jets, we carried out global three-dimensional MHD simulations of a torus threaded by large scale magnetic fields.

We assume that a rotating polytropic torus with constant angular momentum distribution $L = L_0$ is imbedded in a spherical, non-rotating isothermal halo. The gravitational field is assumed to be given by a point mass M. In a cylindrical coordinate (r, φ, z) system, the dynamical equilibrium of the disk is described by

$$-\frac{GM}{(r^2 + z^2)^{1/2}} + \frac{1}{2}L_0^2 r^{-2} + (n+1)\frac{P}{\rho} = const,$$

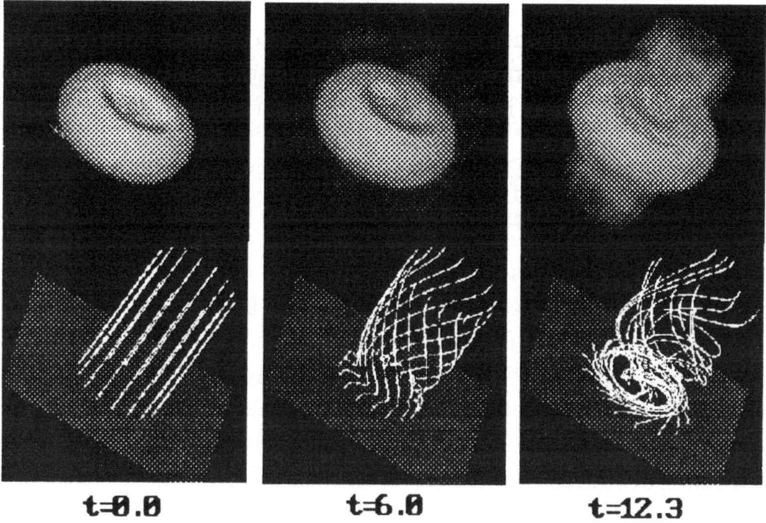

Figure 3. Results of 3D MHD simulation of a torus threaded by large scale magnetic fields. Top panels show volume rendered image of density distribution. The bottom panels show magnetic field lines above the equatorial plane.

where n is the polytropic index. We take the radius of the pressure maximum of the disk $[r_0 = L_0^2/(GM)]$ as the reference radius. The initial magnetic field is assumed to be uniform and vertical ($\mathbf{B} = B_0\hat{z}$). The model parameters for the disk are $A_1 = C_{s0}^2/(\gamma v_{K0}^2)$ and $A_2 = v_{A0}^2/v_{K0}^2$, where γ is the adiabatic index, C_{s0} the sound speed, $v_{A0} = [B_0^2/(4\pi\rho_0)]^{1/2}$ the Alfvén speed, and $v_{K0} = (GM/r_0)^{1/2}$ is the Keplerian rotation speed. We use the normalization $r_0 = v_{K0} = \rho_0 = 1$. The halo parameters are $1/\alpha = C_{sh}^2/(\gamma v_{K0}^2)$ and ρ_h/ρ_0 where C_{sh} and ρ_h are the sound speed and density in the halo at $(r, z) = (0, r_0)$, respectively. We solved the ideal MHD equations in a cylindrical coordinate system.

In Figure 3, we show a result of 3D MHD simulations for a typical model ($n = 3$, $\gamma = 5/3$, $\alpha = 1$, $A_1 = 0.05$, $A_2 = 10^{-3}$, $\rho_h/\rho_0 = 10^{-3}$). The initial plasma β at the pressure maximum of the disk is $\beta_0 = 100$. We initiate the non-axisymmetric evolution by imposing $m = 2$ perturbations for azimuthal velocity. The number of grid points is $(n_r, n_\varphi, n_z) = (101, 32, 101)$.

We confirmed the results of previous 2D axisymmetric simulations (Matsumoto et al. 1996) that the surface layer of the disk accretes faster than the equatorial part like an avalanche because magnetic braking most effectively affects that layer. When the accumulated magnetic twist relaxes, the torsional Alfvén wave entrains the matter in the surface layer of the disk and form bipolar jets. The maximum speed of the jet is of the order of the

Keplerian rotation speed. Due to the self-pinching of the toroidal magnetic field, the outflow is collimated along the rotational axis. The growth of non-axisymmetric modes in the torus bunch the magnetic field lines in the jet into helical bundles. A helical structure can also be seen in the density distribution of the jet.

5. Concluding Remarks

Through direct MHD simulations of accretion disks, we are getting quantitative answers to long standing problems such as the anomalous angular momentum transport and dynamo action in accretion disks. In this paper, we presented results of ideal MHD simulations of magnetohydrodynamic phenomena in accretion disks. As *Yohkoh* observations indicate for the solar atmosphere, magnetic reconnection may be important as a mechanism of releasing magnetic energy in accretion disks. The results of resistive MHD simulations of accretion disks will be presented in subsequent papers.

Numerical computations were carried out on a VPP300/16R at the Astronomical Data Analysis Center of the National Astronomical Observatory, Japan. This work is supported in part by a grant from the ministry of education, science, and culture (07640348).

References

Balbus, S.A., & Hawley, J.F. 1991, ApJ, 376, 214
Balbus, S.A., & Hawley, J.F. 1992, ApJ, 400, 610
Blandford, R.D., & Payne 1982, MNRAS, 199, 883
Brandenburg, A., Nordlund, A., Stein, R. F., & Torkelsson, U. 1995, ApJ, 446, 741
Cannizzo, J. K., Shafer, A. W., & Wheeler, J. C. 1988, ApJ, 333, 227
Galeev, A., Rosner, R., & Vaiana, G.S., 1979, ApJ, 229, 318
Hawley, J. F., Gammie, C. F., & Balbus, S. A. 1995, ApJ, 440, 742
Hayashi, M.R., Shibata, K., & Matsumoto, R. 1996, ApJ, 468, L37
Koyama, K., Hamaguchi, K., Ueno, S., Kobayashi, N. & Feigelson, E. 1996, PASJ, 48, 87
Matsumoto, R., & Tajima, T. 1995, ApJ, 445, 767
Matsumoto, R., Uchida, Y., Hirose, S., Shibata, K., Hayashi, M.R., Ferrari, A., Bodo, G., & Norman, C. 1996, ApJ, 461, 115
Matsuzaki, T., Matsumoto, R., Tajima, T., & Shibata, K., 1997, in this volume
Miyamoto, S., Kitamoto, S., Iga, S., Hayashida, K., & Terada, K. 1994, ApJ, 435, 398
Parker, E. N. 1966, ApJ, 145, 811
Shakura, N. I., & Sunyaev, R. A. 1973, A&A, 24, 337
Shibata, K., & Uchida, Y. 1986, PASJ, 38, 631
Shibata, K., Tajima, T., & Matsumoto, R. 1990, ApJ, 350, 295
Stone, J. M., Hawley, J. F., Gammie, C. F., & Balbus, S. A. 1996, ApJ, 463, 656
Uchida, Y., & Shibata, K. 1985, PASJ, 37, 515
Ueno, S., Mineshige, S., Negoro, H., Shibata, K., & Hudson, H.S. 1997, ApJ, 484, 920

MAGNETIC LOOPS IN THE HOT UNIVERSE

S. TSUNETA
National Astronomical Observatory, Mitaka, Tokyo 181

AND

K. MAKISHIMA
Department of Physics, School of Science,
The University of Tokyo, Tokyo 113

Abstract. There is a large body of evidence that magnetic fields are responsible for the confinement of the high-temperature plasmas (10^{7-8} K) of the Galaxy and Intracluster Medium (ICM). The scaling laws recently confirmed for solar coronal loops are applied to these cosmic magnetic flux tubes: $T[\mathrm{K}] = 2.2 \times 10^2(p[\mathrm{dyncm}^{-2}] \cdot L[\mathrm{cm}])^{0.39}$, and $F_{\mathrm{T}}[\mathrm{ergs}^{-1}\mathrm{cm}^{-2}] = 1.3 \times 10^2 p^{1.37} L^{0.37}$, where T is the maximum temperature of the loops, p a constant pressure, and F_{T} the heating rate per unit area of the flux tubes. Observed parameters of these cosmic plasmas are essentially consistent with those derived from the scaling laws. The temperature for the Galactic plasma predicted by the scaling laws is lower than that observed, implying taht it has a transient nature. The derived volume filling factors of the flux tubes are $\sim o(0.01 - 1)$ for the ICM , and $\sim o(10^{-4})$ for the Galaxy. Magnetic fields may also play important roles for the heating of these hot plasmas.

1. Introduction

There is substantial evidence that magnetic fields play important physical roles for the creation and confinement of cosmic high temperature plasmas. Here, we summarize these pieces of evidence for Galactic and intra-cluster plasmas (Makishima 1994, 1995, 1996, 1997, and references therein).

T. Watanabe et al. (eds.), Observational Plasma Astrophysics: Five Years of Yohkoh and Beyond, 121–130.

1.1. GALACTIC RIDGE EMISSION

Along the Milky Way, an apparently diffuse X-ray emission called Galactic ridge emission is observed. These X-rays have been shown to originate as optically-thin thermal emission from plasmas of electron temperature $(5-10)\times10^7$ K. Because no discrete objects such as stellar coronae or supernova remnants can readily account for the observed X-ray properties, we assume that the ridge emission comes from diffuse plasma distributed along the Galactic plane (Yamauchi et al. 1995, and references therein). The plasma temperature exceeds the escape temperature of the Galaxy ($\sim 5\times10^6$ K), so that it cannot be gravitationally bound. The Galactic ridge plasma has by far the highest pressure among various interstellar components, so that it cannot be contained by external pressure.

If not confined, the ridge plasmas would escape from the Galaxy in $\sim 10^5$ yr, i.e. the sound transit time across the ridge scale height ($\sim$ 0.1 kpc). The rate of supernova (SN) occurrence is too low to sustain the plasma outflow, and the temperature of the plasma produced by SN is too low. The observed nt parameter $\sim 10^{10-11}$ s cm^{-3} suggests that plasma may be confined for a substantial length of time rather than escaping freely.

These considerations lead us to propose that the Galactic ridge hot plasmas are confined within numerous magnetic flux tubes along the Milky way. For this to work, the plasma-β value must be < 1; that is, the *local* magnetic field strength of the tubes must exceed the equi-partition magnetic field strength $B_{\rm eq}$ defined as

$$\frac{B_{\rm eq}^2}{8\pi} = 2nk_{\rm B}T, \qquad (1)$$

where $k_{\rm B}$ is the Boltzmann constant. The parameters for the Galactic ridge plasmas (Table 1) indicate $B_{\rm eq} = 20 - 40\mu$G. This is larger than the *average* field strength of a few μG observed in our Galaxy. However, the equi-partition field strength can be made consistent with the average field strength, if the flux tubes have a low filling factor.

1.2. INTRA CLUSTER MEDIUM

A cluster of galaxies is a self-gravitating system, containing $10^{1\sim4}$ galaxies and a large amount of hot plasma called the Intra-Cluster Medium (ICM, Sarazin 1988 for review). The ICM is thought to be hydrostatically confined within the cluster potential, which is dominated by dark matter. The ICM temperature is roughly the same as the virial temperature of the gravitational potential, and the ICM X-ray luminosity appears to increase with age (Gioia et al. 1990). This indicates that the ICM has primarily been heated by gravitational compression during the cluster formation.

Detailed *ASCA* investigations of the Centaurus cluster of galaxies have revealed the following striking facts (Fukazawa et al. 1994). (1) High ($3-4\times 10^7$ K) and low ($\sim 1\times 10^7$ K) temperature plasmas coexist within 100 kpc ($r < 5'$) of the central dominant (cD) galaxy NGC 4696. (2) The emission measure of the cooler component decreases more rapidly with increasing distance from the center than the hotter component. (3) The metal abundance dramatically increases in the region where the cooler component is significantly seen.

We estimate the conductive equilibrium time between the two ICM components to be 8×10^{14} s $\sim 2\times 10^7$ yr, using $T = 4\times 10^7$ K, $n = 10^{-3}$ cm^{-3}, and $l \sim 50$ kpc to simulate the Centaurus cluster. Since this is much shorter than the typical age ($\sim$ Hubble time) of ICMs, the cooler component would quickly evaporate and disappear (Takahara & Takahara 1979), unless the two plasma components were thermally insulated from each other. The radiative cooling time scale of the cool component is $\sim 10^9$ yr ($<$ Hubble time), so that we need a heat source to maintain the cool component.

The coexistence of hotter and cooler plasmas is a common occurrence in the solar corona (*e.g.* Yoshida & Tsuneta 1996), where plasma is heated in a highly non-uniform way, and the resultant high temperature component is isolated and is maintained by magnetic fields. We suggest that large-scale magnetic fields in the ICM provide these mechanisms, supposing that the cooler plasma is confined within numerous magnetic flux tubes and surrounded by the hotter plasma. The magnetic flux tubes thermally insulate the interior from the surroundings, and provide a stable heating mechanism to the confined plasma, perhaps through magnetic compression and reconnection (Petschek 1964), just as is observed in the solar corona.

Using the same set of parameters as above, we obtain the equi-partition magnetic field of order 20 μG from eq.(1). The field strength estimated from Faraday rotation is ~ 1 μG in most clusters of galaxies, and reaches $6-30\mu$G in Hydra A (Taylor & Perley 1993). Though the observed field strength is again smaller than the equi-partition magnetic fields, small filling factors will be able to reconcile these two values.

2. SCALING LAWS OF SOLAR CORONAL LOOPS

The introduction of magnetic flux tubes qualitatively explains some of the key features provided by the recent new observations of the Galactic ridge emission and ICMs. Since the implied (equi-partition) field strength is fairly high, the flux tubes must be closed within the systems; the field lines have the form of magnetic loops. Since the Galactic plane is located in the center of the ridge emission, and the cD galaxies are usually in the centers of the ICMs, it is natural to assume that the footpoints of these loops are located

TABLE 1. PHYSICAL PARAMETERS OF SOLAR, GALACTIC AND INTRACLUSTER PLASMAS

	Solar	Galactic	Clusters of galaxies (ICM)
T[keV]	0.1–0.5	5–10	1–10
L_x [erg sec^{-1}]	3×10^{16}	$(1-2)\times 10^{38}$	$(0.5\text{–}5) \times 10^{43}$
L[cm]	$0.1-10 \times 10^9$	$\sim 10^{22}$	$(6-60)\times 10^{23}$
Nominal p[dyn cm^{-2}]	0.1–10	$(5-11)\times 10^{-12}$	2×10^{-11}
Nominal n[cm^{-3}]	10^{9-11}	7×10^{-4}	10^{-3}
B [μG]	100 G	2	1–30
Filling factor	–	$2 \times 10^{-4\sim-5}$	10^{-2} (hot)– 1(cool)
Loop p[dyn cm^{-2}]	0.1-10	$7-20 \times 10^{-10}$	2×10^{-10}
Loop n[cm^{-3}]	10^{9-11}	0.05–0.16	10^{-2}
Predicted loop T[KeV]	0.2–0.4	5 – 6	1 – 6
F_T [erg sec^{-1} cm^{-2}]	10^8	0.005–0.05	$3-6 \times 10^{-3}$
B_{eq} [μG]	~ 8 G	132–224	70
B_{mean} [μG]	10 – 100 G	~ 2	7

on the Galactic plane and the central cD galaxies.

This magnetic geometry is quite analogous to solar coronal loops. In this section, we review the key features of magnetic loops in the solar corona, and obtain the simple relation (a scaling law) between properties of coronal loops such as peak temperature, loop length, pressure, and energy loss rate (heating rate), following Rosner, Tucker & Vaiana (1978), Kano (1997), and Kano & Tsuneta (1995, 1996). Table 1 summarizes key parameters of the Galactic, ICM and solar coronal plasmas. Although the physical parameters are quite different between solar and these cosmic plasmas, the scaling laws should hold, if there are steadily-heated magnetic flux tubes in these media.

The *Skylab* and subsequent *Yohkoh* imaging observations of the Sun in soft X-rays revealed that the solar corona is not a single entity, but consists of ubiquitous isolated magnetic flux tubes. Since these flux tubes suppress the classical transport across magnetic fields, the physical properties of single loops depend on their *in situ* present and past heating rates, and do not depend on the properties of nearby loops. In other words, the neighboring loops do not "communicate" with each other except through magnetic reconnection, owing to the fact that the collision frequency is much lower than the gyro-frequency in the solar corona. This is the reason why the scaling laws hold for wide ranges of solar coronal loops.

We consider the situation that coronal loop anchored in the cool photosphere are steadily heated. The input energy is conductively dumped toward the photosphere, while partially losing the energy by radiation. The total input energy is balanced with the total radiative and conductive energy losses. The energy equation in steady-state is

$$E_{\rm H} + E_{\rm R} - \boldsymbol{\nabla} \cdot \boldsymbol{F}_{\rm C} = 0, \qquad (2)$$

where $E_{\rm H}$ is the heat input rate, $-E_{\rm R}$ is the radiative loss and $\boldsymbol{F}_{\rm C}$ is the conductive heat flux along magnetic field. These quantities change only along magnetic field lines. The radiative loss and the heat conduction are, respectively,

$$-E_{\rm R} = n_{\rm e}^2 \,\Lambda(T) = \frac{p^2}{4k_{\rm B}^2}\frac{\Lambda(T)}{T^2}, \qquad (3)$$

$$\boldsymbol{\nabla} \cdot \boldsymbol{F}_{\rm C} = -\frac{\rm d}{{\rm d}s}\left(\kappa_{\|}\frac{{\rm d}T}{{\rm d}s}\right), \qquad (4)$$

where p is a constant pressure, $\Lambda(T)$ is the radiative loss function, s is the coordinate along the loop from the footpoint, and $\kappa_{\|}$ is the thermal conductivity along magnetic field lines; $\kappa_{\|} = 10^{-6}T^{5/2}$ [erg/K/cm/sec] $\equiv \kappa_0 T^{5/2}$.

The scaling law is obtained by integrating equation (2):

$$T[{\rm K}] = (1.1 \sim 1.4) \times 10^3 (p[{\rm dyncm^{-2}}]L[{\rm cm}])^{0.33}, \qquad (5)$$

where T is the peak temperature of the loop, p pressure of the loop, and L the length of the loop. The numerical factor is slightly different, depending on the heat distribution along the loop [heat input at the loop top (1.1), and uniform heat input (1.4)].

The power index of the right hand side of the equation, however, does not depend on the heat input profile along the loop. The total energy input $F_{\rm T}$ [erg/cm^{2}sec^{-1}] is

$$F_{\rm T} = \int E_{\rm H} {\rm d}s = (1.5 \sim 1.7) \times 10^5 p^{1.16} L^{0.17}. \qquad (6)$$

Again only the numerical factor slightly depends on how the heat is applied along the magnetic fields [heat input at the loop top (1.5), and uniform heat input (1.7)]. The two scaling laws are thus robust for wide range of heat deposition profiles along magnetic fields. This is one of the reasons that the scaling laws are useful.

3. SCALING LAWS FOR COSMIC MAGNETIC LOOPS

3.1. COSMIC SCALING LAW

In this section, we apply the scaling laws derived for the solar condition to the cosmic situation. In the cosmic plasma discussed here, the temperatures are generally higher than that of the solar coronal plasma ($T < 6 \times 10^6$ K), and the metal abundance is sometimes lower than the solar value. Thus, the radiative loss function changes from the solar value, affecting the scaling laws. The scaling laws ($T > 2 \times 10^7$ K) is then

$$T[\mathrm{K}] = (1.5 \sim 2.2) \times 10^2 (p[\mathrm{dyncm}^{-2}] L[\mathrm{cm}])^{0.39}, \tag{7}$$

$$F_{\mathrm{T}}[\mathrm{ergcm}^{-2}\mathrm{sec}^{-1}] = (2.3 \sim 2.8) \times 10^2 p^{1.35} L^{0.35} = (1.5 \sim 6.0) \times 10^{-6} T^{3.5} L^{-1}, \tag{8}$$

for the solar abundance. The scaling law for 0.5 times solar abundance is

$$T = (1.3 \sim 1.8) \times 10^2 (pL)^{0.39}, \tag{9}$$

$$F_{\mathrm{T}} = (1.1 \sim 1.4) \times 10^2 p^{1.37} L^{0.37} = (1.6 \sim 6.3) \times 10^{-6} T^{3.5} L^{-1}. \tag{10}$$

The range of the numerical factors corresponds to the case of the delta-function heat input at the loop top and the case of the uniform heat input along the loop. The scaling laws do not change much in the two cases, and we adopt the following scaling laws in the subsequent discussions:

$$T = (1.3 - 2.2) \times 10^2 (pL)^{0.39}, \tag{11}$$

$$F_{\mathrm{T}} = (1.1 - 2.8) \times 10^2 p^{1.37} L^{0.37} = (1.5 - 6.3) \times 10^{-6} T^{3.5} L^{-1}. \tag{12}$$

3.2. PROPERTIES OF THE COSMIC MAGNETIC TUBES

From the X-ray observations, we obtain the temperatures T and emission measures EM. If the volume filling factor of the magnetic flux tubes is f, and the X-ray emitting high temperature plasma is confined to the flux tubes in the apparent volume V of the X-ray emitting plasma, the density of the flux tubes is given by

$$n = \sqrt{\frac{EM}{Vf}}. \tag{13}$$

If the total X-ray luminosity L_{x} balances the heat input rate of the flux tubes, we obtain

$$F_{\mathrm{T}} = \frac{L_{\mathrm{x}}}{Sf}, \tag{14}$$

where S is the apparent cross-sectional size of the high temperature plasma. The filling factor is then given by,

$$f = \frac{L_{\mathrm{x}}}{SF_{\mathrm{T}}} = (0.05 - 0.2) \left(\frac{L_{\mathrm{x}}}{10^{40}[\mathrm{ergsec}^{-1}]}\right) \left(\frac{L}{10^{22}[\mathrm{cm}]}\right) \left(\frac{S}{10^{44}[\mathrm{cm}^2]}\right)^{-1} \left(\frac{T}{10^7[\mathrm{K}]}\right)^{-3.5}, \tag{15}$$

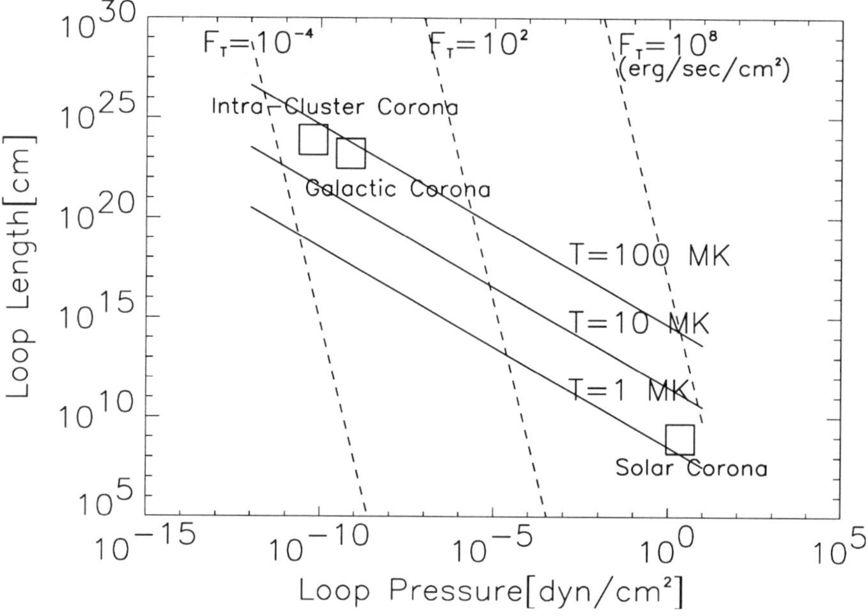

Figure 1. Loop pressure – loop length diagram. The locations of the solar corona, and the Galactic and intracluster coronae are shown by boxes. The solid lines show the scaling law for three different temperatures, and the dashed lines the scaling law for three different heat inputs, F_T. Not only the solar corona, but the Galactic and intracluster coronae also essentially fit with the two scaling laws for proper assumptions of the volume filling factor (see text). Note the pressure of the horizontal axis is not the nominal pressure, but the pressure of the magnetic flux tubes.

where T is the temperature of the loops, and L the length of the loops.

3.2.1. *Galactic Ridge Emission*

The filling factor of the magnetic flux tubes f for the Galactic ridge emission is $\sim 2\times10^{-4} - 2\times10^{-5}$ for $L_x \sim 10^{38}$ erg sec^{-1}, $T \sim 5\times10^7 - 10^8$ K, $S \sim 10$ kpc $\times$ 0.1 kpc, and $L \sim 3$ kpc. The nominal plasma density and pressure are 7×10^{-4} cm^{-3}, and 10^{-11} dyn cm^{-2}, respectively (Table 1). The plasma density and the pressure of the flux tubes are then 0.05–0.16 cm^{-3} and $7\times10^{-10} - 2\times10^{-9}$ dyn cm^{-2}. The heat input rate F_T is 0.005–0.05 erg sec^{-1} cm^{-2}. The equi-partition magnetic field B_{eq} is about 132–224 μG. The mean field strength is $B_{eq}f^{0.5} \sim 2\mu$G, which is consistent with the observations (Ohno & Shibata 1993).

The temperature predicted from the scaling law [equation (11)] is $(5-6)\times10^7$ K for the loop length L of 10 kpc and the filling factor of 10^{-5}. This temperature is lower than the observed highest temperatures. If the filling factor is lower (meaning a higher loop pressure), or the length of the loops is longer than the assumed 10 kpc, the predicted temperature can be higher. In the solar case, the loops which are transiently heated usually have temperatures higher than that predicted by the scaling law (Kano & Tsuneta 1995). The Galactic loops with temperatures higher than 6×10^7 K may deviate from the steady state equilibrium situation. If the Galactic ridge emission is sporadically heated by magnetic reconnection (eg. Tanuma et al. 1997), the conductive cooling time scale is of the order of 2×10^{11} sec $\sim 7 \times 10^3$ yr, which is much shorter than the radiative cooling time. The nt parameter of the magnetic tubes is then about 10^{10} $cm^{-3}s$, which is consistent with observations.

3.2.2. *ICM*

The filling factor of the magnetic flux tubes for the ICM is about 0.01 for $L_x \sim 10^{43}$ erg s^{-1}, $T \sim 7 \times 10^7$ K, and $L \sim 200$ kpc ($S \sim L^2$). The nominal plasma density and the pressure are 10^{-3} cm^{-3} and 2×10^{-11} dyn cm^{-2}, respectively (TABLE 1). The plasma density and the pressure of the flux tubes are thus 10^{-2} cm^{-3} and 2×10^{-10} dyn cm^{-2}. The heat input rate $F_{\rm T}$ is then $3-6 \times 10^{-3}$ erg s^{-1} cm^{-2}. The equi-partition magnetic field $B_{\rm eq}$ is about 70 μG. The mean field strength is $B_{\rm eq} f^{0.5} \sim 7$ μG. This is consistent with observations (Taylor & Perley 1993). The temperature predicted from the scaling law [equation (13)] is 6.7×10^7 K for a loop length L of 200 kpc. This temperatures coincides with the ICM temperatures.

Although the parameters of the ICM high temperature component is consistent with those derived from the scaling laws, the ICM high temperature component is believed to be gravitationally heated. On the the other hand, the cool component located around the central cD galaxy has $T \sim 1-3 \times 10^7$ K, $L_x \sim 10^{43}$ erg s^{-1}, $L \sim 25$ kpc, giving $f \sim 1$ from equation (15). The temperature of the cool component predicted by the scaling law is 16 MK for loops with length 5kpc. This indicates that dense bundles of small and large magnetic flux tubes maintain the low temperature component of ICMs.

Thus there is a possibility that the cool component of ICMs near the central cD galaxy is more closely related to the magnetic activity, whereas there may be no compelling reason to introduce magnetic fields for hot ICMs. However, we point out one interesting conclusion regarding the hot ICMs from the scaling law: There is clear correlation between the ICM temperatures and the the X-ray luminosity; $L_x \propto T^{3.3}$ (e.g. Arnaud 1994). The scaling law shows the relation among the heat input rate, the temperatures, and the loop length; $F_{\rm T} \propto T^{3.5}/L$. If the heat input is balanced by radiative loss, we obtain $L_x \propto F_{\rm T} \propto T^{3.5}$, which is close to the observed correlation.

The locations of the Galactic ridge emission, ICM and the solar corona are shown on a $p - L$ diagram in Figure 1. Note that the pressure here is the pressure of the flux tubes, which is considerably higher than the nominal pressure obtained from the observed emission measure and temperature owing to the small filling factor. The loop length of the vertical axis is not necessarily the size of the system, but can take any value smaller than the size of the system. The solid lines are the scaling law [equation (11)] for the given temperatures, and the dashed lines give the heating rate [equation (12)] for a given pair of loop pressure and loop length. The observed parameters are essentially consistent with those derived from the scaling laws. The predicted temperature for the Galactic plasma is, however, generally lower than the observed highest temperatures, implying that it is transiently heated.

4. CONCLUSIONS

We have assumed that magnetic fields are responsible for the confinement of the Galactic ridge plasmas and the cool component of ICMs. This assumption is supported by the estimated short escape and evaporation times, and by the multiple temperature structures observed with *ASCA*. These features are qualitatively explained by the introduction of magnetic fields, because magnetic fields are effective for the confinement of the high temperature plasma, and for maintaining the multiple temperature structures as demonstrated in the solar corona.

The scaling law developed and verified in the solar context is then applied for these cosmic flux tubes to qualitatively examine this hypothesis. The simple time-scale estimation indicates that Galactic ridge and cool ICM plasmas have to be heated quasi-continuously, thus justifying the key assumption of the scaling law. The filling factor of the high temperature magnetic loops is deduced from the assumption that the energy supply rate is balanced by the loss, due to radiation in X-ray wavelengths, and due to heat conduction. The plasma properties and mean magnetic field strength derived with the scaling laws are essentially consistent with the observations.

Since there is no known heating mechanism for the Galactic ridge emission and the cool component of ICMs, we point out the possibility that magnetic fields play an important role also for the heating of these plasmas (eg. with magnetic reconnection: Tanuma et al. 1997, Raymond 1992, Sturrock & Stern 1980).

References

Arnaud, M. (1995) in *Cosmological Aspects of X-ray Clusters of Galaxies*, ed. Seitter, W. C. (Kluwer Academic Publishers: Dordrecht) p 197.

Fukazawa, Y., Ohashi, T., Fabian, A. C., Canizares, C. R., Ikebe, Y., Makishima, K., Mushotzky, R. F. & Yamashita, K. (1994) *Publ. Astron. Soc. Japan*, **46**, L55.
Gioia, I. M., Henry, J. P., Maccacaro, T., Morris, S. L., Stocke, J. T. & Wolter, A. (1990) *Ap. J.*, **356**, L35.
Kano, R. (1997) Ph. D. thesis, University of Tokyo.
Kano, R. & Tsuneta, S. (1995) *Ap. J.*, **454**, 934.
Kano, R. & Tsuneta, S. (1996) *Publ. Astron. Soc. Japan*, **48**, 535.
Makishima, K. (1994) in *New Horizon of X-ray Astronomy*, eds. Makino, F. and Ohashi, T. (Universal Academy Press: Tokyo) p 171.
Makishima, K. (1995) in *Elementary Processes in Dense Plasmas*, eds. Ichimaru, S. and Ogata, S. (Addison Wesley Publishers: Menlo Park) p47.
Makishima, K. (1996) in *X-ray Imaging and Spectroscopy of Cosmic Hot Plasmas*, eds. Makino, F. and Mitsuda, K. (Universal Academy Press: Tokyo) p137.
Makishima, K. (1997) in these proceedings.
Ohno, H. & Shibata, S. (1993) *MNRAS*, **262**, 953.
Petschek, H. E. (1964) in *AAS-NASA Symposium on Solar Flares*, ed. W. N. Hess (NASA SP-50) p425.
Raymond, J. C. (1992) *Ap. J.*, **384**, 502.
Rosner, R., Tucker, W. & Vaiana, G. (1978) *Ap. J.*, **220**, 643.
Sarazin, C. L. (1988) *X-ray emissions from clusters of galaxies* (Cambridge University Press: Cambridge)
Sturrock, P. A. & Stern, R. (1980) *Ap. J.*, **238**, 98.
Takahara, M. & Takahara, F. (1979) *Prog. Theor. Phys.*, **62**, 1253.
Tanuma, S., Yokoyama, T., Kudoh, T., Shibata, K. & Makishima, K. (1997) *Ap. J. (Letters)*, submitted.
Taylor, G. B. & Perley, R. A. (1993) *Ap. J.*, **416**, 554.
Yamauchi, S. et al. (1995) *Publ. Astron. Soc. Japan*, **48**, L15.
Yoshida, T. & Tsuneta, S. (1996) *Ap. J.*, **459**, 342.

III. Magnetic Reconnection as an Engine of Cosmic Explosions

COMPUTER SIMULATIONS OF TWISTED FLUX TUBES AND MAGNETIC RECONNECTION

T. HAYASHI, M.OZAKI, H.KITABATA, AND T.SATO
National Institute for Fusion Science, Toki, Gifu 509-52 Japan

1. Energy Relaxation of Twisted Magnetic Flux Tubes

To investigate the dynamical problem of continuously twisted flux tubes, elaborate three-dimensional MHD simulations have been performed (Amo, *et al.* 1995). Initially a uniform straight axial magnetic field is assumed. A twisting motion is applied continuously at both ends of the flux tube. An excess magnetic energy is accumulated in the system as the twist goes on. When the flux tube is largely twisted and free energy is sufficiently stored, the structure becomes unstable to kink modes.

Triggered by the instability, the magnetic field lines are largely deformed like a knot-of-tension instability of a bundle of twisted rubber strings. A singular current layer is formed between the deformed twisted flux tube and the surrounding untwisted magnetic field. Reconnection is driven at the singular current layer and the deformed flux tube reconnects with the surrounding straight magnetic field. Thus, the topology of the magnetic field is drastically changed. Through these processes, a large fraction of the stored magnetic energy is converted into thermal and kinetic energy. Because of the temporal change of the nonlinear susceptibility of the system, the energy transfer rate (Poynting Flux) from the continuous twisting motion of the boundary changes with time. When the excess energy is released, the system again absorbs more energy from the twisting motion and a similar relaxation process is repeated. In other words, a burst-like energy relaxation appears intermittently.

Interestingly, after several intermittent burst-like relaxations, the magnetic field structure returns to the initial state. Then, a recurrent behavior is reproduced in the topological change of the magnetic field. The recurrent feature is also seen in the temporal variation of the energy conversion process.

T. Watanabe et al. (eds.), Observational Plasma Astrophysics: Five Years of Yohkoh and Beyond, 133–136.

From these results, we can propose a new scenario of self-organization for an open system where energy is continuously supplied from outside. The initial stable state reaches an unstable state when subjected to a steady energy supply. Nonlinear development brings the system to a marginal (supersaturated) state which is a nonlinear bifurcation point, whereby reconnection is driven and the topology of the magnetic field is changed. Through this relaxation process, stored magnetic energy is rapidly released and a stable state, which is a local minimum energy state, is realized. Upon a continuous energy supply, the energy accumulation and relaxation take place in an intermittent, burst-like fashion. Alternating between a marginal state and a local energy minimum state, the system eventually returns to the initial state, which may be the global minimum energy state. The intermittency and recurrence in the relaxation processes are the common and key features that characterize the self-organization in the open system subject to a continuous energy supply.

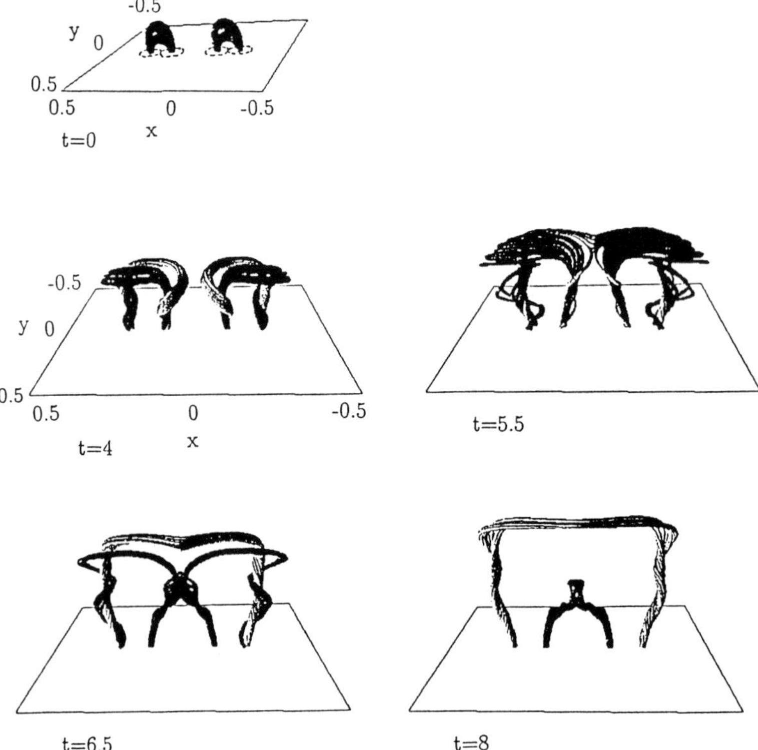

Figure 1. Temporal evolution of the magnetic field lines for a paired loop-like magnetic configuration with attracting force.

Figure 2. Configuration of magnetic field lines before and after reconnection for a paired loop-like magnetic configuration.

2. Nonlinear Behavior of Twisted Magnetic Loops

We have studied temporal evolutions of isolated twisted magnetic loops and sheared magnetic arcades by a three-dimensional MHD simulation (Ozaki, *et al.* 1997) For the initial magnetic field, a current-free (potential) bipolar field emerging from photospheric convection cells is given. Simulations are performed for a single bipolar flux tube and for twin bipolar flux tubes.

In the case of a single magnetic loop or arcade, no drastic topological change of the magnetic field lines is observed. Namely, a single magnetic loop or arcade model may not be able to be a potential candidate for the topological change of the magnetic field and the distinct energy release mechanism. In order to explain the large energy release and drastic reconnection associated with a single magnetic loop or arcade, one may take the overlying flux (ambient field) into account.

In contrast, when two magnetic loops or arcades collide with each other, salient magnetic reconnection is triggered because of the attractive force acting between the loop or arcade currents. Their evolution and resulting reconnection produce a larger overlying expanding loop (arcade) and a smaller lower one (See Fig. 1 and Fig.2). It should be noted that the energy release becomes more intensive and impulsive as resistivity becomes smaller. These features can well be explained by driven magnetic reconnection and appear to be consistent with some important observational features of solar flares and transient X-ray brightenings.

3. Impulsive Nature in MHD Driven Reconnection

We have studied the relaxation process of driven magnetic reconnection by performing a long simulation run. By making a long time scale calculation, we discovered extremely intensive impulsive reconnection and intermittent nature of bursts in the long time behavior of the driven magnetic reconnection (Kitabata, *et al.* 1997). The reconnection rate reaches to an extremely large value, more than ten times that of the driving rate. We used the pure resistive MHD mode without invoking any electron inertia or electron pressure effects in the generalized Ohm's law. The process consists of two phases. In the first phase the reconnection rate is rather slow, and magnetic energy is gradually accumulated in the system. In the second phase, suddenly, the collapsing phenomena occurs, and the magnetic energy is released in quite a short period. This process is repeated. Through this impulsive reconnection, the stored magnetic energy is mainly converted to plasma flow energy. Examination of the magnetic field structure has revealed that the process is directly related with behavior of magnetic islands which are formed along neutral lines. The slow rate of reconnection observed in the first phase corresponds to a period when the islands gradually grow and saturate. The sudden increase in the reconnection rate observed in the second phase is triggered by collapse of the island. These behaviors suggest that very fast reconnection can take place naturally in a driven system (Sato, *et al.* 1978) and that it occurs in a very impulsive and intermittent fashion.

References

H.Amo, T.Sato, A.Kageyama, K.Watanabe, R.Horiuchi, T.Hayashi, Y.Todo, T.H.Watanabe, H.Takamaru, Phys. Rev. E **51**, 3838 (1995).

M.Ozaki, T.Sato, T.Hayashi, R.Horiuchi, K.Watanabe, Y.Todo, A.Kageyama, T.H.Watanabe, H.Takamaru, Astrophys. J., in print (1997).

H.Kitabata, T.Hayashi, T.Sato, J. Phys. Soc. Jpn **65**, 3208 (1996).

T.Sato, T.Hayashi, Phys. Fluids **22**, 1189 (1979); T.Hayashi, T.Sato, J. Geophys. Res. **83**, 217 (1978).

MAGNETIC RECONNECTION AND FLARES IN STAR FORMING REGIONS : FLARES IN PROTOSTARS

MITSURU HAYASHI
Japan Atomic Energy Research Institute (JAERI)
801-1 Mukouyama, Naka-machi, Naka-gun,
Ibaraki, 311-01, Japan

KAZUNARI SHIBATA
National Astronomical Observatory
Mitaka, Tokyo 181, Japan

AND

RYOJI MATSUMOTO
Department of Physics, Faculty of Science, Chiba University
Inage-Ku, Chiba 263, Japan and
Advanced Science Research Center, JAERI
Naka-machi, Naka-gun, Ibaraki, 311-01 Japan

Abstract.

The solar X-ray satellite, *Yohkoh*, has found extensive evidence for magnetic reconnection in solar corona. Here we extend the magnetic reconnection model of solar flares to hard X-ray flares observed in star forming regions. A new ingredient is a protostellar disk which can inject helicity into the magnetosphere if the disk is threaded by the dipole magnetic field of the protostar. We carried out 2.5-dimensional magnetohydrodynamic (MHD) simulations of the disk-star interaction. The closed magnetic loops connecting the central star and the disk are twisted by the rotation of the disk. In the presence of resistivity, magnetic reconnection takes place in the current sheet formed inside the expanding loops. A hot, outgoing plasmoid and hot post flare loops are formed as a result of the reconnection. Numerical results are consistent with the observed plasma temperature ($10^7 - 10^8$K), the length of the flaring loop ($10^{11} - 10^{12}$cm), and the velocity of optical jets ($200 - 400$km/s). We showed by high-resolution numerical simulations using parallel computers that multiple magnetic islands are created in the current sheet due to the growth of the tearing mode instability. The magnetic is-

T. Watanabe et al. (eds.), Observational Plasma Astrophysics: Five Years of Yohkoh and Beyond, 137–142.

lands are ejected quasi-periodically. Intermittent flaring activity continues as long as the disk matter twists the dipole magnetic field.

1. Introduction

Observations by the Hubble Space Telescope (HST) and X-ray satellites have revealed various energetic phenomenon associated with star formation. One remarkable phenomena is the well collimated bipolar plasma outflows (optical jets). Optical observations of HH30 and other protostars by HST confirmed the existence of protostellar disks and showed that optical jets are collimated within 100AU of the central protostar (Burrows et al. 1996).

Although protostars are obscured in optical wavelengths, hard X-rays can peer deep into the core of star forming regions. Observations by the Japanese X-ray satellite ASCA showed that protostars are strong hard X-ray emitting sources. Using ASCA, Koyama et al. (1994,1996) carried out a systematic survey of hard X-ray sources in molecular clouds. A protostar (R1 in Coronet cluster) exhibited a powerful X-ray flare whose spectrum indicates significantly higher temperature ($T \sim 10^8$ K) than that in solar flares and X-ray flares in T Tauri stars which is usually $\sim (1-2) \times 10^7$ K (Koyama et al. 1996). The total energy released by the protostellar flare ($10^{35} - 10^{36}$ erg) is 10^5 times larger than solar flares. The size of the X-ray emitting region is several times larger than the protostar.

X-ray observations of the solar corona by the *Yohkoh* satellite (Tsuneta et al. 1991) gave us the following evidence supporting the idea that magnetic reconnection is responsible for the energy release in solar flares (Tsuneta et al. 1992); (1) After the flare, the loop top region shows a cusp-like structure, implying that a reconnection site is at the top of the flare loop, (2) the height and the footpoint separation of the soft X-ray emitting post flare loop increase with time, (3) the outer loops, which are located closer to the reconnection point, have higher temperatures. Another indication of magnetic reconnection is the detection of a hard X-ray emitting source above the soft X-ray flare loops in impulsive flares (Masuda et al. 1994).

The essential difference between the Sun and protostars is the existence of a rotating disk in protostars. In the innermost region of protostellar disks, the magnetic fields of the protostar may affect the dynamical evolution of the disk. Since the rotation of the disk continuously injects helicity into the closed magnetic fields connecting the central star and the disk, we expect magnetic nonequilibrium and resulting magnetic reconnection similar to those studied in the context of solar flares (e.g. Mikić & Linker 1994; Kusano, Suzuki & Nishikawa 1995). Recently, Hayashi et al. (1996)

carried out nonlinear time-dependent MHD simulations of the interaction between the dipole magnetic field of the central star and the rotating disk threaded by the dipole field. In the following, we show that the outgoing hot rotating spheromak created by the magnetic reconnection can be the origin of optical jets and hard X-ray emission.

2. Numerical Models

We solve the resistive MHD equations in cylindrical coordinate (r, ϕ, z) by applying a modified Lax-Wendroff scheme with artificial viscosity. Axial symmetry is assumed. The effects of radiative cooling and rotation of the central star are neglected. We assume that the magnetic dipole moment is aligned with the rotational axis. The initial state consists of a magnetized central star and a dense disk rotating in a spherical static halo. The disk is assumed to be in rotational equilibrium. The angular momentum distribution of the disk is nearly Keplerian $L/L_0 = (r/r_0)^{0.494}$, where the reference radius r_0 is at the pressure maximum of the disk. The density ratio between the disk and the halo is taken to be $\rho_h/\rho_0 = 2 \times 10^{-5}$ where ρ_0 is the maximum density of the disk. The unit velocity is the Keplerian speed v_{K0} at $r = r_0$. The plasma β in the disk at the pressure maximum is $\beta_0 = 8\pi P_0/B_0^2 = 20$.

We use free boundary conditions at the outer boundaries. At the equatorial plane $(z = 0)$, we assume a symmetric boundary condition. The stellar surface is treated as a thin shell-like damping region where all quantities except magnetic fields are damped to the initial values. The stellar surface is at $r = 0.528 r_0$. We assume anomalous resistivity $\eta(r, z) = \eta_0[\max(v_d/v_c - 1.0, 0)]^2$ where v_c is the critical value above which the anomalous resistivity sets in and $v_d = |\mathbf{J}|/\rho$ (Sato & Hayashi 1979). Here, $\mathbf{J}$ is the current density. We adopt $v_c = 1000$ as a critical value and $\eta_0 = 0.01$ in non-dimensional units.

3. Numerical Results

The results of the simulation of a typical model are shown in Figure 1. Solid curves, gray scale and arrows show the magnetic field lines, temperature, and velocity vectors, respectively. The magnetic field lines connecting the central star and the disk are twisted by the rotation of the disk. As the magnetic twist accumulates, the magnetic loops begin to inflate. They expand quasi-statically in the early stage but after a half rotation they expand dynamically. Magnetic reconnection takes place in the current sheet formed inside the expanding loops. Hot plasmas created by the reconnection are confined in the outgoing magnetic island. The speed of the outflowing hot plasma is the order of the local Alfvén speed $v_A \sim (2-5)v_{K0}$. Crest-shaped fast MHD shock waves are formed in front of the expanding magnetic loops.

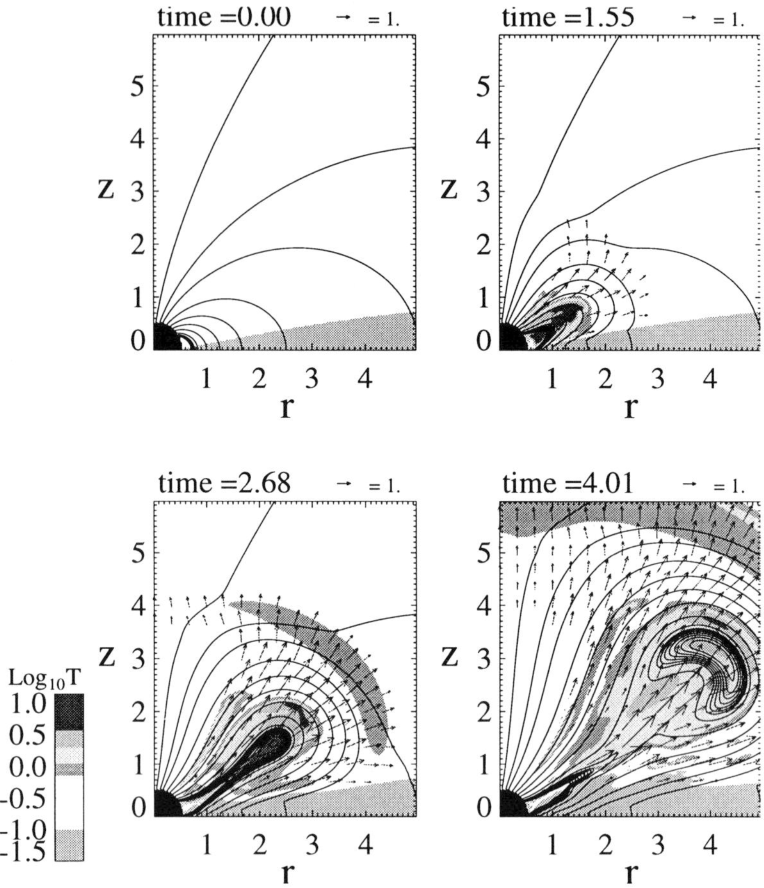

Figure 1. The results of the simulation of a typical model

In the above model, the dipole magnetic field of the central star is not strong enough to support the magnetically braked disk matter. Since the accreting matter drags the footpoints of magnetic field lines anchored to the disk, it deforms the magnetosphere of the central star. The footpoints of the post flare loops created after the magnetic reconnection are removed to the stellar surface before the rotation of the footpoints injects suffient twist into the loops. We carried out another calculation using a stronger magnetic field parameter (five times stronger magnetic field than that in the above model) and truncating the disk at the innermost radius at r=r_0. Although in this model the surrounding disk is not exactly in dynamical equilibrium, the deviation from equilibrium is small because its rotation is nearly Keplerian. The magnetosphere of the star cannot be deformed easily. The results are shown in Figure 2. Solid curves, gray scale and arrows show the magnetic field lines, density, and velocity vectors, respectively. Even after the first reconnection, the inner edge of the disk does not touch the stellar surface. Since the rotation of the footpoints of the post flare loops twists the loops again, they begin to inflate and a current sheet is formed. Magnetic reconnection and plasmoid ejections occur intermittently.

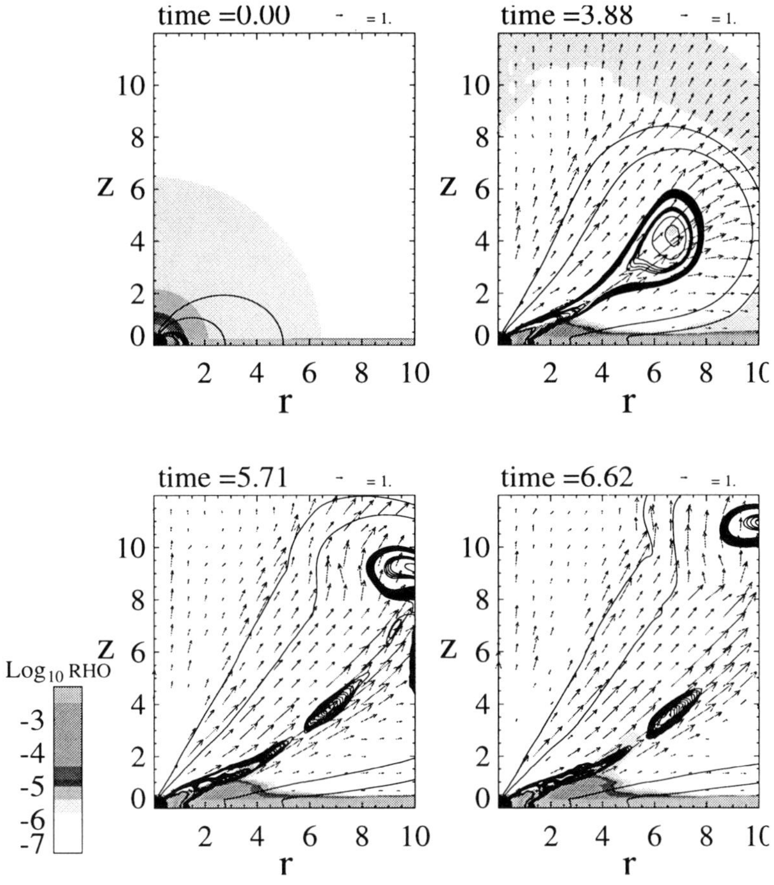

Figure 2. Recursive flaring in protostars

4. Discussion

We demonstrated that if the dipole magnetic field of the central star is connected to the rotating disk, the dipole magnetic field loses equilibrium due to the twist injection and that magnetic reconnection takes place within one rotation period of the disk. We also showed that hot plasmoids are ejected in bipolar directions with velocity 2–5 times the Keplerian rotation speed. The plasma temperature can rise up to the temperature determined by $v_s \sim v_A \sim 2-5 v_K$. Since $v_K \sim 170$ km s^{-1} at $r = 0.03$ AU for 1 $M_{\odot}$ star, the temperature of the plasmoid and the post flare loop can be $10^7 - 10^8$ K. This temperature is consistent with the temperature obtained from the observation of an X-ray flare from a protostar (Koyama et al. 1996).

Our numerical result shows that hot plasmas are created in gigantic loops several times larger than the stellar radius. This is consistent with the observational constraint that the hard X-ray emitting region is larger than the size of the protostar. After the X-ray emitting gas is cooled down (cooling time $\sim$ 1 day), it will be observed as optical jets.

Next, we discuss whether the magnetospheric flares can release the observed total energy of X-ray flares. In our simulation, the initial magnetic energy is thermalized within one rotation period ($\sim 10^5$ sec at 0.03 AU). If the size of the flaring region is $d \sim 10^{11}$ cm and the mean strength of the magnetic field in the flaring region is $B \sim 100 - 200$ G, the released energy $L \sim (B^2/8\pi)\, d^3 \sim 10^{36}$ erg is the order of the observed luminosity.

In conclusion, both hard X-ray flares and optical jets in protostars can be explained by magnetic reconnection driven by the helicity injection from the disk to the magnetosphere. Numerical results show recursive flare-like activities and intermittent mass ejections in a magnetically linked disk-star system. Longer time-scale X-ray observations of protostars may confirm such intermittent activities.

References

Burrows, C. et al. 1996, *ApJ*, **473**, 437
Hayashi, M.R., Shibata, K. & Matsumoto, R. 1996, *ApJ*, **468**, L37
Koyama, K., Hamaguchi, K., Ueno, S., Kobayashi, N., & Feigelson, E. D. 1996, *PASJ*, **48**, L87
Koyama, K. et al. 1994, *PASJ*, **46**, L125
Kusano, K., Suzuki, Y., & Nishikawa, K. 1995, *ApJ*, **441**, 942
Masuda, S., Kosugi, T., Hara, H., Tsuneta, S., & Ogawara, Y., 1994, *Nature*, **371**, 495
Mikić, Z., & Linker, J. A. 1994, *ApJ* , **430**, 898
Sato, T. & Hayashi, T. 1979, *Phys. Fluids*, **22**, 1189
Tsuneta, S. et al. 1991, *Sol. Phys* ,**136**,37
Tsuneta, S.,1992, IAU Colloquim No. 153,"Magnetohydrodynamic Phenomena in the Solar Atmosphere", ed. Kosugi & Uchida, Kluwer
Woltjer, L. 1958, *Proc. Natl. Acad. Sci. U.S.A.*, **44**, 489

MAGNETICALLY DRIVEN JETS FROM ACCRETION DISKS: NONSTEADY 2.5D-MHD SIMULATION

KUDOH, T.
National Astronomical Observatory, Mitaka, Tokyo 181, Japan
MATSUMOTO, R.
Department of Physics, Faculty of Science, Chiba University, Inageku, Chiba 263, Japan
AND
SHIBATA, K.
National Astronomical Observatory, Mitaka, Tokyo 181, Japan

Abstract. We present the results of 2.5-dimensional MHD simulations for jet formation from accretion disks. Although the jets in nonsteady MHD simulations (e.g, Uchida & Shibata 1985) have often been referred to as transient phenomena resulting from a special choice of initial conditions, we found that the basic physics of these nonsteady jets are able to be understood with a theory of steady jets. (1) The velocity dependence on the field strength ($V_{\rm jet} \propto B^{1/3}$ for weak magnetic field) is in rough agreement with that of the one-dimensional steady solutions (Kudoh & Shibata 1995, 1997a, 1997b). The velocity of the jet is of order of the Keplerian velocity of the disk for a wide range of parameters. (2) The ejection point of the jet is determined by the effective potential which results from the gravitational force and the centrifugal force along a field line (Blandford & Payne 1982).

1. Introduction

Astrophysical jets are observed in young stellar objects (YSOs), active galactic nuclei (AGNs), and some X-ray binaries (XRBs). Although the acceleration and collimation mechanism of these jets are still not well understood, these objects are believed to have accretion disks in their central regions. One of the most promising models for the jets is magnetic acceleration from the accretion disk (Fig.1). In this article, we present the results

T. Watanabe et al. (eds.), Observational Plasma Astrophysics: Five Years of Yohkoh and Beyond, 143–148.

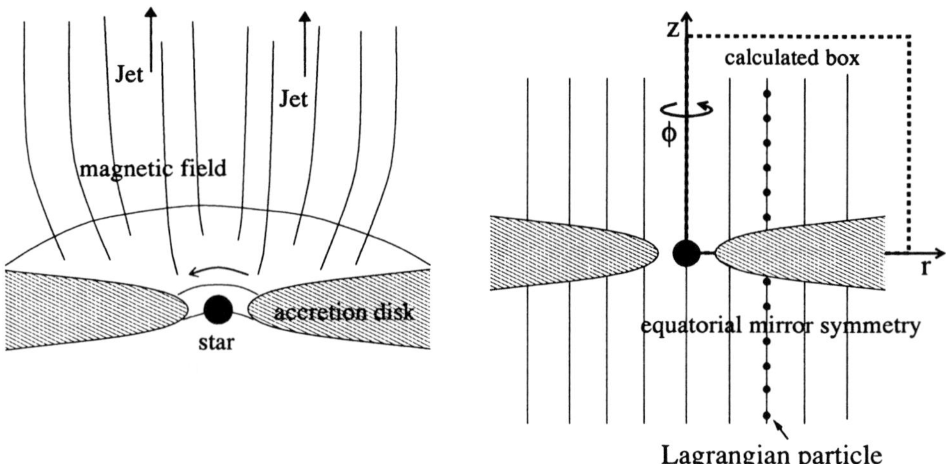

Figure 1. (Left) Schematic picture of jets from accretion disks. (right) A 2-dimensional (2.5D) MHD simulation was performed under the assumption of axial symmetry and equatorial mirror symmetry.

of time-dependent 2.5-dimensional MHD simulations for a magnetohydrodynamical model of jet formation from accretion disks. Time-dependent 2.5-dimensional MHD simulations for jets from accretion disks were first performed by Uchida & Shibata (1985) and Shibata & Uchida (1986, 1990). They showed that magnetically driven jets are ejected from rotating disks. In those papers, however, the physical properties of the jet were not clear because of the complexity of the numerical results. The purpose of this work is to clarify the physical properties of the jets in the time-dependent MHD simulations by comparing with steady MHD jets which have been studied by many authors (eg., Blandford & Payne 1982, Sakurai 1987, Kudoh & Shibata 1995, 1997a).

2. Initial conditions and numerical method

As an initial condition, we assume a rotating disk which is in equilibrium under a central point mass gravitational potential (Matsumoto et al. 1996). The mass distribution outside the disk is assumed to be that of a high temperature corona which is in hydrostatic equilibrium without rotation. The magnetic field in an initial state is assumed to be uniform and parallel to the axis of the rotating disk. Axial-symmetry around the rotation axis is assumed. The numerical methods we use are the CIP method (Yabe & Aoki

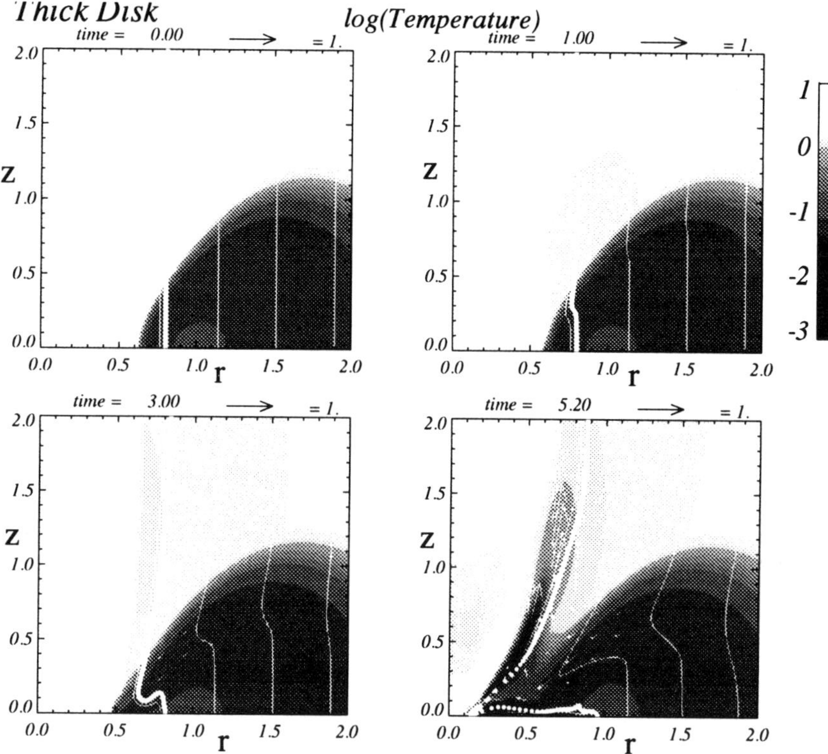

Figure 2. Time evolution of the temperature. Time $t \sim 2\pi \simeq 6.28$ corresponds to one Keplerian orbit at $(r, z) = (1, 0)$. White vertical lines are magnetic field lines. White filled dots are the Lagrangian fluid elements which are initially on a field line at $r = 0.8$. Arrows show the poloidal velocity vectors normalized by the Keplerian velocity at $(r, z) = (1, 0)$.

1991) and the MOC-CT (Stone & Norman 1994) method. The magnetic induction equation is solved by the MOC-CT, and the other equations are solved by the CIP.

3. Numerical result

3.1. EXAMPLES OF RESULTS

Fig. 2 shows the time evolution of the temperature. Time $t = 2\pi \simeq 6.28$ corresponds to one Keplerian orbit at $(r, z) = (1, 0)$. First, a torsional Alfvén wave is generated at the disk surface and propagates up into the corona ($t = 1$). Then the rotating disk loses its angular momentum by the magnetic field and begins to fall into the central region ($t = 3$). When the disk is thick (or when the magnetic field in the disk is weak), the surface

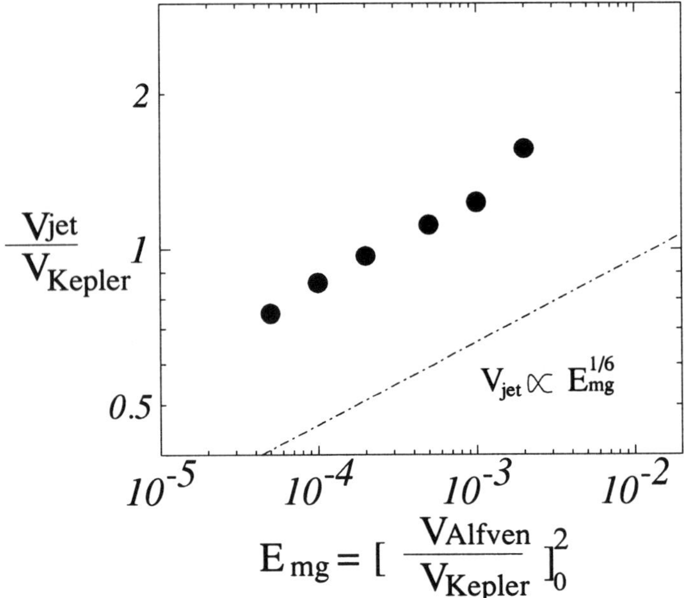

Figure 3. Maximum poloidal velocities of jets as a function of initial magnetic energy in the disk. (The subscript 0 means the initial value.) The dot-dashed line shows the dependence of one-dimensional steady solutions.

layer of the disk falls faster than the equatorial part because the magnetic braking most effectively affects that layer (Matsumoto et al. 1996). Finally, the cold material on the disk surface is ejected as a jet ($t = 5.2$).

3.2. VELOCITIES OF JETS AS A FUNCTION OF MAGNETIC ENERGY

An interesting question regarding magnetically driven jets is how the velocities of the jets depend on the strength of the magnetic field. Fig. 3 shows the maximum velocities of jets as a function of initial magnetic energy in the disk as determined from our simulations. Magnetic energy ($E_{\rm mg}$) is defined as $E_{\rm mg} = (V_{\rm Alfvén}/V_{\rm Kepler})^2$, where $V_{\rm Alfvén}$ and $V_{\rm Kepler}$ are the Alfvén and Keplerian velocities at $(r, z) = (1, 0)$. The maximum velocities are measured along a field line which is initially located at $r = 0.8$. The velocities are normalized by the Keplerian velocity at $(r, z) = (1, 0)$. The velocities of the jets are of the order of the Keplerian velocities for a wide range of magnetic energies in the disk. Its dependence on the magnetic energy is consistent with that of one dimensional steady solutions : $v_{\rm jet} \propto E_{\rm mg}^{1/6} \propto B_0^{1/3}$ (Kudoh & Shibata 1995, 1997a, 1997b).

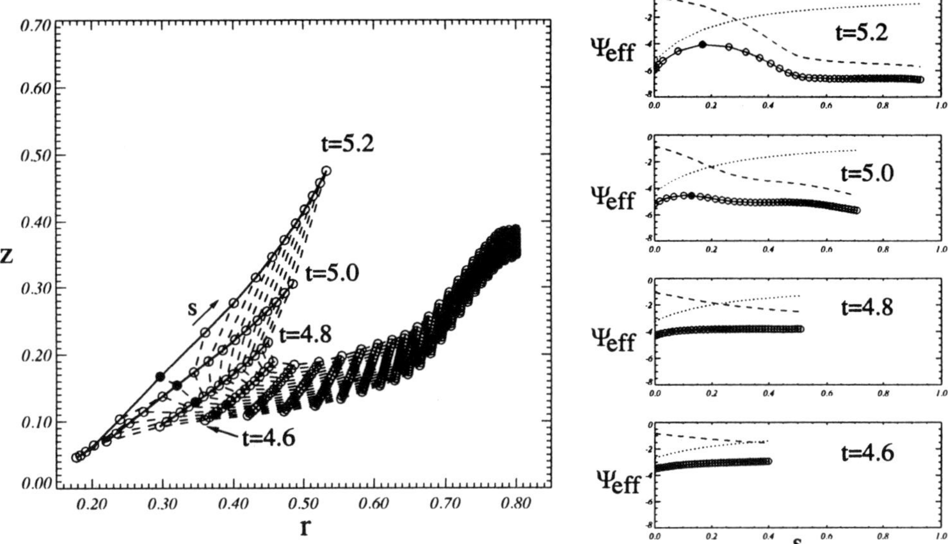

Figure 4. The left side figure is the trajectories of the Lagragian fluid elements which are initially on a field line at $r = 0.8$. The lines connecting fluid elements of the same time represent magnetic field lines. On the right side is the effective potential along the field line. The long-dashed lines and dotted lines show gravitational potential and the centrifugal potentail, respectively.

3.3. THE EJECTION MECHANISM FROM THE DISK

Blandford & Payne (1982) proposed the idea that gas in a rotating disk is flung out by a magnetic field line threading the disk when the centrifugal force is greater than the gravitational force along the field line. We examine the results of our simulation from the view point of this idea. Fig.4 shows the trajectories of the Lagrangian fluid elements which are initially on a field line at $r = 0.8$ (the left side figure), and the effective potential which is the summation of the gravitational potential and the centrifugal potential, along the field line (the right side figure). (A line connecting of fluid elements of the same time represent magnetic field lines.) The effective potential deforms their shape, and the region where $d\Psi_{\rm eff}/ds < 0$ appears as accretion progresses and the field line inclines. $d\Psi_{\rm eff}/ds < 0$ means that the centrifugal force is greater than the gravitational force along the field line. Then, the Lagrangian fluid elements which are in the region of $d\Psi_{\rm eff}/ds < 0$ move out as an outflow (the right side fluid elements represented by filled circles). This result shows that the mechanism proposed by Blandford & Payne (1982) works in our time-dependent model. However, the magnetic pressure of the toroidal component of the magnetic field will also accelerates the jet at the last stage of the acceleration (Uchida & Shibata 1985, Shibata & Uchida 1986).

4. Discussion

The velocities of jets observed in young stellar objects (YSOs), active galactic nuclei (AGNs), and some X-ray binaries (XRBs) are of the order of the escape velocities of the central objects. Our result (Fig. 3) shows that the velocity of the jet is of the order of a Keplerian velocity of the disk. Therefore, these jets are expected to be produced from accretion disks near the central objects. The interaction between the magnetic field of the central object and that of the accretion disk may take place near the central objects. This possibility is studied by Hayashi, Shibata & Matsumoto (1996) and Hirose et al. (1997). Recently, 2.5D-MHD simulation of jets from accretion disks have been performed by Ustyugova et al. (1995), Ouyed, Pudritz & Stone (1997). The differece between thier simulations and ours is the boundary condition at an equatorial plane. They assumed an inflow boundary condition so that they obtained a steady solution from thier time dependent simulations. On the other hand, we assumed a mirror symmetric boundary condition. Therefore, the jets obtained in our simulation are essentially nonsteady, because energy and angular momentum in the disk are finite. We found that the properties of jets in our simulation are able to be understood with a theory of steady jets, although these are essentially nonsteady.

References

Blandford, R. D. & Payne, D. G. 1982, MNRAS, 199, 883.
Hayashi, M.R., Shibata, K., & Matsumoto, R. 1996, ApJ, L37.
Hirose, S., Uchida, Y., Shibata, K., & Matsumoto, R. 1997, PASJ, in press.
Kudoh, T., & Shibata, K. 1995, ApJ, 452, L41.
Kudoh, T., & Shibata, K. 1997a, ApJ, 474, 362.
Kudoh, T., & Shibata, K. 1997b, ApJ, 476, 632.
Matsumoto, R., et al. 1996, ApJ, 461, 115.
Ouyed, R., Pudritz, R. E., & Stone, J. M. 1997, Nature, 385, 409.
Sakurai, T. 1987, PASJ, 39, 821.
Shibata, K., & Uchida, Y. 1986, PASJ, 38, 631.
Shibata, K., & Uchida, Y. 1990, PASJ, 42, 39.
Uchida, Y., & Shibata, K. 1985, PASJ, 37, 515.
Ustyugova, G.V., et al. 1995, ApJ, 439, L39.
Stone, J. M., & Norman, M. L. 1992, ApJS, 80, 791.
Yabe, T., & Aoki, T. 1991, Comp. Phys. Comm., 66, 219.

NUMERICAL SIMULATION OF RELATIVISTIC JET FORMATION IN BLACK HOLE MAGNETOSPHERE

SHINJI KOIDE
Faculty of Engineering, Toyama University, Gofuku, Toyama 930, Japan

AND

KAZUNARI SHIBATA AND TAKAHIRO KUDOH
National Astronomical Observatory, Mitaka, Tokyo 181, Japan

1. Introduction

The radio jets ejected from active galactic nuclei (AGNs) sometimes show proper motions with apparent velocity exceeding the speed of light (e.g., Hughes, 1991). This phenomenon, called superluminal motion, is explained as relativistic jets propagating in a direction almost toward us, and has been thought to be ejected from the close vicinity of hypothetical supermassive black holes powering AGNs (Rees, 1966). Recently, similar superluminal motion has been discovered in some compact radio/X-ray sources (i.e., "microquasars") in our Galaxy, such as GRS 1915+105 (Mirabel & Rodriguez, 1994), which are considered to be black hole candidates. In spite of the vast differences in luminosity and the sizes of microquasar in our Galaxy (e.g., GRS 1915+105, whose luminosity is 3×10^{38} erg/s and size $< 10^6$ cm) and those of AGNs (luminosity $\sim 10^{47}$ erg/s and the size $< 10^{14}$ cm), both objects are believed to be powered by gravitational energy released during the accretion of plasmas onto black holes. The mass of black holes, $M_{\rm BH}$, is estimated to be < 10 solar mass for microquasars, and $\sim 10^8$ solar mass for AGNs. If we normalize the length by the Schwarzschild radius $r_{\rm S} = 2GM_{\rm BH}/c^2$, both objects become similar and thus can be understood in a unified model as in the following.

Since the accreting plasmas have non-zero angular momentum, they form accretion disks orbiting around black holes. Similarly, if the accreting plasmas have non-zero poloidal magnetic flux, the magnetic flux ac-

T. Watanabe et al. (eds.), Observational Plasma Astrophysics: Five Years of Yohkoh and Beyond, 149–154.

cumulates in the inner region of accretion disk to form a global poloidal magnetic field penetrating the disk. Such poloidal fields could also be generated by dynamo action inside the accretion disk. In either case, poloidal fields are twisted by the rotating disk toward the azimuthal direction, and this process extracts angular momentum from the disk, enabling efficient accretion of disk plasmas onto black holes. In addition, magnetic twist generated during this process accelerates plasmas in the surface layer of the disk toward the polar direction by the $\mathbf{J} \times \mathbf{B}$ force to form bipolar relativistic jets which are also collimated by the magnetic force. This magnetic mechanism has been proposed not only for AGN jets (Lovelace, 1976; Blandford & Payne, 1983; Camenzind, 1986; Pelletier *et al.*, 1996) but also for protostellar jets (Pudritz & Norman, 1986; Uchida & Shibata, 1985; Shibata & Uchida, 1986; Ouyed *et al.*, 1997), although no one has yet performed nonsteady general relativistic magnetohydrodynamic (GRMHD) numerical simulations on the formation of jets from the accretion disk around a black hole. It should be noted here that observations of superluminal jet sources show that these are ejected intermittently (Hughes, 1991; Mirabel & Rodriguez, 1994). It has also been theoretically argued recently that accretion disks near black holes can evolve into advection-dominated disks which have large radial accretion flows with possible a time dependence or instability (Abramowicz, 1996; Kato *et al.*, 1996; Balbus & Howley, 1991). Hence a time-dependent approach is essential, even though some basic physics can be understood with steady models (Kudoh & Shibata, 1995). Recently nonrelativistic time dependent MHD simulations showed that intermittent jets can be produced by a magnetically driven mechanism (Ouyed *et al.*, 1997).

2. Numerical Method and Results

We use 3+1 formalism of general relativistic conservation laws of particle number, momentum, and energy and Maxwell equations with the infinite electric conductivity (Thorne *et al.*, 1986; Koide *et al.*, 1996; Koide, 1997). When we take the nonrelativistic limit of these equations ($c \to \infty$), they reduce to Newtonian magnetohydrodynamic (MHD) equations. Therefore these equations are called the GRMHD equations. We solve these equations numerically using the simplified total variation diminishing (TVD) method (Koide *et al.*, 1996; Koide, 1997). The Schwarzschild metric, which provides the space-time around the black hole at rest, is used in the calculation. The simulation is performed in the region $1.2r_S \le r \le 20r_S$, $0 \le \theta \le \pi/2$, with 90×60 meshes, assuming axisymmetry with respect to z-axis and mirror symmetry with respect to the plane $z = 0$. Here r and θ are the radial and polar spherical coordinates, respectively. A free boundary is used at

$r = 1.2r_S$ and a radiative boundary condition is employed at $r = 20r_S$.

Figure 1 shows the time development of the relativistic jet formation in the black hole magnetosphere. These figures show the rest mass density (black-and-white), velocity (vector), and magnetic field (solid lines) in the region: $-7r_S \le x \le 7r_S$, $y = 0$, $-7r_S \le z \le 7r_S$. The black circles at the centers represent the black hole inside the event horizon at Schwarzschild radius r_S. The initial state of the simulation is divided into two parts: a background corona around the black hole and the accretion disk (Fig. 1a). In the corona, plasmas are assumed to be isothermal and in hydrostatic equilibrium, where the proper sound velocity is constant ($v_s = 0.53c$). The accretion disk is located at $|\cot\theta| \le 0.125$, $r \ge r_D = 3r_s$ and rotates around the black hole with Keplerian velocity $v_K = c/[2(r/r_S - 1)]^{1/2}$. The mass density of the disk (white, right and left bands in the figures) is 400 times that of the background corona. In addition, the magnetic field crosses the accretion disk perpendicularly. We use the Wald solution, which provides a uniform magnetic field around the black hole (Wald, 1974). At the inner edge of the accretion disk, the proper Alfvén velocity is $v_A = 0.01c$. Figure 1b shows a snapshot at $t = 10\tau_S$, after the disk rotates one-quarter cycle, where τ_S is defined as $\tau_S \equiv r_S/c$. An Alfvén wave propagates along the magnetic field lines from the rotating accretion disk and the front of the wave reaches at $z \sim 4.5r_S$. The accretion disk drags the magnetic field and loses its angular momentum. The disk accretes into the black hole as shown in the vector plot of Fig. 1b. At $t = 25\tau_S$, a jet begins to be ejected at the inner edge of the accretion disk, around $x \sim 2.5r_S$, $z \sim 1.0r_S$ (Fig. 1c). At this time, the disk rotates almost half a cycle and penetrates into the unstable region ($r \le 3r_S$). The maximum poloidal component of the jet velocity is $\sim 0.4c$. Figure 1d shows the final stage of this simulation at $t = 35\tau_S$, when the accretion disk has rotated almost one cycle around the black hole, and is formed along the magnetic field lines. The jet is ejected from $x \sim 2r_S$, $z \sim 1r_S$. The maximum poloidal component of the jet velocity is 0.88c, which corresponds to a Lorentz factor of 2.1 at $x \sim 3r_S$, $z \sim 2.5r_S$. Figure 2 shows the foot point region of the jet in detail. The velocity plot (vector) shows the inflow to the black hole. Around $r \sim 2r_S$ the plasma of the accretion disk splits into inflow toward the black hole and outflow away from it.

Figure 3 shows a nonrelativistic simulation result ($c = 200v_s$) at $t = 35\tau_S$ to compare the relativistic simulation ($c = 2v_s$) (Fig. 1d). The parameters of the nonrelativistic case are the same as those of the relativistic case except for the ratio of the proper sound velocity in the corona and the velocity of light. This ratio is 100 times smaller in the non-relativistic case (Fig. 3) than in the relativistic case (Fig. 1d). Assuming Keplerian motion for the accretion disk, only a broad jet forms in the nonrelativistic case, whereas a fast, narrow jet forms in the relativistic case (Fig. 1d). This comparison indicates that magnetized plasma with Keplerian motion around a relativistic star (such as a black hole) ejects faster, narrower jets compared to a nonrelativistic star (such as a protostar).

Mirabel & Rodriguez (1995) showed that the velocity of the ejecta from black hole candidates such as microquasar GRS 1915+105 is larger than $0.9c$, whereas it is less than $0.3c$ in neutron star candidates in our Galaxy, such as Cygnus X-3, SS433, and the Crab nebula. Our numerical results of comparing the relativistic and nonrelativistic cases may also explain this observational property.

We thank M. Inda-Koide for her discussion and important comments for this study. We also thank M. Takahashi, T. Yokoyama, K. Hirotani, K.-I. Nishikawa, J.-I. Sakai, and R. L. Mutel for their discussion and encouragement. We thank the National Institute for Fusion Science and the National Astronomical Observatory for the use of their super-computers. This work is partially supported by Sumitomo Foundation.

References

Abramowicz, M. in S. Kato et al. eds, *Physics of Accretion Disks*, 1-14 (1996).
Balbus, S. A. & Hawley, J. F., Astrophys. J. **376**, 214-222 (1991).
Blandford, R. D. & Payne, D. G., Mon. Not. R. Astr. Soc. **199**, 883-903 (1983).
Camenzind, M., Astron. Astrophys. **156**, 137-151 (1986).
Hughes, P.A. eds. *Beams and Jets in Astrophysics* (Cambridge Univ. Press, New York, 1991).
Kato, S., Abramowicz, M., & Chen, X., Publ Astr. Soc. Japan 67-75 (1996).
Koide, S., Nishikawa, K.-I., & Mutel, R. L., Astrophys. J., **463**, L71-L74 (1996).
Koide, S., Astrophys. J. **478** 66-69 (1997).
Kudoh, T. & Shibata, K., Astrophys. J. **452**, L41-L44 (1995).
Lovelace, R. V. E., Nature **262**, 649-652 (1976).
Mirabel, I. F. & Rodriguez, L. F., Nature, **371**, 46-48 (1994).
Mirabel, I. F., & Rodriguez, L. F., Annals of the New York Academy of Sciences, **759**, 21-37 (1995).
Ouyed, R., Pudritz, R. E., & Stone, J. M., Nature **385**, 409-414 (1997).
Pelletier, G., Ferreira, J., Henri, G., & Marcowith, A. in Tsinganos, K. C. eds., *Solar and Astrophysical Magnetohydrodynamic Flows* 643-657 (Kluwer Academic Pub., Dordrecht, 1996).
Pudritz, R. E. & Norman, C., Astrophys. J. **301**, 571-586 (1986).
Rees, M. J. Nature, **211**, 468-470 (1966).
Shibata, K. & Uchida, Y., Publ Astr. Soc. Japan, **38**, 631-660 (1986).
Thorne, Kip S., Price, R. H., & Macdonald, D. A., *Membrane Paradigm* (Yale Univ. Press, New Haven and London, 1986).
Uchida, Y. & Shibata, K., Publ Astr. Soc. Japan **37**, 515-535 (1985).
Wald, R. M., Phys. Rev. D, **10**, 1680-1685 (1974).

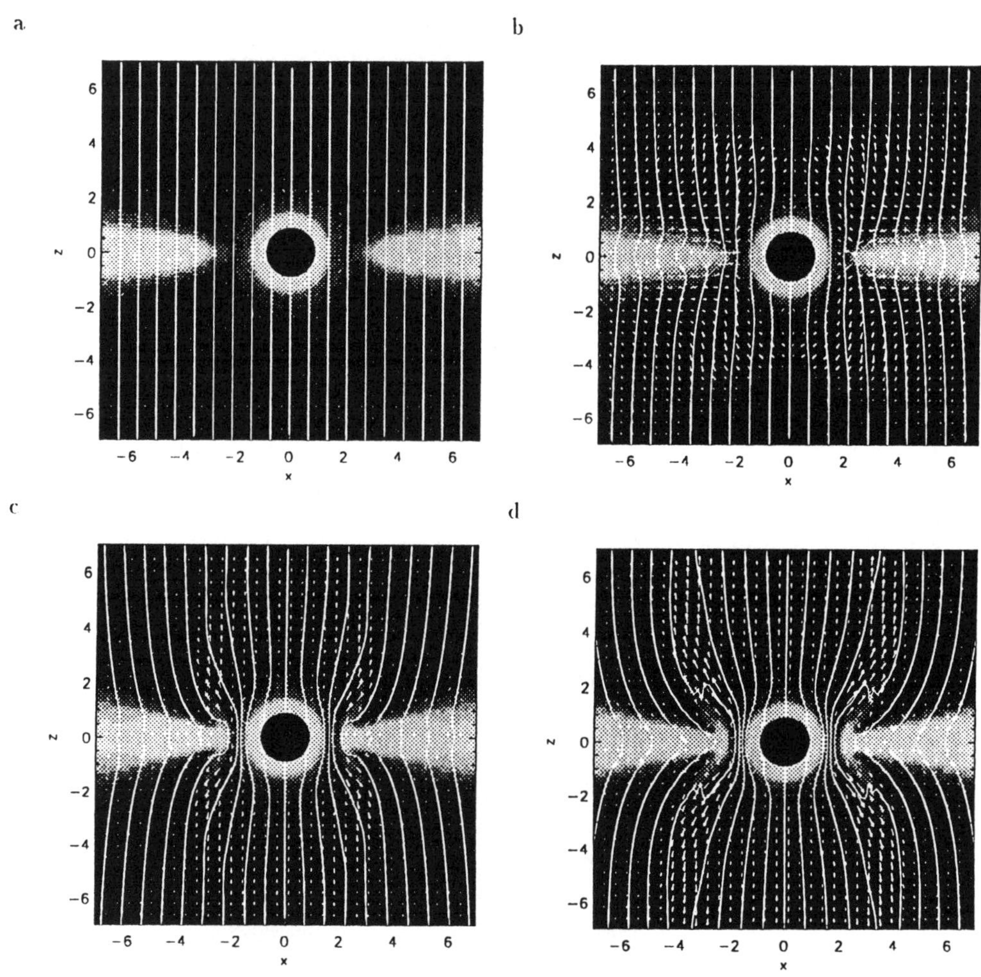

Figure 1. The evolution of the jet formation in a black hole magnetosphere. The black circles indicate the black hole (i.e., the region inside the Schwarzschild radius). The solid lines are the magnetic field lines. The gray-scale plots show the rest mass density (log scale, white indicates the highest value nearly 100). The vector plots show the velocity. (a) The initial condition. (b) $t = 10\tau_S$. (c) $t = 25\tau_S$. The jet begins to be ejected from the inner edge of the disk. (d) $t = 35\tau_S$. The maximum poloidal component of the jet velocity is $0.88c$, which corresponds to Lorentz factor 2.1. The distance and the time are in unit of r_S (Schwarzschild radius) and $\tau_S \equiv r_S/c$, respectively.

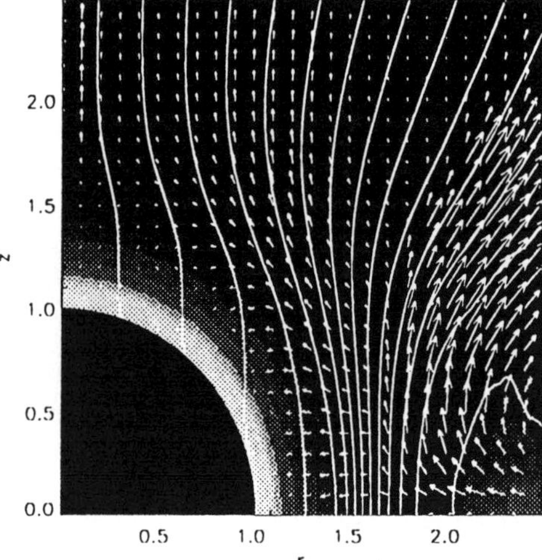

Figure 2. Close-up snapshot of velocity (vector), magnetic field (solid lines), and rest mass density (gray-scale) near the black hole at $t = 35\tau_S$. The circular black region shows the black hole. We see both outflow from the accretion disk and also inflow toward the black hole. The accreting plasma splits into the inflow and the outflow around $r \sim 2r_S$.

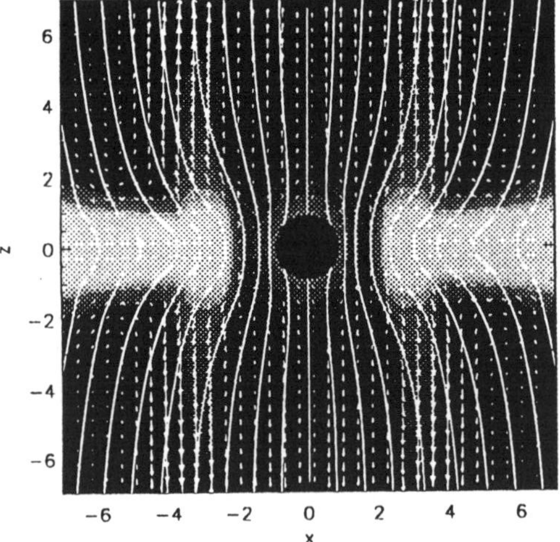

Figure 3. The case of the nonrelativistic jet formation (Fig. 1d) at $t = 35\tau$, where $\tau \equiv r_0/(200v_s)$ and r_0 is the radius of the nonrelativistic star (shown as a black circle). The velocity (vector), the rest mass density (black-and-white), and the magnetic field (solid lines) are shown. The scale of velocity vector of the nonrelativistic case is 100 times smaller than that of the relativistic case (Fig. 1d). Around the accretion disk with the Keplerian motion, the broad jet is formed by the Shibata-Uchida mechanism in the nonrelativistic case. This contracts with the relativistic case, whera a fast, narrow jet formed from the strongly compressed region (Fig. 1d).

PAIR PLASMA PRODUCTION IN A BLACK HOLE MAGNETOSPHERE

KOUICHI HIROTANI AND ISAO OKAMOTO
National Astronomy Observatory
Mitaka, Tokyo 181, Japan

1. Introduction

The model of a stationary, plasma-filled magnetosphere around a rotating black hole has been evoked to explain the energies liberated in the vicinity of a black hole, which possibly form relativistic jets observed in active galactic nuclei (AGNs) (Blandford 1976, 1977; Loveloce 1996). Very close to the black hole, the jet energy will be regarded predominantly electromagnetic, because pair annihilation suppresses the density of material. The source of the electromagnetic power eventually resulting in the γ-ray jet may be the spin of the central black hole (Blandford and Znajek 1976).

It is, however, the electric current that sustains the electromagnetic power in the field-dominated region of the magnetosphere, and it is charged particles, even though inertia-free, that carry the electric current. However, even though under the assumption of massless, it must be real charged particles that constitute the electric current. Therefore, electrons must flow into the horizon for the electric current to flow out, and positrons must flow into there for the electric current to flow in. In other words, there must be a plasma source between these two inflowing and outflowing regions. We thus have to contrive a process of plasma supply in the force-free magnetosphere.

In spite of core significance in feeding the relativistic γ-ray jet from the central engine, the real process of supplying charged particles in the magnetosphere has not been substantiated for a realistic model of radiation fields in AGNs until (Beskin et al. 1992). For typical parameters of the central radiation in AGNs, they showed that the region of pair creation can actually be situated near the 'null' or 'critical' surface separating regions of opposite signs of the so-called Goldreich-Julian charge density, ρ_{GJ}, along field lines.

T. Watanabe et al. (eds.), Observational Plasma Astrophysics: Five Years of Yohkoh and Beyond, 155–160.

The purpose of this paper is to further extend the analysis of Beskin et al. (1992) and to clarify more qualitatively the micro process of interaction of radiation – pair creation under the existence of parallel electric field. In the next section, we formulate basic equations describing pair production cascade in the magnetosphere. We then solve them in §3 and demonstrate that sufficient amounts of plasma is supplied by the cascade so that the hole's rotational energy may be extracted effectively. In the final section, we sum up the results and discuss some implications.

2. Physical Processes and Basic Equations

We assume that the acceleration of $e^{\pm}$'s by $E_{\|}$ is perpendicular to the null surface. In this case, the acceleration works most effectively. We can then write down the Poisson equation

$$\frac{dE_{\|}}{dx} = 4\pi[e(n^{+} - n^{-}) - \rho_{\mathrm{GJ}}], \tag{1}$$

where x is the outwardly increasing coordinate along the field line.

In the potential gap, the $e^{\pm}$'s will lose their perpendicular momentum rapidly owing to Compton scatterings with ambient UV photons. However, their longitudinal motion will be maintained by the the electrostatic acceleration due to the $E_{\|}$ field. Therefore, $e^{\pm}$'s migrate *one*-dimensionally and their motion is *mono*-energetic with a single Lorentz factor. These $e^{\pm}$'s can produce sufficient amounts of high energy γ-ray photons via *inverse Compton scatterings* of background X-ray photons. As a result, $e^{\pm}$'s and γ-ray photons may move one-dimensionally along the magnetic field and the $e^{\pm}$ spectra may be mono-energetic. Under these conditions, assuming stationarity, we next formulate basic equations describing the pair production cascade.

First, let us consider the continuity equations of $e^{\pm}$'s. For simplicity, we shall suppose that positrons migrate outwards while electrons migrate inwards. Then the continuity equations become

$$\pm\frac{d}{dx}\left(n^{\pm}(x)\sqrt{1-1/\Gamma^{2}(x)}\right) = \int_{0}^{\infty} \eta_{\mathrm{p}}(\epsilon_{\gamma})\left[F^{+}(x,\epsilon_{\gamma}) + F^{-}(x,\epsilon_{\gamma})\right] d\epsilon_{\gamma}, \tag{2}$$

where η_{p} is the angle-averaged pair production redistribution function. $F^{\pm}(x,\epsilon_{\gamma})$ are the number density of γ-ray photons propagating in the $\pm\,x$-directions, respectively, at a position x and in the energy interval $\epsilon_{\gamma} \sim \epsilon_{\gamma} + d\epsilon_{\gamma}$. It must be noted that the γ-ray photons, which are produced by the inverse Compton scatterings, are highly aberrated to move in the $\pm x$ direction.

We can easily see that the current, which is carried by $e^{\pm}$'s along a given field line inside the gap, must be conserved along x. From one combination of equations (2), taking acount of $\Gamma \gg 1$, we have

$$[n^{+}(x) + n^{-}(x)] = \frac{j_{0}}{e} \equiv N_{0} \tag{3}$$

where the current density $j_0 \equiv eN_0$ is constant along a field line. In order that the energy and angular momentum may be extracted effectively from a rotating supermassive black hole, j_0 must take the value of

$$j_{0} \sim 10^{-13} \left(\frac{a}{M}\right) \left(\frac{M}{10^{8} M_{\odot}}\right) \left(\frac{B}{10^{4}\mathrm{G}}\right) \frac{\mathrm{abampere}}{\mathrm{cm}^{2}}. \tag{4}$$

in order of magnitude. Another combination of equations (2) gives

$$\frac{d}{dx} \{[n^{+}(x) - n^{-}(x)]\} = 2 \int_{0}^{\infty} \eta_{\mathrm{p}}(\epsilon_{\gamma}) \left[F^{+}(x, \epsilon_{\gamma}) + F^{-}(x, \epsilon_{\gamma})\right] d\epsilon_{\gamma}. \tag{5}$$

Instead of equations (2), we use equations (3) and (5) in the following. The Boltzmann equations for the γ-ray photons become

$$\pm \frac{d}{dx} F^{\pm}(x, \epsilon_{\gamma}) = \eta_{\mathrm{c}} n^{\pm} \sqrt{1 - 1/\Gamma^{2}} - \eta_{\mathrm{p}} F^{\pm}, \tag{6}$$

where $\eta_{\mathrm{c}}(\epsilon_{\gamma}, \Gamma)$ is the Compton redistribution function.

2.1. BOUNDARY CONDITIONS

The purpose of this paper is to elucidate the physical processes of pair production cascade taking place in a magnetosphere surrounding a supermassive black hole in an AGN. To achive this, as a first step, we consider the case in which the functions $E_{||}$, n^{+}, $F^{\pm}$ have symmetric properties so that we may solve the basic equations only in the range $0 \leq x \leq H$.

At the symmetry point $x = 0$, we impose

$$d\Gamma/dx = 0, \quad n^{+} = 0, \quad F^{+} = F^{-} \quad \text{at } x = 0. \tag{7}$$

The outer boundary of the gap is, on the other hand, defined so that $E_{||}$ vanishes at this point. That is,

$$E_{||} = 0 \quad \text{at } x = H. \tag{8}$$

Further, charge density distibution must be continuous at the boundary $x = H$. Outside of the gap, charge density $e(n^{+} - n^{-})$ must coincide with ρ_{GJ}. We therefore impose the condition

$$\frac{1}{4\pi}\frac{dE_{\|}}{dx} = \frac{j_0}{c\sqrt{1-1/\Gamma^2}} - Ax = 0 \quad \text{at } x = H. \tag{9}$$

What is more, any electrons or γ-rays should not come into the gap from the outside ($x > H$). That is, we impose

$$n^- = 0 \quad \text{and} \quad F^- = 0 \quad \text{at } x = H. \tag{10}$$

We shall seek the solutions satisfying the boundary conditions making use of a shooting method.

3. Structure of the Potential Gap

To grasp the rough features, we first show some examples of the solutions of $E_{\|}(x)$ in Fig. 1. For a fixed value of $U_b = 10^6\,\text{ergs/cm}^3$, we depicted the three cases of $\alpha = 1.5$, 2.0, and 2.5 denoted by the dashed, solid, and dotted lines, respectively. This figure indicates that $E_{\|}$ is maximum at the symmetry point, $x = 0$, decreases monotonically with increasing x, and finally vanishes at the boundary, $x = H$. By such $E_{\|}$ fields, the Lorentz factor of $e^{\pm}$ becomes $\Gamma = 10^{2\sim3}$ at the center of the gap, $x = 0$ (Hirotani and Okamoto 1997).

As Fig. 1 indicates, the typical width of the potential gap is $H \sim 0.005 r_H$ for the background radiation field of $U_b = 10^6\,\text{ergs/cm}^3$, where r_H is the hole's radius. Actually, H increases significantly with decreasing U_b. This is mainly because the less luminous background radiation field increases the pair production mean free path, l_p, which essentially determines H.

Because $j_0 = j_{cr}(U_b, \alpha)$ is directly related with H by equation (1) at $x = H$, we can depict $j_{cr}(U_b, \alpha)$ as in Fig. 2. This figure reveals the fact that the pair production cascade provides sufficient current density ($\sim 10^{-13}\,\text{abampere/cm}^2$) for the effective energy and angular momentum extraction from a rotating suppermassive black hole. Moreover, j_{cr}, which sets the maximum extractable energy due to the Blandford-Znajek process, becomes large when U_b decreases.

4. Summary and Discussion

In this paper we qualitatively and self-consistently solved the process of pair production cascade in a thin gap in a rotating black hole magnetosphere. We summarize our results as follows:

1. Once a strong electric field ($E_{\|} = 10^{3\sim6}\,\text{V/m}$) arises along the poloidal magnetic field line, it accelerates migratory or pair-created $e^{\pm}$'s into relativistic regimes with Lorentz factors $\Gamma = 10^{2.5\sim4}$.

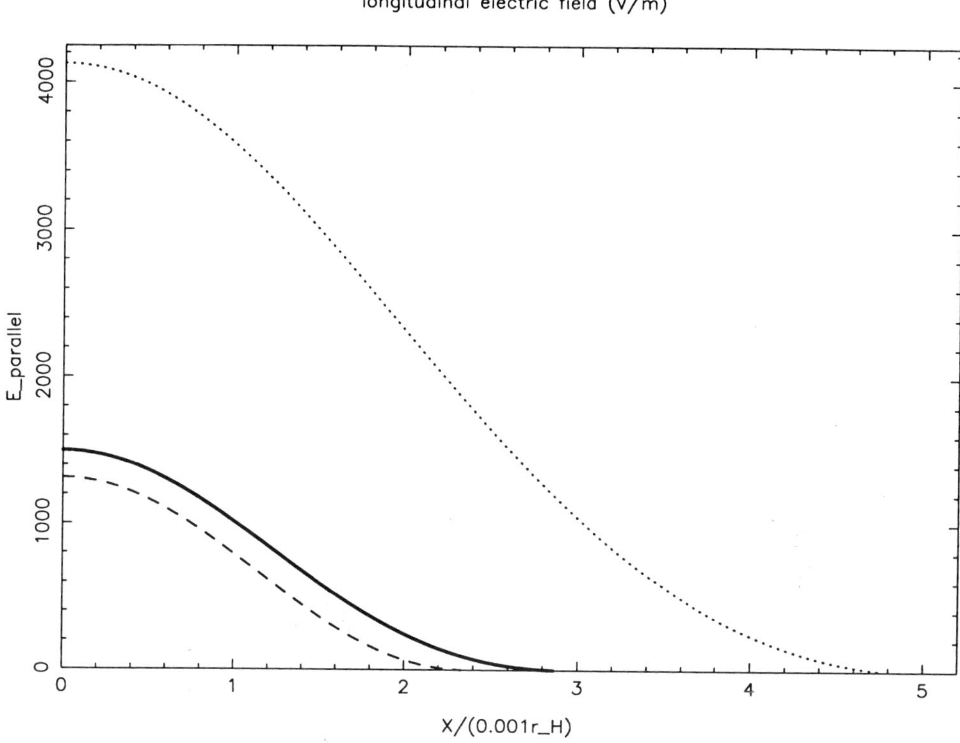

Figure 1. Solutions of $E_{\parallel}$ as a function of the position x. U_{b} is fixed to be 10^6 ergs/s. The solid line corresponds to the case of $\alpha = 2.0$, while the dotted and the dashed lines to $\alpha = 1.5$ and $\alpha = 2.5$, respectively.

2. An accelerated e^+ or e^- scatters background UV and X-ray photons to produce $10^{1\sim3}$ γ-ray photons.

3. Typically, one of such γ-ray photons (produced by a single e^+ or an e^-) collides a background X-ray photon to produce an $e^{\pm}$ pair, which leads to pair production cascade.

4. The cascade supplies sufficient amounts of pair plasma necessary in maintaining the electric current in the magnetosphere and thus ensures efficient extraction of the hole's rotational energy in the form of electromagnetic energy.

5. The width of the pair cascading layer is much less than the hole's radius, r_{H}.

Finally, it might be interesting to investigate whether the same mechanism works around a stellar mass black hole, from which the rotational

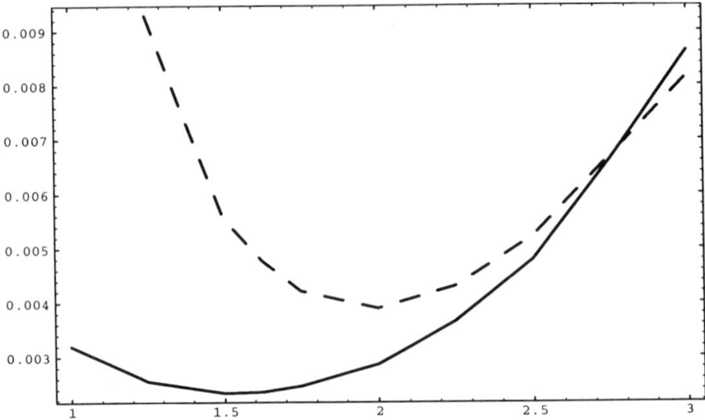

Figure 2. Gap width $H/r_{\rm H}$ vs. photon index α. The solid line corresponds to the case of $U_{\rm b} = 10^6$ ergs/s, while the dashed line to $U_{\rm b} = 10^5$ ergs/s.

energy may be extracted, resulting in a galactic γ-ray jet (cf. (Levinson and Blandford 1996)). Around a supermassive black hole, which is the concern of this paper, the relaxation length l_0 at which $e^{\pm}$'s lose their perpendicular momenta owing to scatterings with background UV photons, is much smaller than $l_{\rm p}$, and H. Therefore, particle motion and γ-ray behaviour can be treated one-dimensionally; in addition, electron/positron spectra are approximated by mono-energetic ones. However, as Beskin et al. (1992) pointed out, around a stellar mass black hole, l_0 becomes comparable with $l_{\rm c}$; therefore, scatterings that lead to pair production can no longer be treated as one-dimensional. This issue will be our next subject of investigation.

References

Beskin, V. S., Istomin, Ya. N., and Par'ev, V. I. 1992, Sov. Astron. 36(6), 642
Blandford, R. D. 1976, MN, 176, 465
Blandford, R. D., 1989, in Theory of Accretion Disks, ed. P. Meyer, W. Duschl, J. Frank & E. Meyer-Hofmeister, Dordrecht: Klewer, 35
Blandford, R. D., and Znajek, R. L. 1976, MN, 179, 433
Hirotani, K., and Okamoto, I. 1997, submitted to ApJ
Levinson, A. and Blandford, R. D., 1996, ApJl, 456, L29.
Lovelace, R. V. E., 1996, Nature, 262, 649

MHD SIMULATIONS OF MAGNETIC RECONNECTION IN THE SOLAR CORONA AND FLARES

YOKOYAMA, T.
National Astronomical Observatory
Mitaka, Tokyo 181 Japan

Abstract. Yohkoh observed many dynamic phenomena in the solar corona, such as flares, microflares, plasma ejections, and jets. These phenomena can be explained by a magnetic reconnection model. This paper is a review of the results of our magnetohydrodynamic (MHD) simulations of magnetic reconnection in the solar corona and flares.

1. Magnetic Reconnection in the Corona

Magnetic reconnection has been thought to play a fundamental role in the occurrence of various magnetic explosive phenomena in the corona. It has, however, not yet been established. One of the reasons is that the magnetic Reynolds number ($R_m = \tau_D/\tau_A$, where τ_D is magnetic diffusion time and τ_A is Alfvén time) is enormously high in the phenomena occurring in the corona (e.g., $\sim 10^{11}$ for solar flares); i.e., the magnetic diffusion (current dissipation) time τ_D is much longer than the Alfvén time τ_A, while the time scale of explosive phenomena (e.g., solar flares) is of order 10 – 100 τ_A. Because of this enormous gap between the diffusion time and the Alfvén time, not only laboratory experiments but also numerical experiments with realistic magnetic Reynolds number are still impossible. This is why the magnetic reconnection model has not yet been fully established as the cause of various explosive phenomena in the corona.

In order to address this issue, Yokoyama & Shibata (1994) have studied the role of the resistivity model for magnetic reconnection driven by the magnetic buoyancy instability (Parker 1966). We showed that the nature of the resistivity model is crucial for the occurrence of fast reconnection and that a key physical process leading to the fast reconnection is the formation and ejection of magnetic islands (plasmoids). We studied two types

T. Watanabe et al. (eds.), Observational Plasma Astrophysics: Five Years of Yohkoh and Beyond, 161–170.

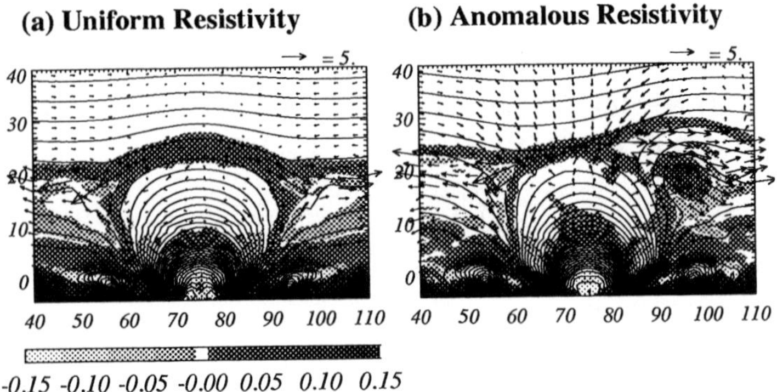

Figure 1. Simulation results of magnetic reconnection driven by the magnetic buoyancy instability for (a) uniform resistivity model and (b) anomalous resistivity model. Two-dimensional distributions of electric current (gray scale), magnetic field lines (lines), and velocity (arrows) are shown.

of resistivity models: (a) a uniform resistivity model and (b) an anomalous resistivity model of functional form $\eta = 0$ for $v_d \geq v_c$ and $\eta \propto (v_d/v_c - 1)^2$ for $v_d \geq v_c$, where $v_d \equiv J/\rho$, is the non-dimensional (relative ion-electron) drift velocity, ρ is the non-dimensional mass density, J is the current density, and v_c is a threshold above which the anomalous resistivity sets in. Figure 1 shows the results. In the anomalous resistivity model (b), several magnetic islands are created in the current sheet by the tearing instability. These islands coalesce with each other by the coalescence instability, and are ejected from the current sheet at the Alfvén speed. After the ejection of the islands, the current sheet suddenly collapses, leading to fast reconnection. There is evidence of a slow mode MHD shock extending from the neutral point (X-point) after the ejection of magnetic islands. (See, e.g., the electric current distribution in Fig. 1, which shows an X-type current sheet along the slow-mode MHD shock.) The length of the current sheet is short and the plasma inflow converges into the X point. Although the dynamics is nonsteady, the situation near the X-point is similar to that predicted by Petschek (1964). In the uniform resistivity model (a), there is no island formation nor slow mode MHD shock formation. The current sheet is long, inflow into the sheet is slow, and hence the reconnection is slow, though the velocity of the outflow along the sheet is about the Alfvén speed. The reconnection in this case is nearly the steady Sweet-Parker type (Sweet 1958; Parker 1963). In fact, the Sweet-Parker scaling holds very well; the reconnection rate $\eta j \propto \eta^{1/2}$, the current sheet thickness $\delta \propto \eta^{1/2}$, the current sheet length $\Delta \propto \eta^0$, and the field strength $B \propto \eta^0$.

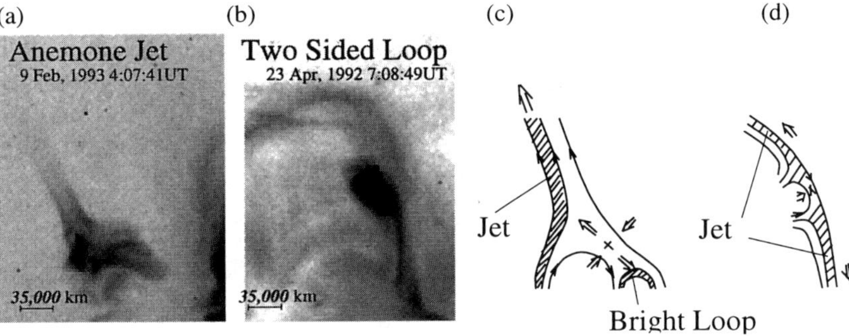

Figure 2. Examples of two types of solar coronal X-ray jets observed by the Yohkoh soft X-ray telescope (after Shibata et al. 1994). In the anemone-jet (a), the active region at the footpoint of the jet looks like a sea-anemone. In the two-sided-loop type (b), a pair of jets appears at both sides of emerging flux along a nearly horizontal field. (c) and (d) are schematic pictures of models for each type jet.

2. X-ray Jets and Microflares

One of the most interesting findings by Yohkoh is the *solar coronal X-ray jets* (Shibata et al. 1992; Strong et al. 1992; Shimojo et al. 1996), which are observed as transitory X-ray enhancements with an apparent collimated motion. Shibata et al. (1994) proposed a phenomenological model of the X-ray jets. They considered that the plasma in the jets is accelerated and heated by magnetic reconnection between emerging flux and pre-existing coronal magnetic fields. There are two types of reconnection. One is the *anemone-jet* type (Fig. 2a), which occurs when the emerging flux appears in a coronal hole where the pre-existing magnetic field is vertical or oblique (Fig. 2c). The other is the *two-sided-loop* type (Fig. 2b), which occurs when the emerging flux appears in a quiet region where the pre-existing magnetic field is almost horizontal. Hot plasma is ejected along the coronal loops away from both sides of the emerging flux (Fig. 2d).

Figure 3 shows the results of simulations of this model (Yokoyama and Shibata 1995, 1996). The emergence of the flux sheet from the convection zone is fully reproduced by simulating the magnetic buoyancy instability (the Parker instability). A uniform oblique field (with inclination angle of 45 degree) is assumed for the initial coronal field, The plasma is assumed to have anomalous resistivity. When the top of the rising loops meets the coronal field, reconnection takes place between the two fields: The plasma at the interface is heated to X-ray temperature ($\approx$ 4–10 MK). At the same time, the field lines of the emerging flux release their stress by straightening themselves out, thereby accelerating hot plasma and creating a pair of reconnection jets (Fig. 3). One of the pair is ejected upward, with velocity reaching about 100 km s^{-1} (which is approximately the Alfvén speed),

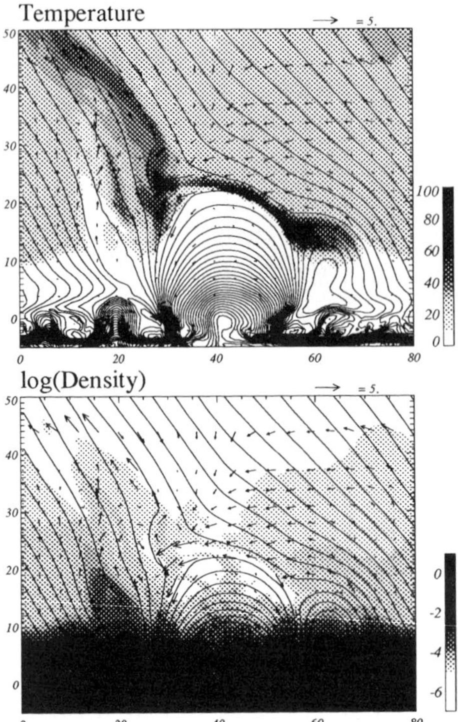

Figure 3. Simulation of a coronal X-ray jet. The top panel shows temperature (T; gray-scale map) in units of 10^4 K and magnetic field lines (**B**; lines). The bottom panel shows density ($\log_{10} \rho$; gray-scale map) in units of 10^{-7} g cm^{-3}, magnetic field lines (**B**; lines), and velocity vectors (**v**; arrows). The length is in units of $H \simeq 200$km.

and total bulk kinetic energy of about 1.2×10^{28} erg. These values are in good agreement with typical observations of X-ray jets. Another interesting feature in this simulation is the high-temperature loops at the right side of the emerging flux in Fig 3. While one of the pair of jets is ejected upward, the other goes down, colliding with the loops next to the emerging loops, generating strong compression and forming very hot loop.

In addition to these hot features, this simulation also reveals a whip-like motion of cool plasma (Fig. 3). This cool plasma originates from the chromosphere. It is carried up with the expanding loops and is ejected by a sling-shot effect (i.e., acceleration by the tension force of disconnected field lines) due to the reconnection, resulting in the whip-like motion. The velocity of this whip-like motion is a few tens km s^{-1}. This cool plasma may be observed as a cool jet, namely an Hα surge (Canfield et al. 1996).

Okubo et al. (1997) studied the effect of the perpendicular component of magnetic field on the ejection of plasma. They found that shear Alfvén waves are important for the acceleration of the cool plasma as first suggested by Shibata & Uchida (1986).

3. Large-Scale Flares

Many pieces of evidence supporting the magnetic reconnection model (Fig. 4) of flares have been found by the Yohkoh X-ray telescopes. In this model, the primary energy release site is high in the corona. The released energy is conducted along magnetic field lines and the hot region becomes a cusp-like structure. Actually, such an X-ray loop was observed with the soft X-ray telescope on board Yohkoh (Tsuneta et al. 1992, Tsuneta 1996). According to the model, a pair of hot plasma jets are ejected from the reconnection site. The observed soft X-ray ejecta associated with flares (Shibata et al. 1995; Ohyama & Shibata 1997; Tsuneta 1997) may be the plasma ejected upward through this reconnection mechanism. The model predicts that the downward jet collides with the closed loop at the base of the reconnection site, and a fast-mode MHD shock is produced at the collision site. The hard X-ray source above a soft X-ray loop found by Masuda et al. (1994) may correspond to this. Magara et al. (1996) performed a two-dimensional MHD simulation of this reconnection model. They found plasma ejection consistent with the observations and a high density region at the top of the flare loops. The latter may have some relation with the observed loop-top hard X-ray source found by Masuda et al. (1994).

When the released energy in the corona through this magnetic reconnection mechanism is transported to the upper chromosphere (Fig. 4), the dense plasma in the chromosphere is suddenly heated-up and expands because of this energy input. The induced pressure-gradient force drives the plasma to go up to the corona along the field lines (right panel of Fig. 4). This backward upflow process is called the chromospheric evaporation. The blue-shifted features of X-ray lines observed by the Bragg Cristal Spectrometer on board Yohkoh is the manifestation of this line-of-sight motion of up-going plasma.

In considering hot plasmas in the corona, it is important to take into account the heat conduction effect because the time scale for heat conduction is almost comparable with or very often faster than the time scale for flows (the Alfvén time scale). There are two important characteristics of the heat conduction. One is nonlinearity: Heat conductivity increases with increasing temperature nonlinearly (e.g. $\propto T^{5/2}$ in collisional plasma). The other is anisotropy: Heat flux can only conduct along the magnetic field direction.

Yokoyama & Shibata (1997a) developed an MHD code which includes

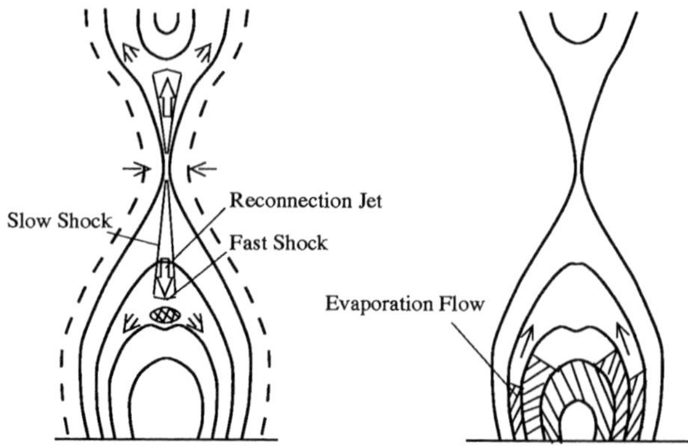

Figure 4. Schematic picture of the magnetic reconnection model for flares.

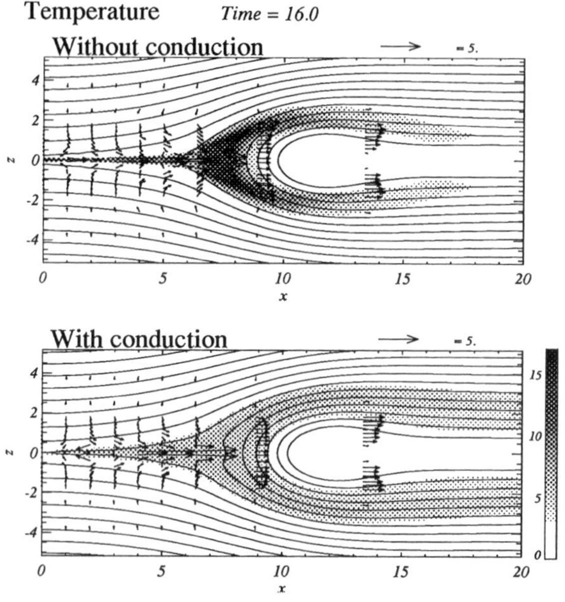

Figure 5. MHD simulation of magnetic reconnection coupled with heat conduction. Two-dimensional distributions of the temperature (gray-scale map), magnetic field lines (solid lines), and velocity vectors (arrows) for the case *without* heat conduction [upper panel] and those for the case *with* heat conduction [lower panel]. The gray-scale map levels are shown to the right. The unit of velocity vectors is shown at the top right of each panel, whose size is $V = 5.0$ in unit of the initial sound velocity. The unit of length is the initial half-thickness of the current sheet. The dimensional units of the length, temperature and density are, respectively, $\delta = 10^9$ cm, $T = 4 \times 10^6$ K, and $n_{\rm normal} = 2 \times 10^9$ cm^{-3}. The initial plasma beta is $\beta = 0.03$ and the Alfvén velocity is $V_{\rm A} \approx 6.5 \approx 1500$ km s^{-1}.

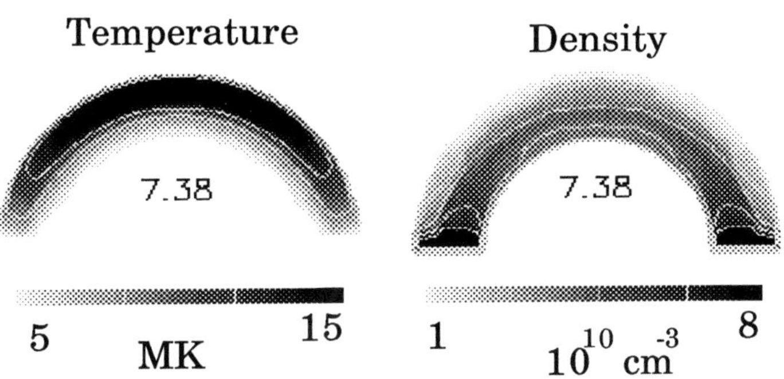

Figure 6. Results of pseudo-two-dimensional hydrodynamic simulation of multiple flare loops by Hori et al. (1997). The left panel shows temperature and the right panel shows density.

nonlinear anisotropic heat conduction. Using this code, they performed a two-dimensional simulation of magnetic reconnection coupled with heat conduction initiated by steady localized resistivity (Fig. 5). They found that the heat conduction introduces the following effects in the magnetic reconnection: (1) An adiabatic slow shock is dissociated into a conduction front and an isothermal slow shock (as had been predicted by Forbes et al. 1989). (2) In the reconnection jet (i.e., slow shock region), the temperature becomes lower (about a factor 2) and the density becomes larger in the case with conduction than without it. (3) The reconnection rate and energy release rate are weakly affected by heat conduction. (4) As a result of the propagation of the heat conduction fronts, the two-dimensional distribution of hot plasma forms a cusp-like shape (Fig. 5).

Hori et al. (1997) have developed a ‘pseudo two-dimensional ’model of flare loops (Fig. 6). It is ‘pseudo ’because they solved only one-dimensional hydrodynamics to construct a two-dimensional system of multiple loops geometrically adjacent. Flare energy is injected succesively from the innermost to the outermost loop, by which an energy source as well as the successive formation of a loop by reconnection is implicitly assumed. By combining the results for several loops, a two-dimensional distribution of physical quantities was derived (Fig. 6).

Yokoyama & Shibata (1997b) performed a two-dimensional MHD simulation of chromospheric evaporation associated with a solar flare. In this simulation, the thermal energy driving the evaporation is first supplied from the coronal magnetic field through the magnetic reconnection mechanism and is transported into the chromosphere by heat conduction. Nonlinear

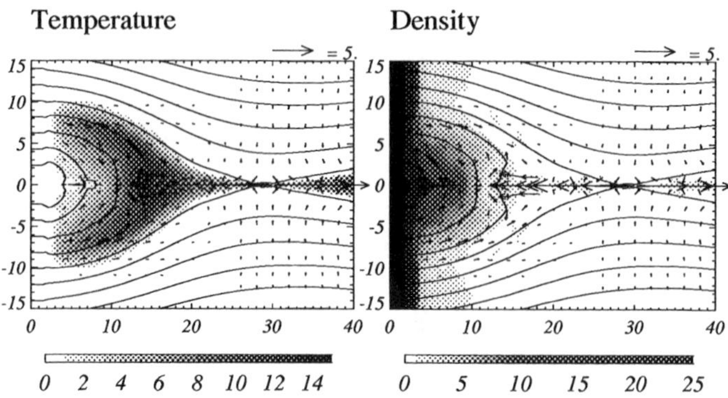

Figure 7. Results of the simulation of chromospheric evaporation driven by magnetic reconnection. The left and the right panel show temperature and density distribution, respectively. The arrows show the velocity, and lines show the magnetic field lines. The unit of length, velocity, time, temperature, and density is 2000 km, 140 km s^{-1}, 15 s, 10^6 K, and 10^9 cm^{-3}, respectively. A dense region is located near the left side of the simulation box in which the density is about 10^5 times that of the other region.

anisotropic heat conduction and radiative cooling of optically-thin plasma are taken into account.

Figure 7 shows the results. Because of the enhanced resistivity around $(x, z) = (30, 0)$, magnetic reconnection starts at this point. The reconnected field lines together with the frozen-in plasma are ejected from this X-point in the positive and negative x-directions due to the tension force of the reconnected field lines. To complement this outflow, an inflow takes place from the positive and negative z-directions of the current sheet. At the boundary between this inflow and the outflow a *slow-mode MHD shock* is formed. Since there are two sets of inflow (from the positive and the negative z-directions), a pair of the shocks is formed, emanating from the neutral point. At these shocks, the plasma is heated by shock heating. A pair of hot plasma jet is thus ejected from the neutral point. The heat conduction front propagates rapidly (Fig. 7) because the time scale of heat conduction becomes shorter as the temperature becomes higher. In the simulation, the temperature of the hot plasma jet reaches about ten times the initial coronal temperature, and so the time scale becomes 3×10^{-3} times shorter. The conduction of heat is only in the direction along the magnetic field line. The outer edge of the conduction front, therefore, traces the magnetic field lines extending from the X-point. This temperature distribution (Fig. 7) is very similar to the cusp-like structure of loops of a so-called long-duration-event (LDE) flare, which are observed by the Yohkoh soft X-ray telescope (e.g., Tsuneta et al. 1992; Tsuneta 1996). In the density distribution, a growing

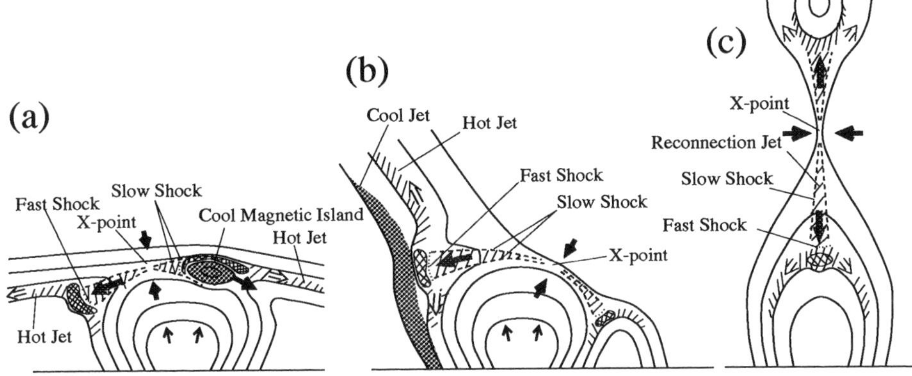

Figure 8. (a) Schematic pictures of the simulation results of the magnetic reconnection between an emerging flux and a coronal field, (b) an X-ray jet, and (c) a large-scale solar flare.

plasma mound can be seen (Fig. 7). This is the direct consequence of the so-called chromospheric evaporation. The chromospheric plasma is heated up and expands suddenly due to the penetration of the heat conduction front. The induced pressure-gradient drives a back-flow toward the corona. This flow carries dense plasma into the corona. An interesting finding of this simulation is the *high density blob above the flaring loops* (around $(x, z) = (13, 0)$ of Fig. 7). This blob is formed by compression at the fast-mode MHD shock at the collision site of the reconnection jet with the flare loops.

4. Conclusion

In this review paper, we described magnetic reconnection models for microflares, jets and large-scale flares (Fig. 8). The key factor for fast reconnection in the coronal plasma with large Reynolds number is the anomalous resistivity which is spatially nonuniform and time dependent. There are plasma ejections accelerated by the tension force of the reconnected field lines. The ejecta is a pair of jets or plasmoids. A fast-mode MHD shocks are produced at the collision site of the jets with the surrounding magnetic loops, The extremely hot plasma behind the fast shock may emit hard X-rays. The thermal energy released by the reconnection mechanism is conducted along the field lines and penetrates into the dense plasma in the chromosphere. The hot plasma evaporates due to the sudden heat-up by this conducted energy.

Finally, we suggest that the magnetic reconnection process discussed in this paper might be applicable to various hot astrophysical plasmas, such as our Galaxy (Tanuma et al 1997), cluster of galaxies (Makishima 1997), and accretion disks.

I thank K. Shibata, M. Ohyama, T. Magara, M. Shimojo, A. Okubo, and K. Hori for the help in preparing this review paper.

References

Canfield R. C. et al., 1996, 464, 1016
Forbes T. G. et al., 1989, Sol. Phys., 120, 258
Hori K. et al., 1997, ApJ, in press
Magara T. et al., 1996, ApJ, 466, 1054
Makishima K., 1997, in these proceedings
Masuda S. et al., 1994, Nature, 371, 495
Ohyama M. and Shibata K., 1997, PASJ, 49, 249
Okubo A. et al., 1997, in these proceedings
Parker E. N., 1963, ApJS, 8, 177
Parker E. N., 1966, ApJ, 145, 811
Petschek H. E., 1964, in *Proc. of AAS-NASA Symp. on the Physics of Solar Flares*, ed. W. N. Hess, (NASA Spec. Pub. 50), 425
Shibata K. and Uchida Y., 1986, Sol. Phys., 103, 299
Shibata K. et al., 1992, PASJ, 44, L173
Shibata K. et al., 1994, in *X-ray Solar Physics from Yohkoh*, eds. Y. Uchida et al., (Tokyo: Univ. Academy Press), 29
Shibata K. et al., 1995, ApJ, 451, L83
Shimojo M. et al., 1996, PASJ, 48, 123
Strong K. T. et al., 1992, PASJ, 44, L161
Sweet P. A., 1958, in *Electromagnetic Phenomena in Cosmical Physics*, ed. B. Lehnert (London: Cambridge Univ. Press), 123
Tanuma S. et al., 1997, ApJ, submitted
Tsuneta S., 1996, ApJ, 456, 840
Tsuneta S., 1997, ApJ, in press
Tsuneta S. et al., 1992, PASJ, 44, L63
Yokoyama T. and Shibata K., 1994, ApJ, 436, L197
Yokoyama T. and Shibata K., 1995, Nature, 375, 42
Yokoyama T. and Shibata K., 1996, PASJ, 48, 353
Yokoyama T. and Shibata K., 1997a, ApJ, 474, L61
Yokoyama T. and Shibata K., 1997b, in preparation

SEVERAL PROBLEMS WITH CLASSICAL "RECLOSING OF OPENED-UP SIMPLE ARCADE" MODELS FOR ARCADE FLARES AND ARCADE FORMATION EVENTS

Y. UCHIDA, K. FUJISAKI, S. MORITA, M. TORII, AND S. HIROSE
Physics Department, Science University of Tokyo
Kagurazaka, Shinjuku-ku, Tokyo

Extended Abstract

Many researchers usually have the so-called classical models by Sturrock and others in mind when they describe the arcade flaring processes. Several new findings derived from the Yohkoh-SXT observations, however, seem to support anti-standard quadruple source model proposed by Uchida (1980).

Observational data derived from the Yohkoh-SXT in some typical arcade flares, including the arcade flares at the limb of Feb 21, 1992 and of Dec 2, 1991, reported by Tsuneta, clearly show that the nucleus of the flaring region has clear high loops connecting the top of the structure back to the photosphere on both sides. Those flares also show a bright feature lying low at the axis of the dark tunnel under the "candle-flame" shaped flaring region in the pre-flare and initial phases of the flares. Those features, clearly not coincidental, do not have any explanation in the above-mentioned classical models. Those very features, however, are *expected* in the quadruple magnetic source model proposed by Uchida (1980).

In this connection, we have examined the process of the same category with larger scale and slower time development, namely, the faint high latitude X-ray arcade formation, as well as its still fainter pre-event coronal magnetic structure by using the SXT data with faint-feature-enhancing techniques. We find that the detailed process of the arcade formation at high latitudes is clearly different from what the classical models predict. Specifically, there appears a bright rising spine-like structure above the polarity-reversal line in the photospheric magnetic field, *much after* the disappearance of the dark filament itself. Contrary to what is expected from the classical models (namely, the appearance of a locus of the reconnecting points in the "reclosing of the once opened-up magnetic arcade"), this rising " spine" consists of *pre-existed threads* that are brightened and

T. Watanabe et al. (eds.), Observational Plasma Astrophysics: Five Years of Yohkoh and Beyond, 171–172.

rising. The "spine" cannot be the locus of reconnecting points because they are multiple ! This "spine" continues to rise, seemingly pulling up the arcade of loops from below into the shape of a tent, just as we reported for the Sep 28, 1991 event. Also, the event of Feb 24-25, 1993 is found to evolve at its western end into a giant cusp type structure. This led us to reexamine the so-called "Giant Cusp" event of Jan 24-25, 1992 (Fujisaki et al. 1997), and found that the event has the remnant of the "spine" extending tens of dgrees eastward, suggesting that the Cusp may not actually be a cusp produced in the reclosing of once opened-up arcade.

The deep examination of the fainter pre-event structure of some polar arcade formation events using faint-feature-enhancing techniques has revealed that the pre-event coronal structure of polar arcade formation events consists of "dual arcades", whose inside legs cross with each others' in the region above the polarity reversal line in the photospheric field (Uchida et al. 1997). The H_α dark filament lies in the region of the crossed legs with a height of several to ten thousand km. The leg crossing, which is quite different from the classical sagged, simply-connected arcade, led us to examine the physical reason for it. We readily found that the photosphere below is *not at all* a clean bipolar magnetic regions, but rather contain a number of opposite polarity patches in the region of the normal field on each side of the polarity-reversal line of the "averaged" field in the photosphere. Surprisingly, the "loops" in the "dual arcades" with crossed legs, coming from one polarity, seem to be connected to the patches of the *same* polarity! This is not possible for magnetic lines of force. A possible explanation Uchida et al. (1997) propose is that the "dual arcades" are not the usual lines of force, but the "separatric surface" in the "quadruple source configuration" proposed by Uchida (1980) by using 2.5 dimension (depending on x and y coordinates, but vector quantities having x, y, and z components) for simplicity.

The details will be presented in the papers given below.

References

Fujisaki, K., Uchida, Y., Torii, M., Morita, S., Hirose, S., and Cable, S., (1997), Polar Arcade formation of February 24-25, 1993 Producing a Large Cusp-like Feature, and the Giant Cusp of Jan 25, 1992, *Publ. Astron. Soc. Japan*, to be submitted.

Uchida, Y., (1980), in *Solar Flares*, ed. Sturrock, P., (Univ. Colorado Press), pp 67, 110.

Uchida, Y., Fujisaki, K., Torii, M., Morita, S., Hirose, S., and Cable, S. (1997), Soft X-ray Coronal Structure Surrounding Polar Region Dark Filaments Observed by Yohkoh, *Publ. Astron. Soc. Japan*, to be submitted.

PLASMOID FORMATION IN ERUPTIVE FLARES

T. MAGARA
Department of Astronomy, Faculuty of Science,
Kyoto University, Sakyo-ku, Kyoto 606-01, Japan

AND

K. SHIBATA AND T. YOKOYAMA
National Astronomical Observatory
Mitaka, Tokyo 181, Japan

Extended Abstract

Mass eruptions are often associated with solar flares, which are sometimes called plasmoid eruptions or filament eruptions (observed in Hα). Theoretically, these eruptive phenomena have been considered to be related to an instability or a loss of equilibrium of the coronal magnetic field. Since the magnetic force is much stronger than both the gas pressure and the gravity force in the corona, coronal structures are mainly controlled by the magnetic field. Therefore, unless a magnetically-driven event occurs, the coronal structures evolve in a series of quasi-static states, but once an instability develops or a loss of equilibrium is achieved, they no longer stay in a static state and enter on a dynamical stage. This scenario has made it important to study the stabilities and natures of the equilibrium configurations of the coronal magnetic field. Recently, the rapid development of computers has enabled us to trace the temporal evolution of the coronal magnetic field directly by means of numerical simulations. Such work is found in, e.g., Mikic, Barnes, and Schnack (1988), Biskamp and Welter (1989), Finn, Guzdar, and Chen (1992), Inhester, Birn, and Hesse (1992), and Choe (1995). These papers studied how the coronal magnetic field evolved under situations where shearing or converging motion was imposed in the photosphere. In the present paper we investigate the resistive processes of plasmoid dynamics in eruptive flares by performing 2.5-D resistive MHD numerical simulations. We start with a linear force-free field

T. Watanabe et al. (eds.), Observational Plasma Astrophysics: Five Years of Yohkoh and Beyond, 173–174.

arcade and impose a localized resistive perturbation on the symmetry axis of the arcade. Then the magnetic fields begin to dissipate, producing inflows toward this region. These inflows make the magnetic fields convex to the symmetry axis, forming a neutral point on this axis and leading to the formation of a magnetic island around the symmetry axis. The magnetic island first rises slowly by the upflow produced by the initial resistive perturbation, and then once the anomalous resistivity becomes effective the magnetic island begins to accelerate. This acceleration stops after the fast MHD shock is formed at the bottom of the magnetic island, which implies that the upflow around the central part of the magnetic island is no longer strong. These three stages in the evolution of the plasmoid are confirmed to exist in the observational results. Moreover, a time-lag between the start time when the magnetic island begins to accelerate and the peak time of the neutral- point electric field, can be explained by the inhibition of magnetic reconnection by the perpendicular magnetic field. We also study the difference of the initial rise motion of the plasmoid between the simulated and the observational results, and we conclude that in actual situations, the initial resistive perturbation proceeds very weakly and at many positions inside the arcade.

A full paper on this work has been submitted to ApJ (Magara, Shibata, and Yokoyama 1997).

References

Biskamp, D. and Welter, H. (1989), *Sol. Phys.*, **Vol.120**, pp.49
Choe, G. S. (1995), Ph. D. Thesis, Univ. Alaska
Finn, J. M., Guzdar, P. N., and Chen, J. (1992), *ApJ*, **Vol.393**, pp.800
Inhester, B., Birn, J., and Hesse, M. (1992), *Sol. Phys.*, **Vol.138**, pp.257
Magara, T., Shibata, K., and Yokoyama, T. (1997), *ApJ*, in press
Mikic, Z., Barnes, D. C., and Schnack, D. D. (1988), *ApJ*, **Vol. 328**, pp.830

PLASMA MOTION IN SOLAR FLARES OBSERVED BY THE NOBEYAMA RADIOHELIOGRAPH

K. SHIBASAKI
Nobeyama Radio Observatory
Minamimaki, Minamisaku, Nagano 384-13, Japan

1. Introduction

Large scale and high density plasma cloud motions were detected in the rise phase of flares by the Nobeyama Radioheliograph. The Nobeyama Radioheliograph is a radio interferometer dedicated to solar observations. It has a capability to synthesize one full disk solar image every second at 17 GHz and to observe continuously during 8 hours every day. The spatial resolution is about 10 arc seconds. Fast motions and changes of radio brightness caused by activity anywhere on or near the solar disk can be detected with high probability due to its wide field of view and its continuous observation with high cadence.

2. Flares

Three flares in which ejections were detected are listed in Table 1. They were located on the solar limb where the ejected radio sources could be found easily against dark sky background. Events 1 and 2 occurred when the *Yohkoh* satellite was not observing the Sun. *Yohkoh* observed Event 3 only after the peak of the flare, when the radio source ejection had ended. All events lasted more than an hour in radio and are categorized as gradual flares.

3. Plasma Motions

The peak flux of radio sources were comparable to that estimated from the plasma parameters deduced from GOES two channel soft X-ray data. The peak brightness temperatures of these sources were higher than or equal to 10^4 K and the degrees of circular polarization were lower than one percent.

T. Watanabe et al. (eds.), Observational Plasma Astrophysics: Five Years of Yohkoh and Beyond, 175–178.

TABLE 1. Event list

Event ID	Event 1	Event 2	Event 3
Date	2-Jan-93	4-Jan-93	20-Nov-93
Time(Dur)	04:20(1.5h)	00:45(1h)	00:20(2h)
Peak Flux	11 sfu	7 sfu	4 sfu
Flare Class	M 1.1	C 6.6	C 5.2
Location	W limb	E limb	E limb
Active Region	AR 7376	AR 7391	AR 7622
Others(Time)		Type III(00:45)	BATSE(00:20)

No hard X-ray events were associated with these motions. Taking these facts into account, we infer that the emission mechanism of these sources were thermal free-free emission from dense plasma clouds. Motions of these radio sources are interpreted as plasma cloud motions.

Plasma cloud parameters were deduced from the sequences of radio images of moving radio sources. They are listed in Table 2. The column emission measure is calculated from the radio brightness temperature assuming that the plasma temperature is the same as that deduced from the GOES soft X-ray two channel data and that it is uniform over the source. We assume the emission is optically thin. This assumption may not be valid everywhere in the source, but it is valid as a whole. The total flux calculated from the plasma parameters deduced from GOES data are comparable to that of the observed radio flux, at least in the rise phase of the flares. The filling factor of the plasma is assumed to be unity and the size in the line of sight direction is assumed to be comparable to the width of the radio source.

TABLE 2. Plasma parameters of ejected clouds

Event ID	Event 1	Event 2	Event 3
Speed(km/s)	500-700	280	200
Duration(min)	4	10	10
Size(10^9 cm)	1.5 x 1.5	4.5 x 13	1.5 x 20
Mass(10^{13} g)	4	400	140
Kinetic En.(10^{29} erg)	1	20	3
Form	trapped	explosive	explosive

4. Morphology

The plasma cloud ejections were preceded by the appearance of bright sources at lower altitude (on the disk) from where the ejection started. This location was one end of a bright microwave loop which were formed later in each flare. The period when the ejections were going on coincided well with the increasing phase of the GOES soft X-ray flux.

In Event 1, the ejected source was trapped at the top of the loop, brightened, and eventually became part of the bright flare loop. The kinetic energy content of an individual ejection is smaller compared to the other two events, but at least a couple of ejections could be identified. The ejection speed reached 700 km/s.

In Events 2 and 3, ejections lasted about 10 minutes with velocities of 280 and 200 km/s respectively. Ejected plasma clouds continued to ascend up to several arc minutes above the limb. A metric Type III event and a BATSE X-ray event were reported respectively for these events coinciding with the onset of the ejections, but not during the ejections.

5. Conclusions and Discussion

In this study, we have shown that large fraction of total flare energy is contained in the form of plasma cloud motion during the rise phase of flares. Concentrated energy release is needed in the chromosphere to heat and propel the high density and high temperature plasma cloud. This plasma cloud motion is different from plasmoid ejection which is associated with the process of opening up closed magnetic field before the onset of flares found by *Yohkoh* SXT (Shibata *et al.,* 1995), which are not dense enough to be detected by radio. The initial energy release in these flares will not be at the top of the loops, but rather at one end of the loops. The present results suggest that the emerging flux flare model by Heyverts *et al.* (1977) is working in these flares. Similar event was also observed by *Yohkoh* SXT (Ohyama and Shibata 1997) in the 1993 November 11 flare. It is interpreted mainly in the frame of opening and reclosing magnetic field flare model they refer to as the CSHKP model. They pointed out inconsistencies between the observations and the theory based on the CSHKP model.

References

Heyverts, J., Priest E.R. and Rust D.M. (1977) *Ap. J.*, **216**, 123

Shibata, K., Masuda, S., Shimojo, M., Hara, H., Yokoyama, T., Tsuneta, S., Kosugi, T. and Ogawara, Y. (1995) *Ap. J.*, **451**, L83

Ohyama, M and Shibata, K (1997) *PASJ*, **49**, 249

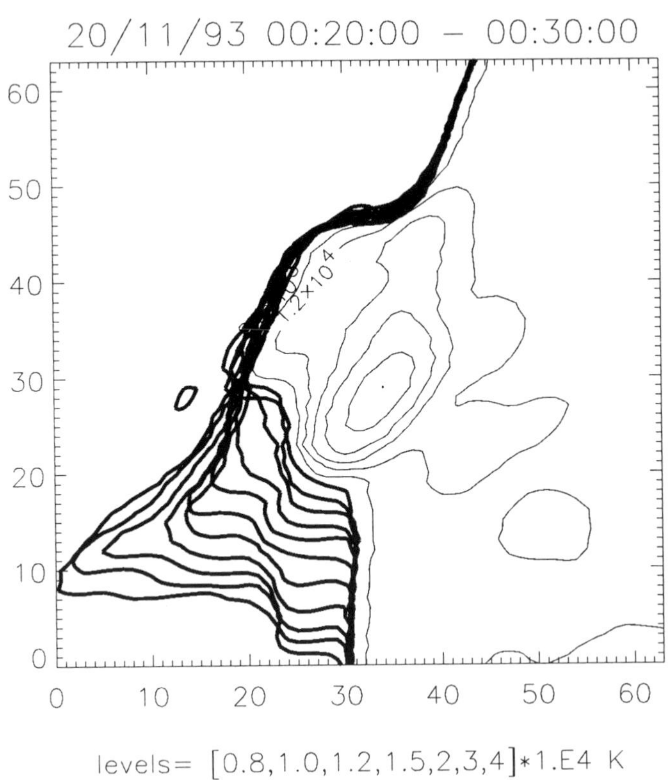

Figure 1. Plasma ejection in Event 3. The thick coutours are at the time of the event's start: 00:20 UT. The contour levels are [0.8, 1.0, 1.2, 1.5, 2.0, 3.0, 4.0] $\times 10^4$ K. Thin lines are countour levels of 0.8 $\times 10^4$ K successive every minute starting from 00:20 UT until 00:30 UT. The field of view is 5.3 $\times$ 5.3 arcmin.

CURRENT SHEET FORMATION AND RECONNECTION ON SEPARATOR FIELD LINES

D.W. LONGCOPE
Montana State University
Department of Physics
Bozeman, MT 59717-3840, USA

Abstract. Magnetic reconnection is believed to play a fundamental role in many types of coronal activity such as flaring and transient loop brightenings. Reconnection occurs when flux is transferred between neighboring coronal flux systems. A single flux system consists of coronal magnetic field lines originating in a common pair of photospheric features (i.e. flux tubes). Changing the flux in such a system requires a substantial electric field in the corona, parallel to a magnetic field line in at least one place. Such an electric field is not consistent with the highly conductive nature of the hot corona.

We present a model for the evolution of a complexly structured coronal magnetic field driven by motion of photospheric flux tubes. Barring reconnection, motion of photospheric flux tubes leads to the development of three dimensional current sheets (current ribbons) along particular field lines called separators. A separator is a field line which lies at the boundaries of four distinct flux systems simultaneously. The amount of current in a given ribbon is an increasing function of flux tube displacement. The presence of current implies free magnetic field energy, and reconnection must occur eventually if the separator current and stored energy are not to grow indefinitely. When it occurs, reconnection adjusts the fluxes in each system to temporarily obviate the need for current ribbons, thereby liberating the stored free energy. Such instances of reconnection occur sporadically during a typical evolutionary scenario. The frequency of these events, and the energy released, is consistent with transient brightenings observed by Yohkoh (Shimizu *et al.*, 1992).

T. Watanabe et al. (eds.), Observational Plasma Astrophysics: Five Years of Yohkoh and Beyond, 179–184.

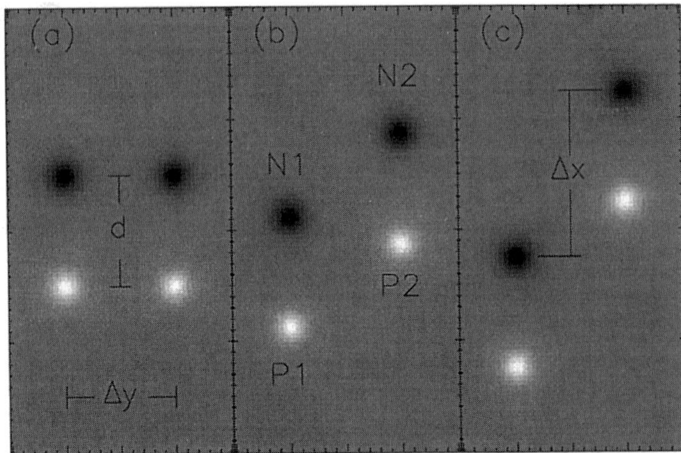

Figure 1. Synthetic magnetograms of the two bipoles shown with x running vertically and y horizontally. Here $\Delta y = d$. Three cases are shown: (a) $\Delta x = 0$, (b) $\Delta x = 0.75\,d$ (c) $\Delta x = 1.5\,d$.

1. Quantifying flux interlinkage

To introduce our model we will consider a case of two interacting magnetic bipoles. This case, considered in many previous investigations (Sweet, 1958; Baum and Bratenahl, 1980; Gorbachev and Somov, 1988; Longcope, 1996), is one of the simplest with enough geometric complexity to manifest generic features of three dimensional magnetic fields such as a separator. The configuration consists of four magnetic poles ("charges") $P1$, $P2$, $N1$ and $N2$ each with flux ψ_0, arranged in two bipoles (see Fig. 1). The poles interact with one another as the distance between bipoles, Δx, changes due to some relative photospheric motion.

The magnetic poles represent the tops of magnetic flux tubes. These are confined to a tube below the canopy by flows and plasma pressure, but in the corona the field expands into a space-filling equilibrium (see e.g. Gabriel 1976). Almost all of the coronal field lines will connect a positive (P) charge to a negative (N) charge. The set of field lines connecting, for instance, pole $P2$ to $N1$ will have some total net flux Ψ_{21}. This flux can be calculated by integrating the vector potential around a path $\mathcal{C}$ which encloses only those field lines

$$\Psi_{21} = \oint_{\mathcal{C}} \mathbf{A} \cdot d\mathbf{l} \quad . \tag{1}$$

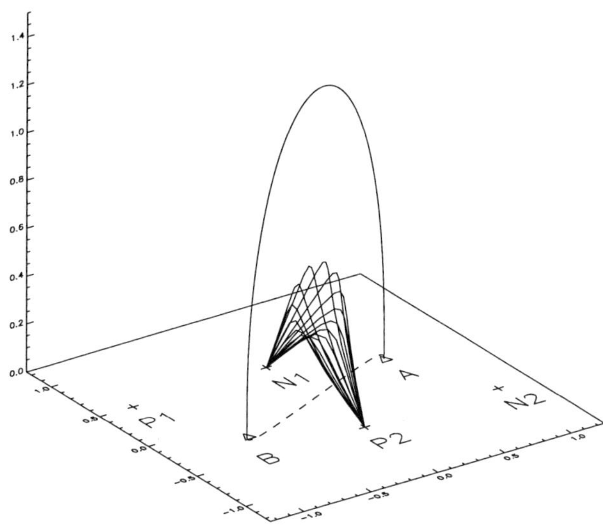

Figure 2. The separator σ connecting null points B and A (solid), along with a possible closure (dashed) formed by projecting σ onto $z = 0$. Also shown are some of the $P2$–$N1$ field lines.

This path must lie at the boundary separating those field lines which connect $P2$ to $N1$ from all other field lines. There is a single field line, called the separator σ, with this property (Longcope, 1996). We will take $\mathcal{C}$ to consist of the separator field line and some curve lying just below the $z = 0$ surface (see Fig. 2).

The value of Ψ_{21} will depend on the configuration of poles, and on the form of the coronal magnetic field. Taking the coronal field to be current-free (its state of lowest energy) yields a curve $\Psi_{21}^{(\mathrm{v})}(\Delta x)$ shown in Fig. 3. The flux is zero for $\Delta x = 0$, due to symmetry about the x axis, and increases up to $\Delta x \simeq 1.33\, d$ where $P2$ and $N1$ are very close to one another. For still larger Δx the bipoles become increasingly isolated and $\Psi_{21}^{(\mathrm{v})} \to 0$.

2. Response to flux motion — Current ribbons

Changing the flux Ψ_{21} would require an electric field along $\mathcal{C}$. Since the bottom of this path lies within a conductor ($z = 0$) this electric field must exist along the separator field line in the corona. Using typical values for ψ_0, d and $d(\Delta x)/dt$ gives an average field of 0.1 V/m, which is much larger than the highly conductive corona could support parallel to the magnetic field. We must conclude that the corona cannot remain current-free in the face of flux tube motion. Instead the coronal field is a sum of the vacuum

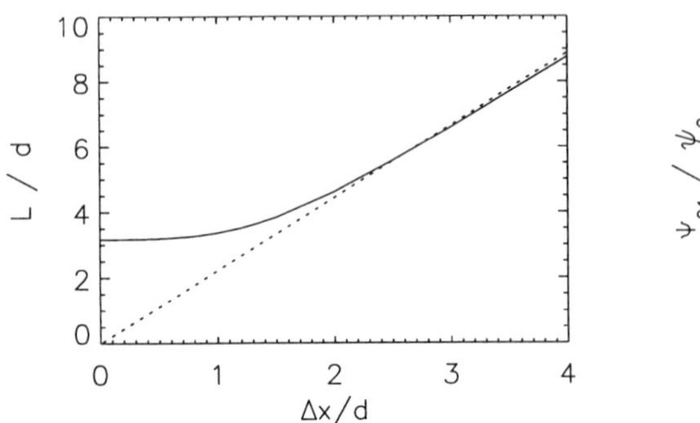

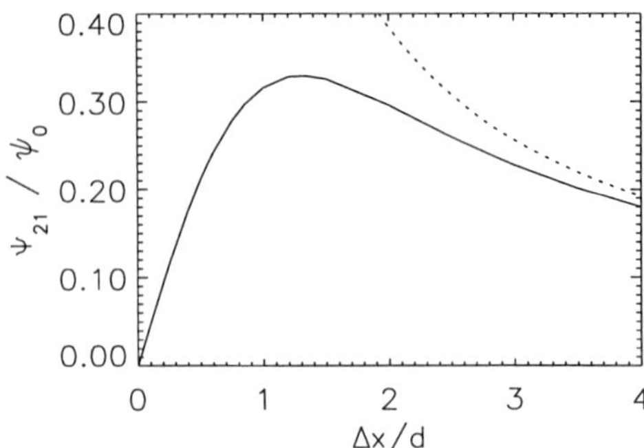

Figure 3. (Left) The separator length L scaled to the bipole separation d, and (Right) the flux $\Psi_{21}^{(\mathrm{v})}$ scaled to ψ_0, the flux of a single pole. Its maximum value is 0.34; one-third of the flux from $P2$ links to $N1$, the remining two-thirds links to $N2$.

field from the point charges and a contribution, vanishing at $z = 0$, from coronal currents. The change in vacuum flux $\Delta\Psi_{21}^{(\mathrm{v})}$ must be canceled by self-flux from the coronal currents. A careful analysis of this problem by Longcope and Cowley (1996) lead to the conclusion that the coronal current forms a singular ribbon paralleling the separator. The current will flow from one separator footpoint to the other, through the corona, and return in a distributed current layer just beneath the canopy ($z = 0$). Since the separator lies at the boundary of flux systems, current can close through the canopy without crossing any field lines, in contrast to models with current flowing on flux tubes themselves (Melrose, 1995).

The formation of a current ribbon on a separator is directly analogous to the formation of a current sheet at an X-point in a two-dimensional magnetic field (Syrovatskii, 1971). In the case of a generic three-dimensional field the separator field line plays the same role as an X-point in two-dimensions (Greene, 1988; Lau and Finn, 1990).

The net coronal current is determined by the requirement $\Psi^{(\mathrm{cr})}(I) = -\Delta\Psi_{21}^{(\mathrm{v})}$. The relationship between current and self-flux is essentially one of self-inductance $\mathcal{L}$. The (differential) self-inductance will have the form

$$\mathcal{L} = \frac{1}{c}\frac{d\Psi^{(\mathrm{v})}}{dI} \sim \frac{2L}{c^2}\ln(L/w) \quad , \tag{2}$$

where w is the breadth of the current sheet which is a property of the detailed structure of the singular magnetic equilibrium. For small I the current will distort the equilibrium only locally and $w \sim \sqrt{|I|}$ by analogy to two dimensions (Longcope and Cowley, 1996). The constant of proportionality depends on the local gradient of the vacuum magnetic field at the separator. This gradient can be used to define a function $I^{\star}(\Delta x)$ according to which the approximate current-flux relationship is

$$\Psi^{(\mathrm{cr})}(I) = \frac{LI}{c} \ln\left[\frac{I^{\star}}{|I|} \frac{e^3}{256}\right] . \tag{3}$$

The exact definition of $I^{\star}$, depending only on the current-free field for Δx, is given in Longcope (1996).

3. Effects of reconnection — microflaring

Changing Δx will give rise to a coronal current I flowing roughly parallel to the separator. This will store free energy (i.e. in excess of the vacuum field energy) as for any inductor. This energy is not localized to the sheet; it corresponds to the extended magnetic field due to the coronal current. We propose that some upper limit exists to the current in the ribbon. This limit might be due to instability at large or small scales, but can be phrased as a threshold: $|I| \leq \theta I^{\star}$, where θ is some number probably less than unity. When this threshold is exceeded we assume that a large electric field occurs along the current ribbon (stick-slip reconnection). This changes the flux Ψ_{21}, bringing it closer to the present $\Psi_{21}^{(\mathrm{v})}(\Delta x)$ and thereby reducing I. Thus, some of the flux linking $P2$–$N2$ is transported across the separator and remapped so as to link $P2$–$N1$: *this is reconnection.* Complete reconnection takes $\Psi_{21} \rightarrow \Psi_{21}^{(\mathrm{v})}$, eliminating I entirely, thereby releasing all the free magnetic energy it stores. This, we propose, is the underlying cause of sudden coronal energy releases, such as transient X-ray brightenings (Shimizu *et al.*, 1992).

Applying this model to the above configuration gives the time history shown in Fig. 4. The threshold parameter has been set at $\theta = 0.15$, and physical values are those of relatively small bipoles, $\psi_0 = 3 \times 10^{20}$ Mx, $d = 30,000$ km, which are moving moderately rapidly $d(\Delta x)/dt = 300\,\mathrm{m/s}$. In the resulting evolution there are a series of small ($\Delta E \sim 10^{27}$ ergs) events occurring at intervals of 3–6 hours while the bipoles interact. After this initial phase there is a lull, followed by a longer period of smaller, less frequent events as the bipoles become increasingly isolated.

4. Discussion

The example above shows how a coronal field originating in multiple distinct magnetic sources will evolve. We have chosen a simple case, with only two sources of each polarity, which contains only a single separator. Motion of the sources leads to coronal current which flows in a thin ribbon along the separator. Cases with more magnetic sources have more separators, forming

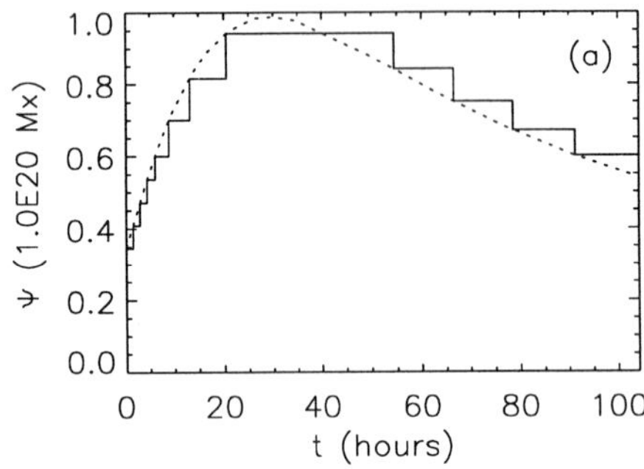

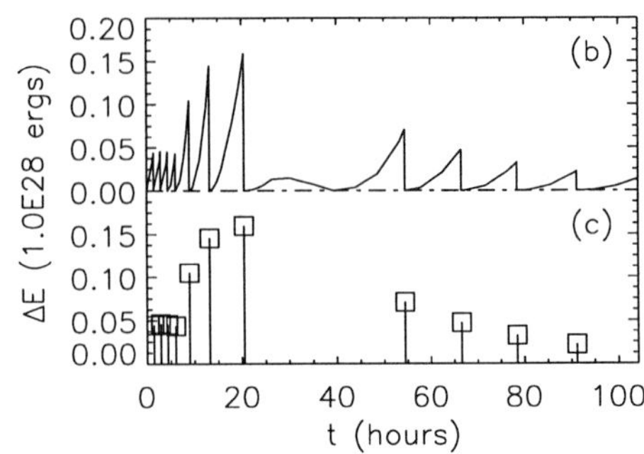

Figure 4. The time history of the evolving bipoles. (a) The flux Ψ_{21} (solid) which attempts to follow $\Psi_{21}^{(v)}$ (dotted) through a series of reconnection events. (b) The free-energy from the current. This drops to zero with each reconnection event. (c) The amplitude of the drops in energy. This is the energy liberated by reconnection.

a rich array of coronal current paths. Motion of these sources would result in sporadic reconnection on many of the separators. Each reconnection event would liberate energy causing a transient loop brightening there. The resulting picture, an array of loops sporadically brightening, is reminiscent of the now familiar Yohkoh SXT movies.

References

Baum, P. J. and Bratenahl, A. (1980), Flux linkages of bipolar sunspot groups: a computer study, *Solar Phys.* **67**, 245–258.

Gabriel, A. H. (1976), A magnetic model of the solar transition region, *Phil. Trans. R. Soc. Lond.* **A281**, 339–352.

Gorbachev, V. S. and Somov, B. V. (1988), Photospheric vortex flows as a cause for two-ribbon flares: a topological model, *Solar Phys.* **117**, 77–88.

Greene, J. M. (1988), Geometrical properties of three-dimensional reconnecting magnetic fields with nulls, *JGR* **93**, 8583–8590.

Lau, Y.-T. and Finn, J. M. (1990), Three-dimensional kinematic reconnection in the presence of field nulls and closed field lines, *ApJ* **350**, 672–691.

Longcope, D. W. (1996), A model for current ribbon formation and reconnection in a complex three-dimensional corona, *Solar Phys.* **169**, 91–121.

Longcope, D. W. and Cowley, S. C. (1996), Current sheet formation along three dimensional magnetic separators, *Phys. Plasmas* **3**, 2885–2897.

Melrose, D. B. (1995), Current paths in the corona and energy release in solar flares, *ApJ* **451**, 391–401.

Shimizu, T., Tsuneta, S., Acton, L. W., Lemen, J. R., and Uchida, Y. (1992), Transient brightenings in active regions observed by the soft X-ray telescope on Yohkoh, *Proc. Astron. Soc. Japan* **44**, L147–L153.

Sweet, P. A. (1958), The neutral point theory of solar flares, in Lehnert, B., editor, *Electromagnetic Phenomena in Cosmical Physics*, pp. 123–134, Cambridge University Press, Cambridge, U.K.

Syrovatskii, S. I. (1971), Formation of current sheets in a plasma with frozen-in strong magnetic field, *Sov. Phys. JETP* **33**, 933–940.

BIFURCATION-TRANSITION DYNAMICS IN SOLAR CORONAL PLASMA

K. KUSANO

Faculty of Science, Hiroshima University
Higashi-Hiroshima 739, Japan

1. Introduction

Solar flares must be a sort of catastrophic process, in which the free energy stored in the coronal magnetic field is explosively released into the heat and kinetic energy of coronal plasmas. However, the supply of free energy from photospheric motions is believed to be a greatly slower process compared to flares. Therefore, it is worthwhile to consider the problem of why flares abruptly appear and proceed as an explosive process.

We have theoretically considered this problem based on the minimum energy principle, and recently proposed a new theory, the *'bifurcation-transition flare model'*. According to minimum energy theory, when magnetic helicity is increased, the solution of the linear force free field equation $\nabla \times \vec{B} = \alpha \vec{B}$ bifurcates into two solutions; a loop-type and a plasmoid-type solution [1]. We found that the loop-type solution becomes linearly unstable to a coronal loop instability (CLI). The bifurcation-transition flare model asserts that the transition from the loop-type solution into the plasmoid-type solution could cause flares [2].

However, we also found that the growth rate of the CLI is proportional to $1/\sqrt{S}$, where S is the magnetic Reynolds number [3]. This means that the growth time of the CLI must be much longer than the magnetohydrodynamic time-scale when S is very large, as expected in the solar corona. Therefore, it is quite important to investigate the growth time for the nonlinear growth of the CLI. This paper is a brief report on the simulations we have carried out in this regard.

T. Watanabe et al. (eds.), Observational Plasma Astrophysics: Five Years of Yohkoh and Beyond, 185–186.

2. Model

The simulation model is almost the same as in [3]. The initial state is given by the loop-type solution of the linear force free field, which is unstable to the CLI. In this study, however, the evolution of the plasma pressure and the density is taken into account. Also we employ the grid-size refinement technique, in which the numerical resolution is dynamically improved as the finer structure is nonlinearly generated.

3. Result

Two simulations are carried out for $S = 2 \times 10^3$ and 10^4, where the initial pressure is distributed uniformly. The maximum β value in the initial state is 10^{-4}. In both cases, the CLI can grow, so that the total kinetic energy K increases exponentially in the initial phase. We found that the initial growth rate, which is given by $\gamma = (1/K)(dK/dt)$, is reduced when S increases, as expected from linear analysis [3]. However, in the case for $S = 10^4$, the growth rate is enhanced as K increases, so that the instability grows faster than an exponential function in time.

4. Discussion

The mechanism of the escalation of γ can be understood from the fact that the current sheet, which is generated as the result of the growth of the CLI, becomes unstable to the resistive tearing mode. Since the growth rate of the tearing instability is larger than that of the original coronal loop instability, the growth rate is accelerated.

Since the escalation of the growth rate is not observed in the case for $S = 2 \times 10^3$, it must be a specific feature to the high S regime. These results suggest that, when S is as large as it is in the solar corona ($\sim 10^{10}$), the instability might grow in a much shorter time scale than that predicted by linear stability analysis.

References

1. Kusano, K., Suzuki, Y., and Nishikawa, K. (1995) A Solar Flare Triggering Mechanism Based on the Woltjer-Taylor Minimum Energy Principle, *Astrophys. J.* **441**, pp. 942–951
2. Kusano, and Nishikawa, K. (1996) Bifurcation and Stability of Coronal Magnetic Arcades in a Linear Force-Free Field, *Astrophys. J.* **461**, pp. 415–423
3. Kusano, and Nishikawa, K. (1996) Simulation Study of Magnetic Reconnection in Solar Corona, *Magnetic Reconnection in the Solar Atmosphere.* Astronomical Society of the Pacific, San Francisco, pp. 375–378

A UNIFIED MODEL OF SOLAR FLARES

KAZUNARI SHIBATA
National Astronomical Observatory
Mitaka, Tokyo 181, Japan

Abstract: A unified model of flares, which we call the *plasmoid-induced-reconnection model*, is presented. This model explains various observed features (e.g., homologous flares, time scale, energy release rate) of impulsive flares, LDE flares, giant arcades associated with filament eruptions or CMEs, and microflares in a unified scheme. This model is an extention (a unified version) of the CSHKP model and the emerging flux model.

1. Introduction

Before *Yohkoh*, solar observers had long thought that solar flares could be classified into two types, such as LDE flares vs impulsive flares, or eruptive vs confined, or two ribbon vs simple loop (or compact), etc. (e.g., Priest 1982). The former has often been thought to be explained by the so called "CSHKP" (Carmichael-Sturrock-Hirayama-Kopp-Pneuman) reconnection model, whereas the latter has been attributed to different models, such as the emerging flux reconnection model (Heyvaerts et al. 1974).

Yohkoh, however, has revealed that there are many common features in both types of flares, e.g., the ejection of hot plasmas (possibly plasmoids) (Shibata et al. 1995, Shibata 1996, Tsuneta 1997, Ohyama and Shibata 1997a,b), x-type or y-type morphology suggesting the presence of current sheets or neutral points (Tsuneta et al. 1992a,b, Masuda et al. 1994, 1995, Tsuneta 1996, 1997), change of field configuration, etc. Even microflares have sometimes shown hot plasma ejections or jets (Shibata et al. 1992b, Shimojo et al. 1996) and change of morphology (possibly as a result of reconnection) (Shibata et al. 1994). It is now not easy to classify flares into two types, and a unified view of flares has emerged on the basis of *Yohkoh* observations (Shibata 1996, 1997, Kosugi and Shibata 1997). This view is also consistent with the statistical properties of many solar flares, such as

T. Watanabe et al. (eds.), Observational Plasma Astrophysics: Five Years of Yohkoh and Beyond, 187–196.

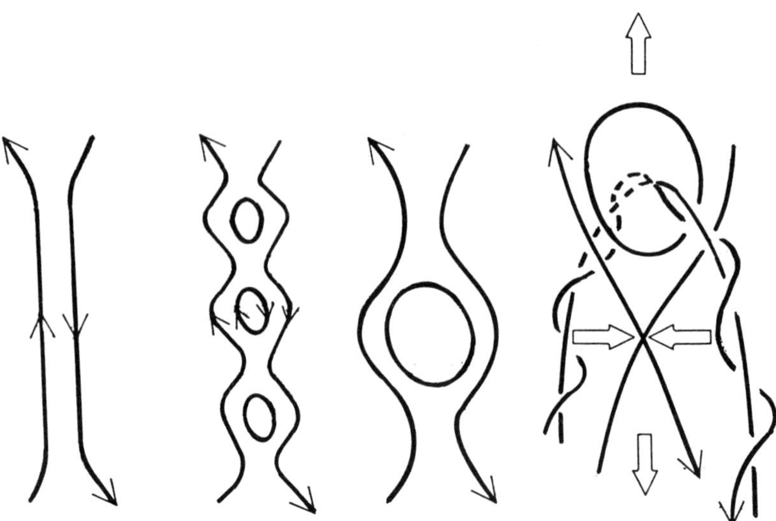

Figure 1. Schematic drawings showing current sheet formation, tearing, coalescence, plasmoid ejection, and triggered fast reconnection at high magnetic Reynolds number, all of which have been found in numerical simulations. In three-dimensional space, the plasmoid would be seen as a helically twisted loop or filament. Eruptive prominences observed in Hα are a kind of plasmoid ejection.

the frequency distribution of flares as a function of X-ray intensity (or flare energy), which shows the same power-law distribution not only for major flares but also for microflares (e.g., Shimizu 1995), and the relationship between the temperature and the emission measure for major and micro flares (e.g., Watanabe 1994).

On the other hand, recent numerical simulations of magnetic reconnection with high spatial resolution have revealed that there are common fundamental (evolutionary) features in fast reconnection at high magnetic Reynolds number ($R_m > 10^3$); once a current sheet is formed, magnetic reconnection proceeds intermittently with nonsteady processes (see Fig. 1), such as tearing in the current sheet, coalescence of magnetic islands (plasmoids), and ejection of plasmoids (e.g., Lee and Fu 1986, Ugai 1989, Shibata et al. 1992a, Yokoyama and Shibata 1994, Karpen et al. 1995, Kusano et al. 1995, Kitabata et al. 1996). It has been found that even reconnection driven by emerging flux (i.e., the emerging flux model) results in the ejection of plasmoids (Shibata et al. 1992a, Yokoyama and Shibata 1994, 1995, 1996) similar to those seen in CSHKP-type reconnection (e.g., Magara et al. 1996, Yokoyama and Shibata 1997). In this sense, there is no fundamental difference in the physics of reconnection in the emerging flux model and in the CSHKP model. These numerical simulation results (see Yokoyama

1997 for a review), especially the common occurrence of plasmoid ejections and associated intermittent nonsteady (bursty) reconnection at high magnetic Reynolds number, are quite similar to actually observed behavior of solar flares as discussed above. Since the magnetic Reynolds number in the active region corona is enormously high $\sim 10^{13}$, it is natural to think that the observed impulsive nature and plasmoid ejections are a consequence of fast reconnection at high magnetic Reynolds number (Shibata 1997).

In this paper, taking account of these recent developements in both observations and numerical simulations, we propose a unified model of solar flares, which we call the *plasmoid-induced-reconnection model* since plasmoid ejections play a key role in triggering fast reconnection in the model. This model unifies not only LDE flares and impulsive flares, but also microflares. In a previous paper (Shibata 1996), we call the former unification the "unified model", and the latter the "grand unified model". However, in this paper, we simply use the term "unified model," since it has become clear that there is no essential difference in our previously defined unified and grand unified models.

2. The Plasmoid-Induced-Reconnection Model

2.1. BASIC FEATURES OF THE MODEL

Observations show that strong acceleration of plasmoids occurs just before the peak of the impulsive phase of flares (Ohyama and Shibata 1997a,b). If the intensity of hard X-rays is a measure of the electric field (i.e., reconnection rate) at the X-point, this suggests that the ejection of high speed plasmoids induces fast inflow, i.e., fast reconnection. Similar behavior has also been found in numerical simulations of reconnection (e.g., Ugai 1989, Yokoyama and Shibata 1994, 1996, Magara et al. 1997).

Hence we adopt the hypothesis that *impulsive energy release due to fast reconnection is induced by the fast ejection of plasmoids.* In this case, the velocity of inflow into the X-point is estimated to be

$$V_{inflow} \sim V_{plasmoid} \sim 50 - 400 \text{ km/s}, \tag{1}$$

from the conservation of mass, assuming that plasma density does not change much during the process. Then, the Alfven Mach number of the inflow becomes

$$M_A = V_{inflow}/V_A \sim 0.02 - 0.1 V_A. \tag{2}$$

This is comparable to the inflow speed expected from Petschek theory. We call this the *plasmoid-induced-reconnection model* (see Fig. 2).

The merit of this model is that it can easily explain *homologous flares*, as illustrated in Figure 3. In either the long current sheet case (e.g., Kitabata

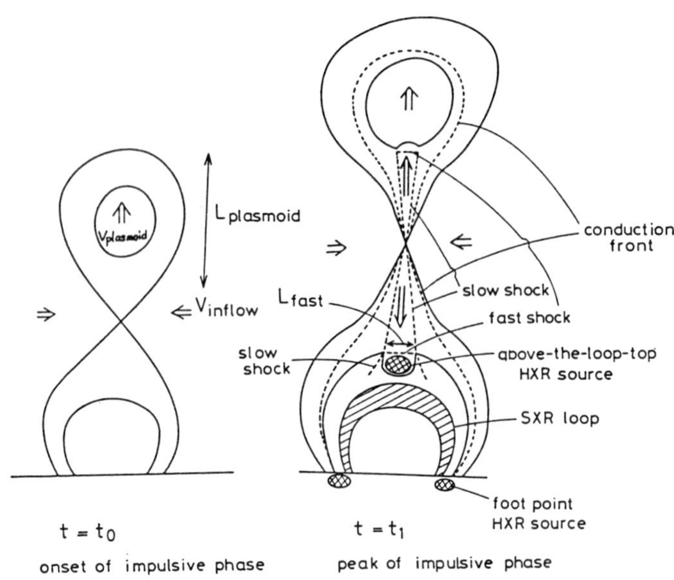

Figure 2. The *Plasmoid-Induced-Reconnection Model* for flares.

et al. 1996) or the sheared arcade case (e.g., Kusano et al. 1995, Magara et al. 1997), if there is a continuous driving motion to compress the current sheet or arcade, the following cycle is repeated: *current sheet or arcade → formation of plasmoids → ejection of plasmoids → recovery of the initial current sheet or arcade*. Note that although the evolution of the system is driven by external agents, the reconnection rate or the energy release rate is determined by the local condtions (i.e., anomalous resistivity and plasmoid ejection). The highly intermittent and recurrent behavior is the basic feature of this model.

We shall discuss the role of plasmoid ejections in more detail. Often people have argued that filament (a kind of plasmoid) ejection is the source of flare energy. In our model, however, the energy source is not in a filament (plasmoid) itself, but is stored in a volume surrounding the filament (plasmoid). The role of the plasmoid ejection is simply to trigger fast reconnection (i.e., fast inflow into an X-point). Ohyama and Shibata (1997a,b) observationally confirmed that the kinetic energy of an ejected plasmoid is much smaller than the total flare energy in the case of some typical flares.

Let us consider a current sheet with a plasmoid (magnetic island) inside it (Fig. 4). Apparently, *the plasmoid inhibits the reconnection.* Only after the plasmoid has been ejected out of the current sheet, will reconnection become possible. Once the reconnection has begun, the released energy helps to accelerate the plasmoid. If the plasmoid's speed increases, then the speed of inflow into the neutral point will also increase. Since the inflow speed determines the reconnection rate, this means that the ultimate origin

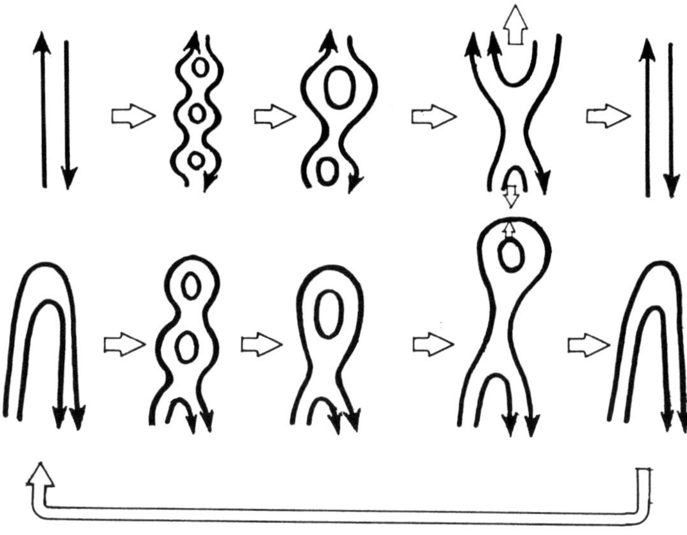

Figure 3. Recurrent behavior of plasmoid-induced-reconnection process, which naturally explains *homologous flares.*

of the fast reconnection is the fast ejection of the plasmoid.

This process can be understood as the collapse of a current sheet after the ejection of a plasmoid (Fig. 4). An important point is that the collapse itself is an ideal MHD process and the energy released during the collapse is non-negligible; indeed, it amounts to one-fourth of the magnetic energy stored in the original current sheet in the case of a simple model illustrated in Figure 4. In this case, once a large plasmoid has been ejected, at least one-fourth of the magnetic energy contained around the plasmoid will be released even if reconnection is inhibited after the ejection. Actually, the strong inflow due to the collapse induces fast impulsive reconnection, followed by a slow reconnection phase corresponding to the growth phase of a plasmoid, and vice versa, leading to intermittent fast reconnection (and plasmoid ejection). Each cycle (ejection of a plasmoid and triggered fast reconnection) may correspond to a different impulsive HXR/microwave peak in a flare with multiple peaks (Aschwanden et al. 1996).

Hence, as larger plasmoids are formed in the system, the stored magnetic energy becomes larger. Consequently, if a larger plasmoid is ejected, a larger energy release will occur.

2.2. APPLICATION TO IMPULSIVE FLARES

Magnetic reconnection theory predicts the existence of two oppositely directed high speed jets from the reconnection point with velocity at the

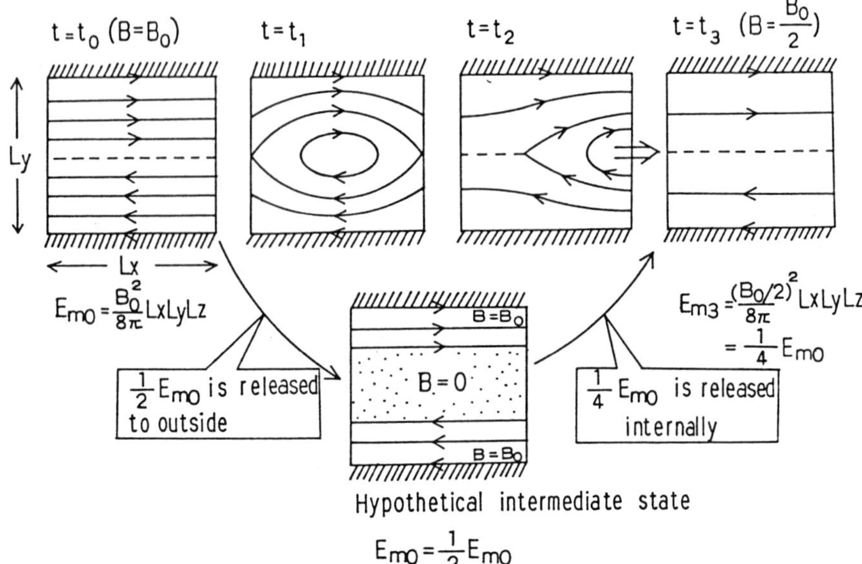

Figure 4. Detailed processes during the collapse of a current sheet accompanied by the ejection of a plasmoid. It is assumed here that half of the initial magnetic flux is reconnected to form a plasmoid ($t = t_1$). After the ejection of the plasmoid ($t = t_2$), the current sheet collapse occurs; i.e., the remaining magnetic flux expands to compress the current sheet. At the final stage ($t = t_3$), the magnetic field strength decreases to half of the initial value, and the magnetic energy in the box (E_{m3}) is one-forth of the initial magnetic energy ($E_{m3} = E_{m0}/4$, where E_{m0} is the initial magnetic energy contained in the box). This process and the effect of the collapse of the current sheet is more easily understood if we consider a hypothetical intermediate state between the ejection of the plasmoid and the collapse; i.e., a state in which half of the initial magnetic flux is removed out but the remaining flux has not yet expanded. After this hypothetical state, the remaining flux expands to fill the box. This process (i.e., the current sheet collapse) releases *one-forth of the initial magnetic energy.* It is interesting to note that this is an ideal MHD process. Hence even if the reconnection is inhibited after the ejection of the plasmoid, we can expect that one-forth of the initial magnetic energy is released by the collapse of the current sheet.

Alfven speed,

$$V_{jet} \sim V_A \simeq 2000\Big(\frac{B}{100\text{G}}\Big)\Big(\frac{n}{10^{10}\text{cm}^{-3}}\Big)^{-1/2}\text{km/s}. \tag{3}$$

The downward jet collides with the top of the SXR loop, producing an MHD fast shock. We consider that a very hot region heated by the fast shock corresponds to a loop-top impulsive HXR source. The temperature just behind the fast shock becomes

$$T_{fast} \sim m{V_{jet}}^2/(6k) \simeq 1 \times 10^8 \text{ K } \Big(\frac{B}{100\text{G}}\Big)^2\Big(\frac{n}{10^{10}\text{cm}^{-3}}\Big)^{-1}, \tag{4}$$

where m is the proton mass and k is the Boltzmann constant. This explains the observationally estimated temperature of the loop-top HXR source (Masuda 1994).

The extent of the fast shock, L_{fast} (see Fig. 2), is of order of

$$L_{fast} \sim \frac{V_{inflow}}{V_A} L_{plasmoid} \simeq M_A L_{plasmoid} \simeq 1000\mathrm{km}\left(\frac{M_A}{0.05}\right)\left(\frac{L_{plasmoid}}{2\times 10^9\mathrm{cm}}\right). \tag{5}$$

The volume emission measure of the very hot region behind the fast shock becomes

$$EM_{fast} \sim n^2 {L_{fast}}^3 \simeq 10^{44}\mathrm{cm}^{-3}\left(\frac{n}{10^{10}\mathrm{cm}^{-3}}\right)^2\left(\frac{M_A}{0.05}\right)^3\left(\frac{L_{plasmoid}}{2\times 10^9\mathrm{cm}}\right)^3. \tag{6}$$

Here, we assumed that the extent of the fast shock in the direction perpendicular to the plane of Figure 2 is comparable to that parallel to the plane. This value is roughly comparable to the actually observed emission measure of loop-top impulsive HXR sources, $10^{44} - 10^{46}$ cm^{-3} (Masuda 1994).

The time scale of the impulsive phase is determined from the duration of the strong inflow, which may be comparable to the travel time of the plasmoid across its size, i.e.,

$$t_{imp} \sim \frac{L_{plasmoid}}{V_{plasmoid}} \sim 20 t_A \left(\frac{M_A}{0.05}\right)^{-1}$$

$$\sim 200 \text{ sec} \left(\frac{M_A}{0.05}\right)^{-1}\left(\frac{B}{100\mathrm{G}}\right)^{-1}\left(\frac{n_e}{10^{10}\mathrm{cm}^{-3}}\right)^{1/2}\left(\frac{L_{plasmoid}}{2\times 10^9\mathrm{cm}}\right), \tag{7}$$

where $t_A = L_{plasmoid}/V_A$. This is roughly consistent with the observed duration (1 – 3 min) of one impulsive peak. (The total duration of the impulsive phase ranges from 1 min to 10 min. Longer impulsive phases usually include multiple impulsive peaks.)

The magnetic energy stored around the current sheet and the plasmoid is suddenly released through reconnection into kinetic, thermal, and non-thermal energies after the plasmoid is ejected. The magnetic energy release rate at the current sheet (with the length of $L_{cs} \sim L_{plamoid} \simeq 2 \times 10^4$ km) is estimated to be

$$\frac{dE}{dt} = 2 \times L_{plasmoid}^2 B^2 V_{inflow}/4\pi$$

$$\sim 4 \times 10^{28} \text{ erg/s} \left(\frac{V_{inflow}}{100 \text{ km/s}}\right)\left(\frac{B}{100 \text{ G}}\right)^2\left(\frac{L_{plasmoid}}{2\times 10^9 \text{ cm}}\right)^2. \tag{8}$$

This is comparable with the energy release rate during the impulisve phase, $4 - 100 \times 10^{27}$ erg/s, estimated from the footpoint HXR source (Masuda 1994), assuming the thick target model and the lower cutoff energy of non-thermal electrons as 20 keV.

Table I Application of Plasmoid-Induced-Reconnection Model to Various Flares

flare	B (G)	n (cm^{-3})	$L_{plasmoid}$ (cm)	T_{fast} (K)	EM_{fast} (cm^3)	t_{imp} (sec)	$\dot{E}$ (erg/s)
impulsive	100	10^{10}	2×10^9	10^8	10^{44}	200	10^{28}
LDE	30	$10^{9.5}$	4×10^9	$10^{7.5}$	10^{44}	700	2×10^{27}
giant arcade	10	10^9	6×10^9	10^7	$10^{43.5}$	1700	3×10^{26}
microflares	100	10^{10}	4×10^8	10^8	10^{42}	40	4×10^{26}

3. Application to LDE Flares, Giant Arcades, and Microflares

Observations show many common features in impulsive flares, LDE flares, giant arcades associated with filament eruptions or CMEs, and even microflares. For example, Shimojo, Yaji, and Shibata (1995) found an X-ray jet ejected from an M-class impulsive flare.

We can apply the above equations to LDE flares, giant arcades, and microflares. The results are summarized in Table I, where the magnetic field strength B is assumed to be comparable to the average photospheric field strength in the flare-occurring regions. These numbers are applicable to the impulsive phase or the rise phase. From this table, we predict that the loop-top HXR source in LDE flares has lower temperature ($\sim$ 30 MK) than that of impulsive flares, and its time scale is longer. We also predict that microflares have loop-top HXR source of temperatures similar to those of impulsive flares, although the emission measure of the source is much less and is not easily detected. We also expect it to be difficult to observe plasmoid ejections in microflares, because the plasmoids would collide and reconnect with ambient fields and disappear in a short time scale ($<$ 100 sec).

4. Discussion

Ground based observations suggest that emerging flux plays an important role in driving flares (e.g., Kurokawa 1987, Zhang et al 1997, Nishio et al. 1997, Hanaoka 1997). For example, a famous X-class impulsive flare, the 15 Nov 1992 flare (e.g., Sakao et al. 1992), was driven by a moving satellite spot (or emerging flux). Even the 21 Feb 1992 LDE flare (e.g.,

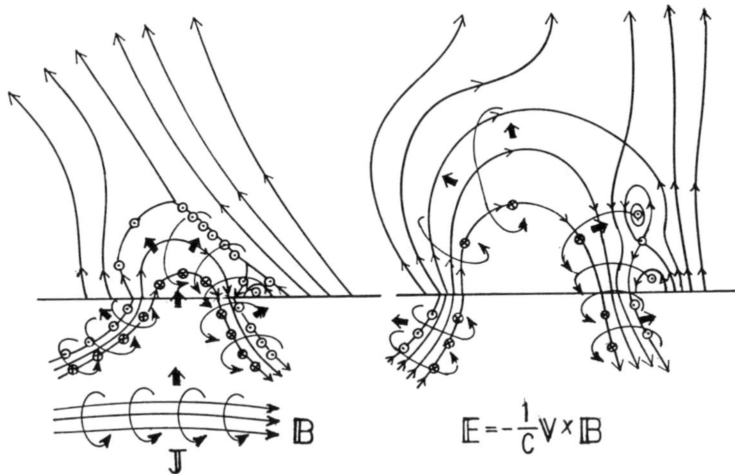

Figure 5. Current circuit and origin of electric field in a unified model (*plasmoid-induced-reconnection model*). This model unifies both the CSHKP model and the emerging flux model. Since the origin of electric field is emerging flux, this scheme may be called also the generalized emerging flux model.

Tsuneta 1996), and a homologous to it on 24 Feb 1992 (Morita et al. 1997) seem to be driven by growing flux (or emerging flux) (Zhang et al. 1997). Nevertheless, these flares clearly show filament or plasmoid ejections as well as the morphology predicted by the CSHKP model. Thus there is a need to unify the CSHKP and the emerging flux models. Such a unification is indeed possible in our *plasmoid-induced-reconnection model* as illustrated in Figure 5, in which it is shown that the driving electric field may originate from emerging flux.

The author would like to thank T. Yokoyama, K. Kusano, T. Hayashi, M. Ohyama, and S. Tsuneta for stimulating discussions.

References

Acshwanden, M. J., et al. 1996, ApJ, 464, 985.
Hanaoka, Y. 1997, in these proceedings.
Heyvaerts, J., Priest, E. R. and Rust, D. M. 1977, ApJ, 216, 123.
Karpen, J. T., Antiochos, S. K., and DeVore, C. R. 1995, ApJ, 450, 422.
Kitabata, H., Hayashi, T., Sato, T., et al. 1996, J. Phys. Soc. Japan, 65, 3208.
Kosugi, T., and Shibata, K. 1997, in Proc. Chapman Conference, Magnetic Storms, Geophysical Monograph 98, AGU, p. 21.

Kurokawa, H. 1987, Space Sci. Rev. 51, 49.
Kusano, K., Suzuki, Y., and Nishikawa, K. 1995, ApJ, 441, 942.
Lee, L. C., and Fu, Z. F. 1986, J. Geophys. Res., 91, 6807.
Magara, T., et al. 1996, ApJ, 466, 1054.
Magara, T., Shibata, K., Yokoyama, T. 1997, ApJ, in press.
Masuda, S. 1994, Ph. D. Thesis, U. Tokyo.
Masuda, S., et al. 1994, Nature, 371, 495.
Masuda, S., et al. 1995, PASJ, 47, 677.
Morita, S., et al. 1997, in these proceedings.
Nishio, M., et al. 1997, ApJ, in press.
Ohyama, M., and Shibata, K. 1997a, PASJ, 49, 249.
Ohyama, M., and Shibata, K. 1997b, ApJ, submitted.
Priest, E. R. 1982, Solar Magnetohydrodynamics, Reidel.
Sakao, T., et al. 1992, PASJ, 44, L83.
Shibata, K., Nozawa, S., and Matsumoto, R. 1992a, PASJ, 44 , 265.
Shibata, K., et al. 1992b, PASJ, 44, L173.
Shibata, K., et al. 1994, ApJ, 431, L51.
Shibata, K., et al. 1995, ApJ, 451, L83.
Shibata, K. 1996, Adv. Space Res., 17, (4/5)9.
Shibata, K. 1997, Proc. of Hitachi Solar Flare Workshop, T. Sakurai et al. (eds.) in press.
Shimizu, T. 1995, PASJ, 47, 251.
Shimojo, M., et al. 1996, PASJ, 48, 123.
Shimojo, M., Yaji, K., Shibata, K. 1995, ASJ fall meeting abstract (in Japanese).
Tsuneta, S., et al. 1992a, PASJ, 44, L63.
Tsuneta, S., et al. 1992b, PASJ, 44, L211.
Tsuneta, S. 1996, ApJ, 456, 840.
Tsuneta, S. 1997, ApJ, 483, 507.
Ugai, M. 1989, Phys. Fluids B, 1, 942.
Yokoyama, T., and Shibata, K. 1994, ApJ, 436, L197.
Yokoyama, T., and Shibata, K. 1995, Nature, 375, 42.
Yokoyama, T., and Shibata, K. 1996, PASJ, 48, 353.
Yokoyama, T., and Shibata, K. 1997, ApJ, 474, L61.
Yokoyama, T. 1997, in these proceedings.
Zhang, H. Q., et al. 1997, in these proceedings.
Watanabe, T. 1994, in Proc. Kofu Symposium, p. 99.

Activity Caused by Interacting Loops and Evolution of Magnetic Field

YOICHIRO HANAOKA
Nobeyama Radio Observatory, NAOJ
Nobeyama, Minamimaki, Minamisaku, Nagano 384-13, Japan

Abstract.
Interaction between newly emerging flux regions and overlying loops is a well-known cause of various types of activity – flares, jets, and surges. We studied magnetic configuration which consists of an emerging loop and an overlying loop, and the relation between the evolution of emerging flux regions and various active phenomena. The main conclusions are as follows.
(1) We reveal the topology of an emerging loop and an overlying loop using microwave and soft X-ray observations of the flares, along with their magnetograms. Two of the footpoints of the loops, one from the emerging loop and the other from the overlying loop, are included in a single magnetic polarity patch. Therefore, the footpoints of the two loops do not form a quadrapole structure, but they are distributed in three magnetic patches. The two loops form a 'three-legged' structure, and the magnetic field has a 'bipolar + remote unipolar' structure.
(2) Emergence of parasitic polarities near the main preceding-polarity region or the following-polarity region in active regions are observed. The life-time of the parasitic polarities are a few days. Not only flares, but microflares, jets, and surges occur at the parasitic polarities, and they repeat homologously during the evolution of the parasitic polarities. Consequently, emerging flux activity produces a double-loop configuration with an overlying loop, and this magnetic configuration is a source of various active phenomena.

1. Introduction

Emerging flux plays a key role in causing various types of activity in the solar atmosphere. Recent observations with *Yohkoh* and the Nobeyama Radioheliograph give new insights into the emerging flux-related activity (Hanaoka, 1996 and 1997; hereafter Papers I and II). We review here the results obtained from our analyses on various aspects related to emerging flux activity.

Flux emergence inevitably causes collisions between the emerging loop and overlying loops. Interactions between newly emerging flux and overlying magnetic fields is one of the well-known typical conditions to cause flares (e.g. Zirin, 1983; Kurokawa, 1989). The existence of two loops is essential to release magnetic energy. In this paper we call the loop topology consisting of an emerging loop and an overlying loop, a 'double-loop configuration'.

We survey flares which are considered to occur under the double-loop configuration. Some of flares which show a compact brightening in X-rays/microwaves at the main flare site also show a remote brightening in microwave images taken with the Nobeyama Radioheliograph. Such flares are presumed to occur under the double-loop configurations; the compact brightening at the main flare site suggests

T. Watanabe et al. (eds.), Observational Plasma Astrophysics: Five Years of Yohkoh and Beyond, 197–206.

TABLE I
List of selected flares and their active regions

Date	Time	GOES class	location	NOAA	slope[a]	flux emergence[b]
1992 Dec 15	0113	C6.9	S20W37	7360	yes	yes
1992 Dec 16	0325	C4.6	S20W55	7360		yes
1992 Dec 16	0605	C4.7	S19W57	7360		yes
1992 Aug 21	0018	M1.3	N18W34	7260	yes	yes
1993 Feb 06	0527	C5.6	S09E67	7420	yes	
1993 Apr 10	2335	C9.1	S08W51	7469	yes	yes
1993 Apr 11	0030	C6.2	S13W54	7469	yes	yes
1993 Jun 07	0549	C4.1	S09W30	7518	yes	yes
1993 Sep 30	2243	C1.7	S09E43	7590	yes	yes
1993 Apr 11	0616	C8.9	S09W61/N02W58	7469/7472		

[a] slope of the degree of polarization at the main source
[b] flux emergence related to the small loop

the existence of a small flaring loop, and the remote brightening suggests that a large loop connects the main and the remote sources. First, we selected flares which have two bright points in microwave images observed with the Nobeyama Radioheliograph. Since the spatial resolution of the Radioheliograph is about 13″, two radio sources are well separated in the selected flares. Next, among these, we select flares which show a remarkable compact brightening in X-rays and/or Hα at one of the microwave sources. The flares which are confirmed to have the double-loop configuration are listed in Table I.

As a result of analyses of the flares and the active regions listed in Table I, we construct a new view of the magnetic structure of emerging flux regions, as shown in Figure 1. Footpoints of the two loops distributed in three magnetic patches, or the two loops form a 'three-legged' structure. Such a structure is formed by evolution of a parasitic polarity region. Flares, microflares, and various other active phenomena occur recurrently during the evolution of the parasitic polarity due to the interactions of two loops.

The configuration of loops revealed from the analyses of the flares are described in Section 2. In Section 3, the various types of activity that occurred in the active regions listed in Table I are described. The relation between evolution of magnetic field and occurrence of flares and microflares are described in Section 4.

2. Configuration of Flare Loops

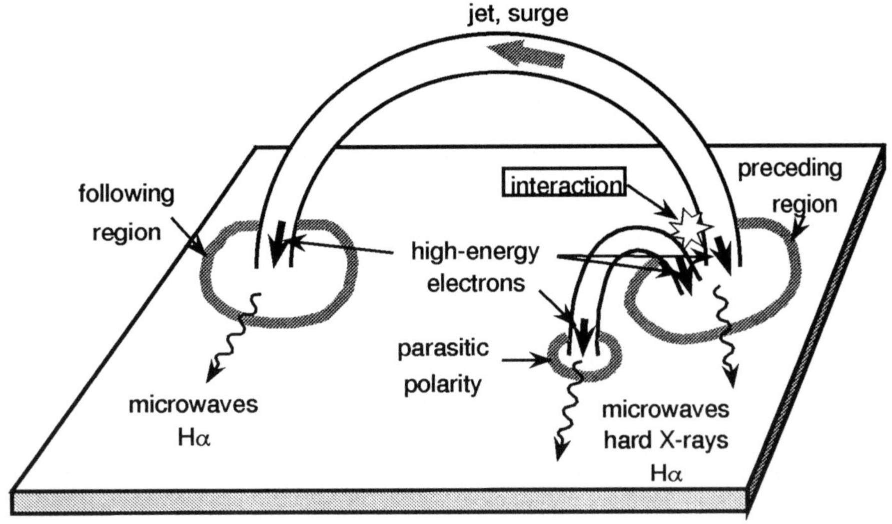

Fig. 1. Schematic drawing of a 'three-legged' structure of two loops. Interaction between a pre-existent large loop and a newly emerging loop causes not only flares but also jets and surges.

Images of a flare on 1993 June 7 are shown in Figure 2, a typical example of a 'three-legged' structure. The flares listed in Table I show similar morphology, and their detailed descriptions appear in Papers I and II. There are two bright points in microwaves (Figure 2(a)), which are connected by a large soft X-ray loop (Figure 2(c)). Intense and compact brightenings in soft and hard X-rays occur at the western footpoint of the large loop, and therefore probably there is a flaring small loop at this location (which we call the 'main source'). The large loop becomes bright later in soft X-rays. In Hα (Figure 2(e)), there is a strong brightening at the main flare site and a faint brightening at the remote site. This flare shows a remote brightening both in microwaves and Hα.

The main source is observed as a single microwave source, but we presume that the main source actually includes two opposite-polarity microwave sources based on the microwave images and the magnetogram. The degree of polarization in microwaves of the main source has a slope (Figure 2(b)). Such a slope cannot be produced by a single source. Two unresolved microwave sources, which have opposite polarities, can produce a slope in the degree of polarization, even if they merge into a single source in the intensity map. If the two opposite polarity sources have equal intensity, the composite polarization varies from right- to left-handed gradually. However, if their intensities and/or degree of polarizations are not equal, the composite polarization shows an asymmetry, as seen in Figure 2(b). The opposite polarizations correspond to the opposite magnetic polarities. Therefore the

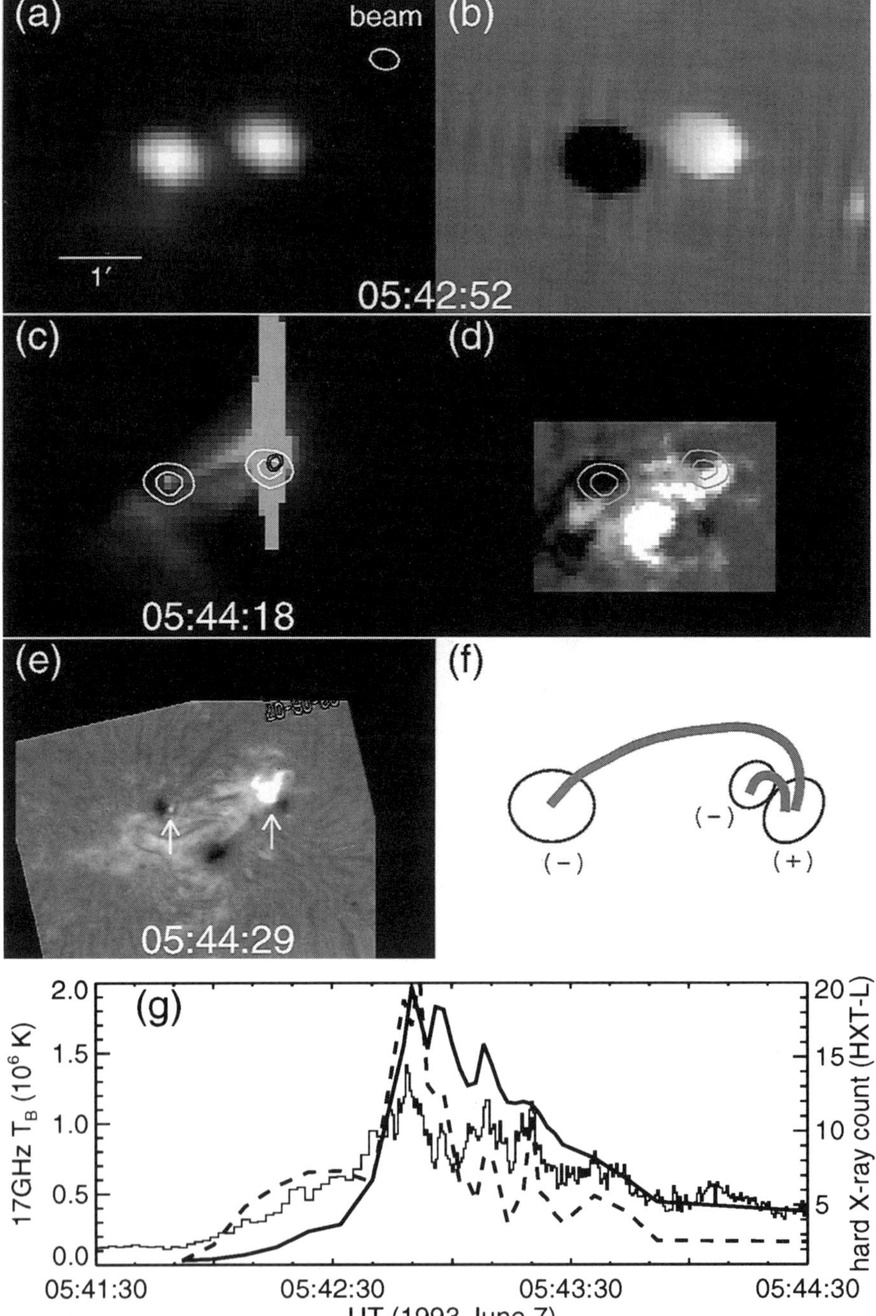
(a)
beam
1′
(b)
05:42:52
(c)
05:44:18
(d)
(e)
05:44:29
(f)
(−)
(−)
(+)
(g)
17GHz T_B (10^6 K)
hard X-ray count (HXT-L)
2.0
1.5
1.0
0.5
0.0
20
15
10
5
05:41:30
05:42:30
05:43:30
05:44:30
UT (1993 June 7)

Fig. 2. Images of the C4.1 flare of 1993 June 7 in NOAA 7518. Each image has a field of view of $5.2' \times 3.9'$, and solar north is to the top. (a)(b) Images of the intensity and the degree of polarization at 17 GHz taken with the Nobeyama Radioheliograph. In the map of the degree of polarization, the right- and left-handed polarizations are displayed in white and black, respectively. The beam shape at half of the maximum is shown. (c) Soft X-ray images taken with the SXT on board *Yohkoh* around the peak of the flare, overlaid by white contours showing the microwave image and black contours showing the hard X-ray image of the M1-band (23–33 keV) at 05:43:06 taken with the HXT on board *Yohkoh*. (d) Magnetogram taken with the Stokes Polarimeter of Mees Observatory at 0h UT on June 7. Positive and negative polarities are displayed in white and black, respectively. The contours of the microwave image are overlaid. (e) Hα picture of the flare taken with the Domeless Solar Telescope of Hida Observatory. Hα bright points at the footpoints of the loops are denoted by arrows. (f) Schematic drawing of the relation between the footpoints of the loops and the magnetic polarities. (g) Time variations of the brightness temperature of the western radio source (thick line), that of the eastern source (thick dashed line), and total hard X-ray counts of the L-band (14–23 keV) (thin line).

slope of the degree of polarization indicates microwave brightening from the two footpoints of a small loop.

The magnetogram in Figure 2(e) shows that a parasitic region with negative polarity touches a positive polarity region at the main source. The direction of the positive-negative polarity complex is consistent with the gradient of the degree of polarization. Therefore the microwave observations are consistent with the magnetographic information. Since there is a bipolar structure at the main source, and the remote source corresponds to a unipolar magnetic field of negative polarity, the magnetic field of the flare has a 'bipolar + remote unipolar' structure. Therefore, although there are four footpoints associated with the two loops, the magnetic field is not a quadrapole structure. The footpoints are distributed in three magnetic patches, as shown in Figure 2(f). The small loop and the large one form a 'three-legged' structure. As a result of analyses, we concluded that all the flares in Table I involve 'bipolar + remote unipolar' magnetic field structures.

Two-dimensional reconnection models by Heyvaerts, Priest, and Rust (1977), Forbes and Priest (1984), and Shibata, Nozawa, and Matsumoto (1992) assumed reconnection between anti-parallel magnetic fields. A quadrapole structure, which is similar to the above models, was proposed by Machado *et al.* (1983, 1988) to account for the flares in active regions 2372 and 2779, and it was confirmed by Mandrini *et al.* (1991) and Bagalá *et al.* (1995) with separatrix analyses. However, in the 'three-legged' structure, the directions of the magnetic fields of two loops are not anti-parallel. Démoulin, Hénoux, and Mandrini (1994) have also pointed out that some flares occur without the neutral points which are created by anti-parallel magnetic loops. Nevertheless, our result is not necessarily in conflict with the reconnection model. In three-dimensions, two magnetic loops, which are not anti-parallel, can reconnect. Even if the loops are nearly parallel, twists of loops can be a source of energy release (see Priest, 1992). The current-loop interaction (reviewed by Sakai and de Jager, 1996; references therein) can occur between two nearly parallel loops. Recently, a generalized reconnection model in three-dimensions, which can be applied to not-anti-parallel loops, has been developed by Démoulin *et al.* (1996). Based on their method, Mandrini *et al.* (1996) proposed a magnetic configuration for an XBP flare, which was described by van Driel-Gesztelyi *et al.* (1996), in their Figure 2. The proposed configuration is similar to our observed 'three-legged' structure, and the change of magnetic configuration described by Mandrini *et al.* (1996) is applicable to our cases.

The time variations of our observed two radio sources coincide very well, and they show good agreement with the hard X-ray counts of the L-band (14–23 keV) of the HXT(Figure 2(g)). Several impulsive brightenings of the two microwave sources coincide within a few seconds, and they also show good agreement with the hard X-ray counts. Since the distance between the two microwave sources is about 10^5 km, the energy must be transferred by high-energy electrons. Therefore both the main and the remote sources emit microwaves produced by high-energy electrons. The existence of two loops, as shown in Figure 2(f), suggests that the flare is caused by interactions between the loops. The high-energy electrons are presumed to be produced at the interaction region of the two loops and precipitate into the footpoints of the loops, but we cannot specifically define the acceleration site from our analyses. On the other hand, Aschwanden *et al.* (1996a) find that the acceleration site in the flare shown in Figure 2 is not far from the hard X-ray source, or just above the top of the small loop (see their Figure 3), on the basis of electron time-of-flight measurements of hard X-ray data taken with the Burst and Transient Source Experiment (BATSE) on board the Compton Gamma Ray Observatory (CGRO). This supports the conclusion that the interaction between the two loops causes the flare. Furthermore, Aschwanden *et al.* (1996a) show that some of the flares analyzed by Nishio *et al.* (1997), which are also proposed to be caused by interacting loops, have electron acceleration sites just above small loops.

The magnetic configuration of the flares in Table I, which show an impulsive nature, is much different from the cusp-type configuration of the impulsive flares analyzed by Masuda *et al.* (1995). However, the electron time-of-flight analyses by Aschwanden *et al.* (1996b, c) show that the electron acceleration sites of cusp-type flares are also just above the loop tops. Therefore, although flares caused by interacting loops and cusp-type flares have much different magnetic configurations, they are similar to each other from the viewpoint of electron acceleration. To elucidate the differences and the common properties of the flares caused by interacting loops and the cusp-type flares is very important to understanding impulsive solar flares.

3. Various Types of Activity Related to the Double-Loop Configurations

Generally, activity of parasitic polarities is well-known to be a source of various active phenomena. Shimojo *et al.* (1997) studied the relationship between jets and evolution of magnetic field, and concluded that in many cases jets are caused by the activity of parasitic regions. Roy (1973) revealed the close relationship between the activity of satellite sunspots and surges. The fact that most jets accompany microflares and flares was pointed out by Shibata *et al.* (1992). Yokoyama and Shibata (1995, 1996) showed that both jets and surges occur due to reconnection between a pre-existing loop and emerging flux, and Canfield *et al.* (1996) studied jets and surges commonly originating from the activity of satellite spots. These results lead to the conclusion that the emergence of a parasitic polarity is a common source of various types of activity.

In most of the active regions in Table I, the emergence of the parasitic polarity is observed at the main flare site (see Table I). Various types of active phenomena are also observed in those regions. In region NOAA 7360, of which the loop configuration is shown in Figure 3, many microflares occur recurrently in a small loop (Paper I). More than half of the microflares cause plasma flow into a large loop. The typical velocity of the plasma flow is about 1000 km s^{-1}. Such plasma flow with a typical velocity of 1000 km s^{-1} is very similar to the thermal wave front observed by Rust, Simnett, and Smith (1985). Furthermore, they have the same characteristics as the jets analyzed by Shibata *et al.* (1992) and Shimojo *et al.* (1996). The velocity of about 1000 km s^{-1} is the highest in the jets statistically analyzed by Shimojo *et al.* (1996). Therefore, the observed plasma flow and the thermal wave front are considered to be jets occurring in active region loops.

Recurrent occurrences of microflares at the main flare site are observed in some of the other active regions in Table I as well as region 7360. Furthermore, the 1993 April 10 C9.1 flare and the 1993 April 11 C6.2 flare in region NOAA 7469 show surges ejected from the main source into the large loop during the flares.

Consequently, the active regions in Table I show typical active phenomena, which are caused by emerging flux activity - not only flares, but also microflares, jets, and surges. The double-loop configuration is the common source of various active phenomena. Therefore, the 'three-legged' structure shown in Figure 1, which is a typical loop configuration for flares, is also one of the basic magnetic field configurations which cause various other types of activity.

4. Double-Loop Configuration and Evolution of the Magnetic Field

As shown in Figure 1, in most of the active regions analyzed the large loop connects the major preceding and following polarities of the active regions, and a parasitic region near the preceding or following polarity produces the double-loop configuration. We have checked the evolution of the parasitic polarities in the

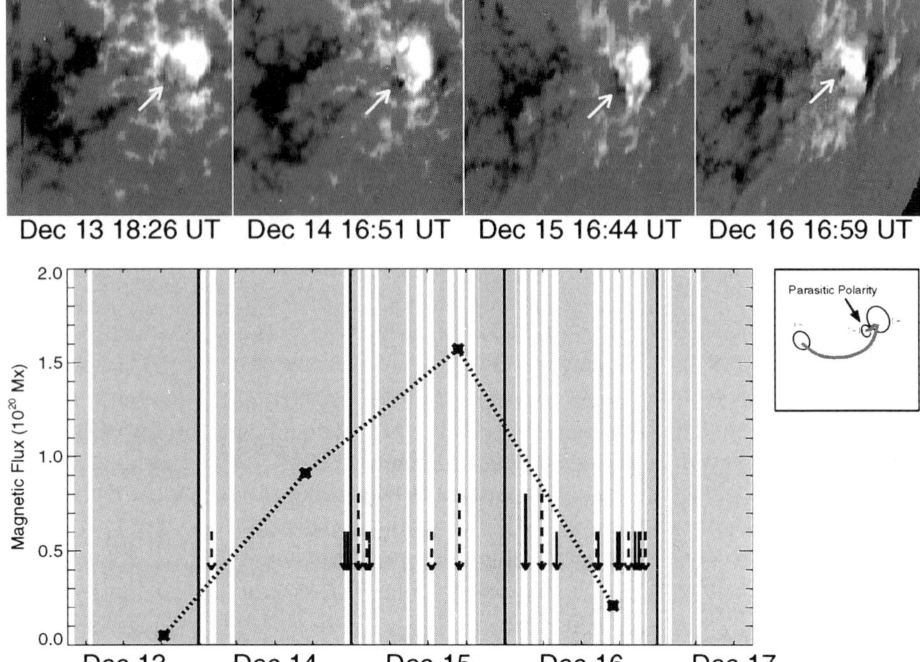

Fig. 3. Evolution of magnetic field of NOAA 7360. NSO/Kitt Peak magnetograms of four days are shown in the upper panels. The field of view is $3.8' \times 3.8'$, and solar north is to the top. A parasitic polarity region is pointed by arrows. The lower-right cartoon shows the loop configuration. The lower-left panel shows the variations of magnetic flux of the parasitic region, and occurrence times of flares (long arrows) and microflares (short arrows) found in SXT images. Solid line arrows denote (micro)flares accompanied by jets. The SXT does not observe this region in periods shown in shaded areas.

active regions. For example, region NOAA 7518 shown in Figure 2, shows both the development and the decay of a parasitic polarity. The relation between magnetic flux of the parasitic polarity measured on the Mees magnetograms and occurrences of flares and microflares observed with the SXT is shown in Table II.

The lifetime of the parasitic region is only about one day, and no (micro)flares occur before the appearance of the parasitic polarity or after its disappearance are observed. It is obvious that the (micro)flares are caused during the evolution of the parasitic polarity.

TABLE II
Magnetic flux and flare occurrences in NOAA 7518

	Date	Time (UT)	Magnetic flux	Flares
1993	June 6	17h	-3×10^{19} Mx	
	June 6	1844		microflare
	June 6	21h	-11×10^{19} Mx	
	June 7	00h	-13×10^{19} Mx	
	June 7	0051		microflare
	June 7	0549		flare
	June 7	17h	-0.2×10^{19} Mx	
	June 7	20h	-0.1×10^{19} Mx	

While only 3 events are observed in region 7518, many flares and microflares occur in region NOAA 7360, as described in Section 3. The relation between magnetic flux of a parasitic region and (micro)flares in region NOAA 7360 is shown in Figure 3, which is a similar diagram to those by Shimojo *et al.* (1997), showing the relation between magnetic flux and jets. A parasitic region of negative polarity appears near the preceding region, and an emerging loop collides with an overlying loop. It grows and decays in three days. Before the birth and after the disappearance of the parasitic region, there are no (micro)flares in the vicinity of the parasitic polarity. Figure 3 clearly shows the close relationship between the parasitic polarity and the flare occurrences. Both the regions 7518 and 7360 show (micro)flare occurrences during both the developing and the decay phases of the parasitic polarities, as in the regions analyzed by Shimojo *et al.* (1997). Therefore, the word 'evolving flux' is more appropriate than the word 'emerging flux' as a cause of active phenomena.

As shown above, emergence of a parasitic polarity is a typical cause of double-loop configurations. However, in the case of the 1992 August 21 flare, the double-loop configuration was produced by the development of a δ-type spot, and the double-loop configuration of the 1993 April 11 C8.9 flare involves two active regions (see Paper II). This means that the double-loop configuration can be produced under various situations, and that it is common on the solar surface. The reason why the emergence of a parasitic polarity is a typical source of a double-loop configuration is that it is a commonplace event in active regions.

Acknowledgements

The author expresses his hearty thanks to the National Solar Observatory/Kitt Peak and Mees Solar Observatory of the University of Hawaii, for providing the magnetograph data. The NSO/Kitt Peak magnetograms are produced cooperatively by NSF/NOAO, NASA/GSFC and NOAA/SEL. The magnetograms of Mees Solar Observatory were produced with the support of NASA grant NAGW-1542 and NASA contract NAS8-37334. Hida Observatory of Kyoto University provided the Hα pictures.

References

Aschwanden, M. J., Kosugi, T., Hudson, H. S., Wills, M. J., and Schwartz, R. A.: 1996a, *Astrophys. J.* **470,** 1198.

Aschwanden, M. J., Hudson, H., Kosugi, T., and Schwartz, R. A.: 1996b, *Astrophys. J.* **464,** 985.

Aschwanden, M. J., Wills, M. J., Hudson, H. S., Kosugi, T., and Schwartz, R. A.: 1996c, *Astrophys. J.* **468,** 398.

Bagalá, L. G., Mandrini, C. H., Rovira, M. G., Démoulin, P., and Hénoux, J. C.: 1995, *Solar Phys.* **161,** 103.

Canfield, R. C., Reardon, K. P., Leka, K. D., Shibata, K., Yokoyama, T., and Shimojo, M.: 1996, *Astrophys. J.* **464,** 1016.

Démoulin, P., Hénoux, J. C., and Mandrini, C. H.: 1994, *Astron. Astrophys.* **285,** 1023.

Démoulin, P., Hénoux, J. C., Priest, E. R., and Mandrini, C. H.: 1996, *Astron. Astrophys.* **308,** 643.

Forbes, T. G. and Priest, E. R.: 1984, *Solar Phys.* **94,** 315.

Hanaoka, Y.: 1996, *Solar Phys.* **165,** 275 (Paper I).

Hanaoka, Y.: 1997, *Solar Phys.* in press (Paper II).

Heyvaerts, J., Priest, E. R., and Rust, D. M.: 1977, *Astrophys. J.* **216,** 123.

Kurokawa, H.: 1989, *Space Sci. Rev.* **51,** 49.

Machado, M. E., Somov, B. V., Rovira, M. G., and de Jager, C.: 1983, *Solar Phys.* **85,** 157.

Machado, M. E., Moore, R. L., Hernandez, A. M., Rovira, M. G., Hagyard, M. J., and Smith, Jr., J. B.: 1988, *Astrophys, J.* **326,** 425.

Mandrini, C. H., Démoulin, P., Hénoux, J. C., and Machado, M. E.: 1991, *Astron. Astrophys.* **250,** 541.

Mandrini, C. H., Démoulin, P., van Driel-Gesztelyi, L., Schmieder, B., Cauzzi, G., and Hoffman, A.: 1996, *Solar Phys.* **168,** 115.

Masuda, S., Kosugi, T., Hara, H., Tsuneta, S., and Ogawara, Y.: 1995, *Nature,* **371,** 495.

Nishio, M., Yaji, K., Kosugi, T., Nakajima, H., and Sakurai, T.: 1997, *Astrophys, J.* in press.

Priest, E. R.: 1992, in Z. Švestka, B. V. Jackson, and M. E. Machado (eds.), 'Eruptive Solar Flares', Springer Verlag, Heidelberg, p.15.

Roy, J.-R.: 1973, *Solar Phys.* **28,** 95.

Rust, D. M., Simnett, G. M., and Smith, D. F.: 1985, *Astrophys. J.* **288,** 401.

Sakai, J.-I. and de Jager, C.: 1996, *Space Sci. Rev.* **77,** 1.

Shibata, K., Ishido, Y., Acton, L. W., Strong, K. T., Hirayama, T., Uchida, Y., McAllister, A. H., Matsumoto, R., Tsuneta, S., Shimizu, T., Hara, H., Sakurai, T., Ichimoto, K., Nishino, Y., and Ogawara, Y.: 1992, *Publ. Astron. Soc. Japan,* **44,** L173.

Shibata, K., Nozawa, S., and Matsumoto, R.: 1992, *Publ. Astron. Soc. Japan,* **44,** 265.

Shimojo, M., Hashimoto, S., Shibata, K., Hirayama, T., Hudson, H. S., and Acton, L. W.: 1996, *Publ. Astron. Soc. Japan,* **48,** 123.

Shimojo, M., Shibata, K., and Harvey, K. L.: 1997, submitted to *Solar Phys.*

van Driel-Gesztelyi, L., Schmieder, B., Cauzzi, G., Mein, N., Hofmann, A., Nitta, N., Kurokawa, H., Mein, P., and Staiger, J.: 1996, *Solar Phys.* **163,** 145.

Yokoyama, T. and Shibata, K.: 1995, *Nature,* **375,** 42.

Yokoyama, T. and Shibata, K.: 1996, *Publ. Astron. Soc. Japan,* **48,** 353.

Zirin, H.: 1983, *Astrophys. J.* **274,** 900.

LOOP-LOOP INTERACTION IN IMPULSIVE SOLAR FLARES INFERRED FROM MICROWAVE AND X-RAY IMAGES

M. NISHIO, K. YAJI, T. KOSUGI AND H. NAKAJIMA
Nobeyama Radio Observatory
Minamisaku, Nagano 384-13, Japan

AND

T. SAKURAI
National Astronomical Observatory
Osawa 2-21-1, Mitaka 181, Japan

We have analyzed fourteen solar impulsive flares simultaneously observed with microwave and X-ray imaging instruments (the *Nobeyama Radioheliograph* and the *Yohkoh* X-ray telescopes). All of the events analyzed are short-lived, consisting of a single or multiple spikes with each spike duration of about 10 s and showing quite hard, hard X-ray spectra with relatively small photon counts in the HXT L-band. Hence, we believe that these events reveal most clearly the behavior of energetic electrons without severely contamination by thermal emission from super-hot ($\sim$ 30MK) flare plasma.

Figure 1 shows coaligned microwave, hard X-ray, and soft X-ray images of the 1994 Jan. 6 event. Figure 1(a) is a hard X-ray image (in contours) overlayed on the microwave image (in gray scale) and Figure 1(b) a soft X-ray image. The field of view is $2'.6 \times 2'.6$. Figure 1(c) is a cartoon showing the relative locations of microwave, hard X-ray, and soft X-ray sources. Light curves in microwaves at 17 GHz and in the HXT M1-band are shown in Figure 1(d).

The relative locations of microwave, hard X-ray, and soft X-ray sources, in at least ten cases out of the fourteen flares analyzed, clearly reveal that two loops, at least, are involved in a single event. The sizes of the two loops are different from each other; typically one is $\leq 20''$ and the other $30''$–$80''$. Microwave emission is detected from both loops, while hard X-ray emission is preferentially radiated from the shorter of the two loops. The shorter loop is brighter than the longer loop in soft X-rays. The intensity variations of

T. Watanabe et al. (eds.), Observational Plasma Astrophysics: Five Years of Yohkoh and Beyond, 207–208.

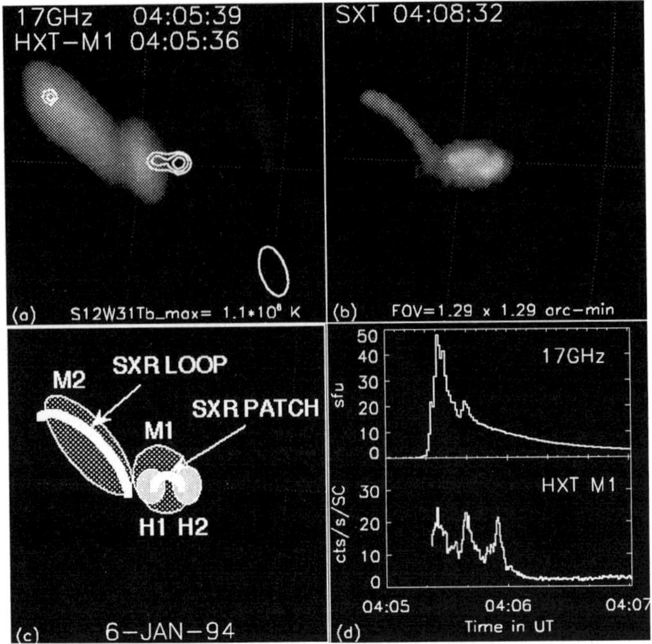

Figure 1. Relative locations of microwave, hard X-ray, and soft X-ray sources.

microwaves from the two loops resemble each other. From comparisons of the microwave/X-ray images with photospheric magnetograms, we find that a remote microwave/HXR source seen at a footpoint of the longer loop is located above a region with uniform polarity of relatively strong magnetic field, while a site involving a close HXR double source seen at the footpoints of the shorter loop lies above an asymmetric bipolar magnetic field region, i.e. a region which consists of a sunspot with strong magnetic field and a compact sunspot or penumbra with weak opposite magnetic field.

These observations suggest that in the majority of impulsive flare events two loops interact with each other, releasing magnetic energy and producing energetic electrons. The observations are therefore consistent with the emerging flux model of flares (e.g., Heyvaerts et al. 1977).

Details of this work are fully given in Nishio et al. (1997).

References

Heyvaerts, J., Priest, E. R. and Rust, D. M. (1977): *Astrophys. J.*, **216**, pp. 123–137

Nishio, M., Yaji, K., Kosugi, T., Nakajima, H. and Sakurai, T. (1997): *Astrophys. J.*, **488** (in press)

HORSESHOE-SHAPED Hα FLARES ON 13 OCT. 1995 AND AN EMERGING FLUX MODEL OF FLARES

Y. SUEMATSU, Y. NISHINO, Y. YAN, AND J. SATO
National Astronomical Observatory
Mitaka, Tokyo 181, Japan

Abstract. We present a study of two horseshoe-shaped impulsive Hα flares that occurred in active region NOAA 7912. Soft X-ray images show that the main body of the flare occupied a small part of the Hα horseshoe structure, implying a confined energy release location. We present some observational evidence, such as sunspot proper motion and hard X-ray structures, that the flares are due to an emerging flux model process.

1. Introduction

It is known that the energy for solar flares can be stored and released in coronal magnetic fields, and then flow down to the chromosphere along magnetic field lines in the form of high-energy particle beams and thermal conduction. As a result, an enhancement of chromospheric line emission is observed in the flare. Therefore, coordinated observations of X-rays originating in the corona, Hα images, and photospheric magnetic fields are very useful for studying the coronal magnetic topology responsible for flares, and hence their energy storage and release mechanisms. We obtained a good set of such observational data for homologous flares that occurred on 13 Oct. 1995. Here we describe and discuss the evolution of these events to gain further insight into one possible flare model: an emerging flux model of flares.

2. Observational Data and Flare Evolution

Two unusual horseshoe-shaped Hα flares occurred on 13 Oct. 1995 in a flare-productive active region, NOAA 7912. These two flares are regarded to be homologous and to have a similar origin; the initial flare was small

T. Watanabe et al. (eds.), Observational Plasma Astrophysics: Five Years of Yohkoh and Beyond, 209–214.

(GOES X-ray class C7.7) and the following one, which occurred four hours later, was larger (M4.8).

For these flares, the Solar Flare Telescope (Sakurai et al. 1995) provided Hα filtergrams in the line center and at ± 0.6 Å from line center, G-band (4305 Å) filtergrams, and vector magnetograms in the Fe I 6302 Å line. With these data, we studied the evolution of the chromospheric flares, velocity fields in the chromosphere, proper motions of sunspots, and magnetic field distributions in the photosphere.

Soft and hard X-ray images for the M4.8 flare were provided by the YOHKOH satellite (Kosugi et al. 1991; Tsuneta et al. 1991); obsevation of the C7.7 flare occurred during YOHKOH night. Soft and hard X-ray images were carefully coaligned with the Hα images.

2.1. EVOLUTIONAL CHARACTERISTICS OF THE C7.7 FLARE

We show in Figure 1 Hα and G-band images for this flare. It can be seen that flare kernels form a horseshoe shape, and two of the kernels are located at pores indicated by the letters ‘p’ and ‘f’ in the top right panel of Fig. 1. Since the magnetograms indicate that the ‘p’ pore has negative polarity and the ‘f’ pore has positive polarity, and because the separation of the pores and the magnetic flux of the pores increased rapidly in the day, the pores represent an emerging flux region (EFR). We note that the ‘p’ pore was surrounded by a positive polarity region, which is a suitable situation for a magnetic reconnection.

In the initial phase of the flare, we see an ejection to the north of the flare site (top left panel in Fig. 1). Later in the flare, we see a pair of bright points at the leading pore (p) (indicated by arrows in the bottom-left panel of Fig. 1). As discussed below, the ejection and formation of the bright point pair seem to be caused by magnetic reconnection near the ‘p’ pore between the emerging flux and preexisting coronal fields.

2.2. EVOLUTION OF THE M4.8 FLARE AND THE EMERGING FLUX MODEL OF FLARES

For this flare, YOHKOH data are available, and hence detailed comparisons between soft X-ray, hard X-ray, and Hα images are possible. Figure 2 gives a time series in Hα and YOHKOH soft X-ray images and Figure 3 gives a light curve and images in hard X-rays.

In the preflare (top in Fig. 2), the corona above the ‘p’ pore was already bright. In the impulsive phase, this bright region flared up in soft X-rays as well as in Hα images. At the same time, an explosive ejection heading east from the bright region was seen in the Hα and soft X-ray images (middle panels of Fig. 2). The ejection lasted about one min and coincided with the

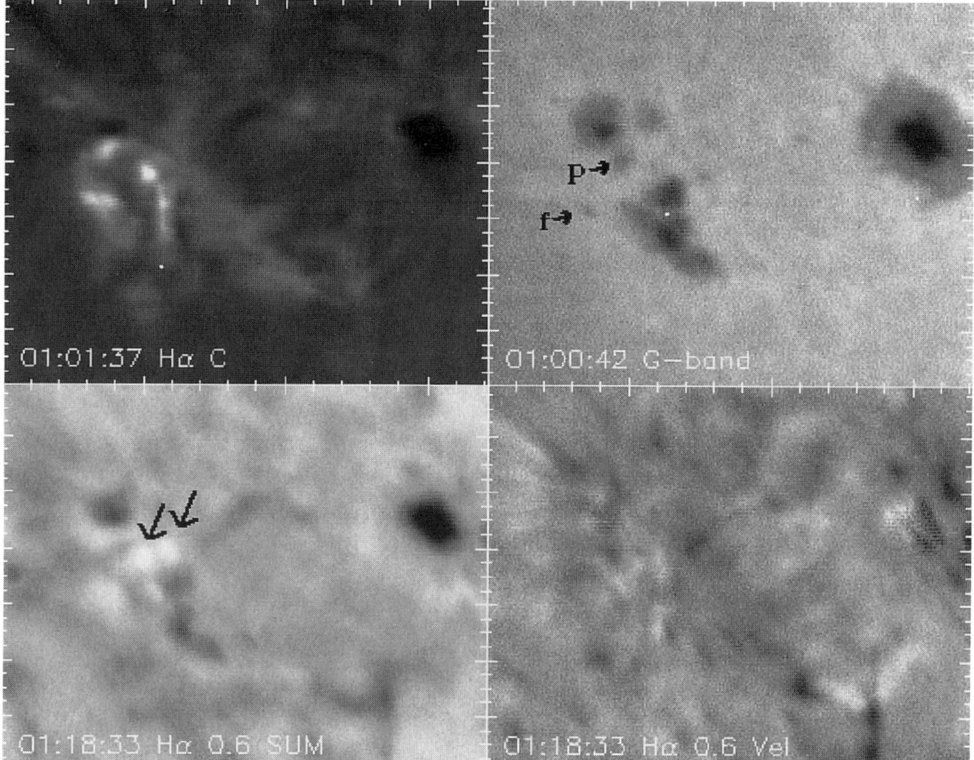

Figure 1. Hα and G-band filtergrams for the GOES X-ray class C7.7 flare which started at 01:00 UT on 13 Oct. 1995. The top two images were taken during maximum phase; the left is in Hα center and the right is in G-band. The bottom two images were taken during the decay phase; the left is the sum of Hα+0.6 with -0.6Å images and on the right is the -0.6 image subtracted from the +0.6Å image, showing velocity fields; bright is upward and dark is downward. The field of view is 163"×138"; sky north is at the top and east is at to left.

time span of the hard X-ray burst (see Fig. 3), and its velocity was about 500 km s^{-1}. In hard X-rays, two distinct sources coinciding with the Hα horseshoe formed during this phase. In the maximum phase (bottom panels in Fig. 2), the flare formed a horseshoe shape in Hα. Interestingly, however, the soft X-ray bright structure was located only around the initial bright region: the west part of the Hα horseshoe.

In the decay phase, we see an arcade structure in the soft X-ray image and a two-ribbon structure in the Hα image around the 'p' pore similar to the bright point pair in the preceding C7.7 flare. A downward moving feature was seen near the 'f' pore.

From the magnetic field distribution around the flare region and the proper motion of sunspots, an emerging flux model (e.g. Heyvaerts et al. 1977; Yokoyama and Shibata 1995) is basically fit for this flare. According to this model, the X-ray bright region in the preflare phase could be a sign of current sheet formation near the 'p' pore where the magnetic fields are antiparallel, and the ejection and hard X-ray burst in the impulsive phase can be explained by an explosive reconnection caused by the onset of anomalous resistivity. The arcade formation and the two-ribbon structure are good evidence of closed field formation resulting from the reconnection. Judging from the distance between the two ribbons (about 10000 km), we estimate the height of the initial reconnection to be less than 5000 km above the photosphere.

From the measurement of the proper motion of sunspots and pores around the flare, we found that the colliding motion of the 'p' pore with the preceding positive sunspot increased drastically about one hour before the flare onset. Hence, it is likely that the magnetic energy storage started one hour before the flare. We estimated the magnetic free energy as a function of time, assuming that the magnetic field configuration above the observed region is force-free over a semi-infinite space above the surface (e.g. Semel 1988, Yan 1995), and found that the free energy increased by about 10^{33} ergs between two and one hour before the flare, and then monotonously decreased at an average rate of 10^{32} ergs hour^{-1}. The thermal energy content of the flare was estimated to be about 5×10^{31} ergs, using the temperatures and emission measures derived with the SXT filter ratio method (Hara et al. 1992). Although we should be cautious in this result because the free energy has a large uncertainty (of the order of 10^{33} ergs), it is interesting to note that the timing of the enhancement in free energy coincides with that of the abrupt increase in the sunspot proper motion.

3. Discussion

Our observations suggest that the horseshoe-shaped homologous flares on 13 Oct. 1995 were caused by magnetic reconnection in the low corona ($z < 5000$ km) between emerging flux and the preexisting coronal fields. By extending the usual 2-D reconnection model into 3-D, taking into account the actual magnetic field distribution in the photosphere, we find that the footpoints of emerging flux should form the horseshoe shape.

It is a mystery, however, why the soft X-ray flare did not form a horseshoe shape similar to that of the Hα flare. This could be explained by different magnetic topologys on the east and west side of the horseshoe. The west side is the main energy release region for the reconnection, and as a result, a closed magnetic topology is formed, which is suitable for the confinement of heated plasma. In contrast, the east side consists of open-up magnetic fields because the closed fields of the emerging flux change into open fields after the reconnection. The chromosphere on the east side could

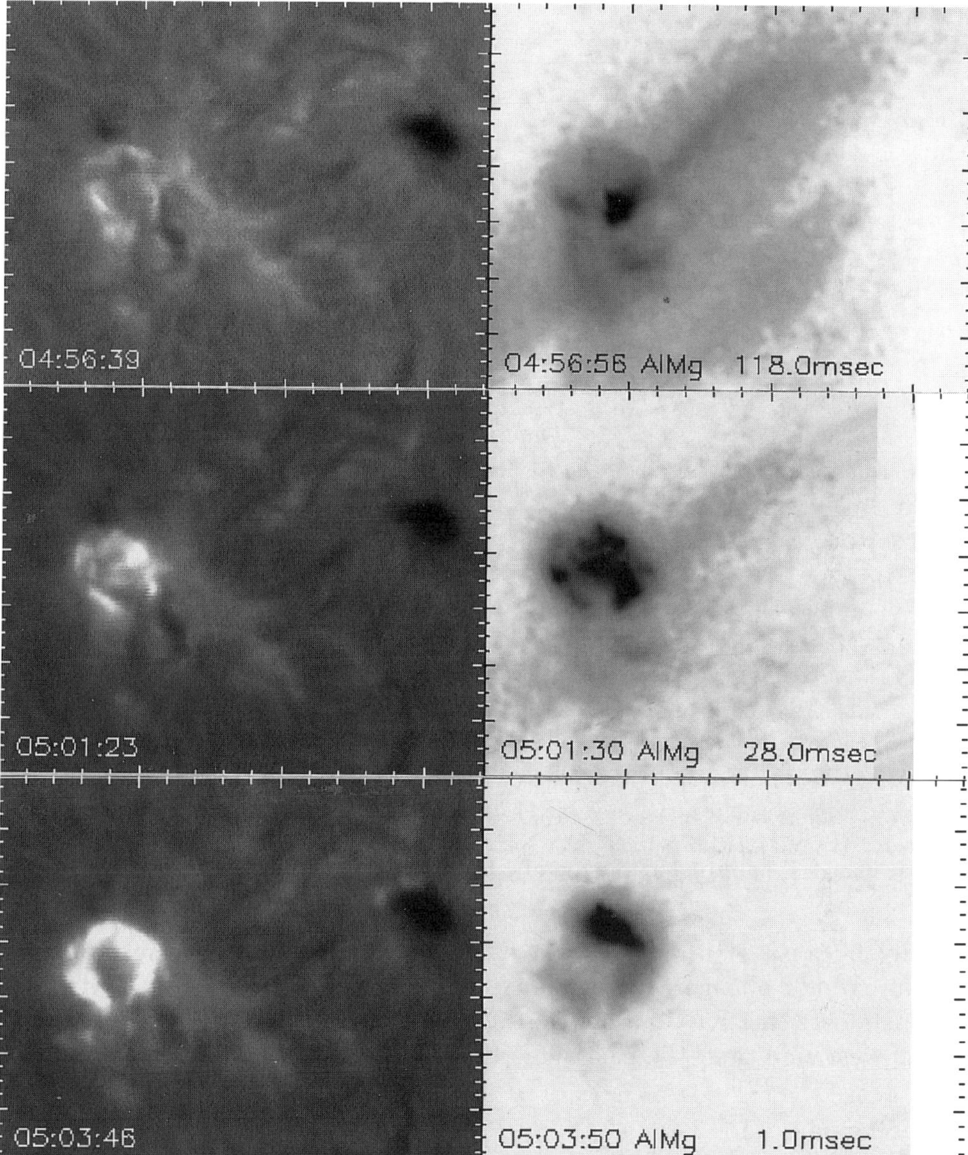

Figure 2. Time series of Hα and YOHKOH soft X-ray images (negative) for the GOES X-ray class M4.8 flare of 13 Oct. 1995. The field of view is the same as that in Figure 1. Explosive ejections heading east (left) are seen in both Hα and soft X-ray images at 05:01:30 UT. Note that the flare structure in soft X-ray is quite different from that in Hα.

be heated by energy transport along the closed field lines from the reconnection region, as indicated by two hard X-ray sources in Fig. 3, but the resultant hot plasma would evaporate into the open-up fields, cooling down

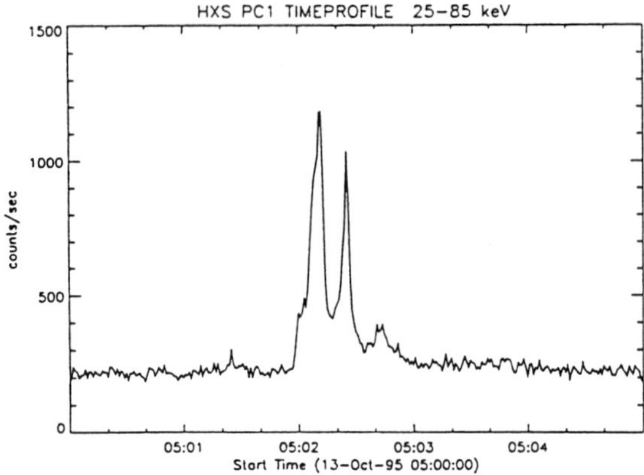

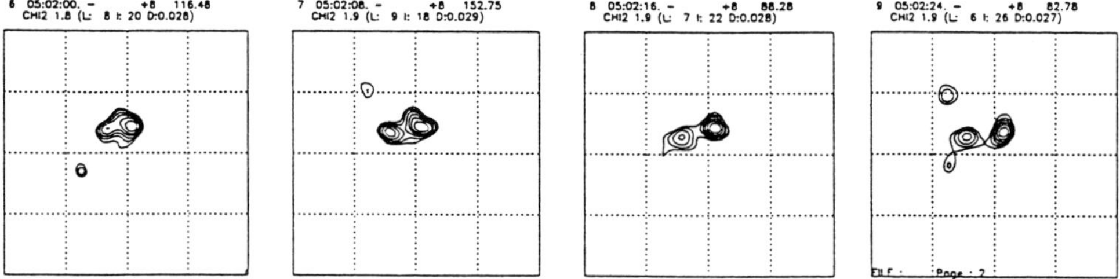

Figure 3. Light curve in hard X-rays (25–85 kev) and time evolusion of hard X-ray structures (FOV is 155.6"×155.6") in the M2 channel (33–53 kev). The time of each contour images is indicated at the top of each panel.

quickly because of large outward expantion. In conclusion, the emission measure of hot plasma (higher than about 3.5 MK, to which the YOHKOH SXT is most sensitive) and hence soft X-ray intensity could be much larger at the west side than the east side of the horseshoe.

References

Hara, H. et al.: 1992, Publ. Astron. Soc. Japan, **44**, L135.
Heyvaerts, J., Priest, E.R., and Rust, D.M.: 1977, Astrophys. J. **216**, 123.
Kosugi, T. et al.: 1991, Solar Phys. **136**, 17.
Sakurai, T. et al.: 1995, Publ. Astron. Soc. Japan **47**, 81.
Semel, M.: 1988, Astron. Astrophys. **198**, 293.
Tsuneta, S. et al.: 1991, Solar Phys. **136**, 37.
Yan, Y.: 1995, Solar Phys. **159**, 97.
Yokoyama, T. and Shibata, K.: 1995, Nature **375**, 42.

< Joint Observation of Yohkoh and SoHO >

SOHO - YOHKOH SCIENCE COLLABORATION

PETRUS C. MARTENS
ESA/SSD at NASA/GSFC, Greenbelt, MD 20771, USA

1. Introduction

I present an overview of joint SOHO - Yohkoh observing programs carried out since the beginning of SOHO science operations in April 1996. This is followed by an synopsis of SOHO results to date that are of interest for coordinated SOHO - Yohkoh data analysis.

In the second part of this review I will summarize those areas that, in my view, are most promising for joint observations and joint analysis. This is followed by the outline for a series of informal joint SOHO - Yohkoh data analysis workshops, the first of which has been completed at the writing of this review, to the great satisfaction of both the Yohkoh and SOHO participants.

Finally I present a summary on the ways to get access to the SOHO data for those that are not directly involved with the SOHO mission.

2. Completed SOHO - Yohkoh Joint Observing Programs

The SOHO catalogs are accessible to anyone interested via the World Wide Web. The Universal Resource Locator (URL) for the campaign catalog is

http://sohodb.nascom.nasa.gov/catalog/campaign.html.

Through the catalogs it will be possible to retrieve the data obtained during the campaigns listed, for those that are in the public domain. (More about data availability is said at the end of this paper.)

As of late March 1997, 29 SOHO campaigns (out of a total of 201) involving Yohkoh-SXT are listed. This list may not be complete as the SOHO catalogs are still being verified and updated. These campaigns have subjects ranging from "X-ray Bright Points," "Filling Factors of Hot Coro-

T. Watanabe et al. (eds.), Observational Plasma Astrophysics: Five Years of Yohkoh and Beyond, 217–225.

nal Loops," "The Diffuse Corona," to "Observations of Polar Jetlets," and "Filament Movies." Since the number of SOHO - Yohkoh joint observations increases on a weekly basis, rather than producing a table here, I refer the reader to the URL listed in the previous paragraph.

3. SOHO Results from the First Year Relevant for Yohkoh

SOHO experiments and operations are described in a dedicated Solar Physics issue (Fleck, Domingo, and Poland, 1995).

SOHO has three helio-seismology experiments on-board to study the interior of the Sun. While their results are fascinating and fundamental, they are not of prime interest for comparison with Yohkoh data. The one exception is MDI's maps of convective flows below the Sun's surface, obtained with the technique of time-distance helioseismology (Duvall *et al.*, 1997, Kosovich and Duvall, 1997) – the first glimpse ever of convective flows below a stellar surface. Figure 1a shows rapid down flows confined in narrow channels, and relatively mild upflows in the regions in between. This has clear relevance for the generation and transport of the magnetic flux tubes that eventually rise above the solar surface and determine the morphology and brightness of the soft X-ray corona seen in Yohkoh-SXT images.

SOHO's solar visible light and UV imagers and spectrographs (MDI, SUMER, CDS, EIT, LASCO, and UVCS) are producing a wealth of data, the analysis of which has just barely begun. Among the first results, immediately clear from inspection of time-lapse sequences, is that there is no such thing as the quiet sun: SOHO's UV detectors show structure at all observable length scales and time scales in regions that are quite unremarkable in soft X-ray images. One example are the so-called polar plumes, columns of high temperature and density plasma in the open field regions of the poles (Fig. 1b)

The SOHO/LASCO coronographs have found that Coronal Mass Ejections (CME's) are more frequent and carry more material than previously thought. In fact, the concept is emerging now (Sheeley *et al.*, 1997), that the "slow" solar wind, which emanates from the Sun's equatorial regions, consists of an almost uninterrupted series of small and large "puffs", resulting from tops of helmet streamers being torn away by the isothermal solar wind. The high cadence synoptic sequence of LASCO observations is perfectly suited to verify this hypothesis. Figure 1c shows a remarkable LASCO image of a CME simultaneous with the comet Hyakutake approaching its perihelion.

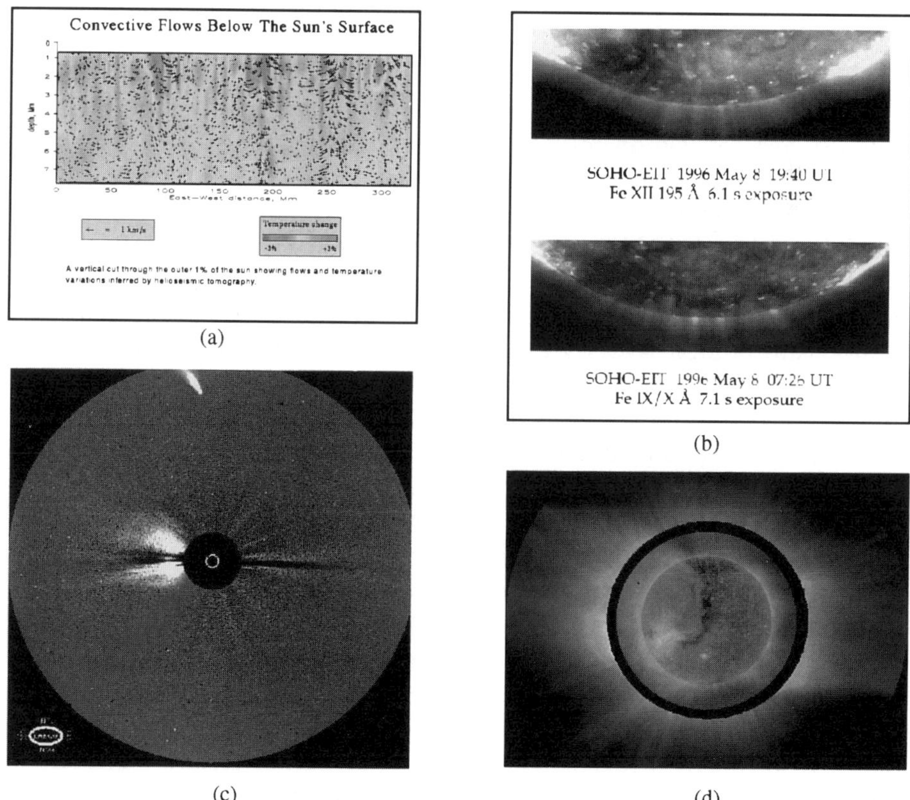

Figure 1. (a) A vertical cut through the upper convection zone showing the subsurface flows and sound speed inhomogeneities observed by MDI The flow field is shown as vectors (longest arrow 1.5 kms^{-1}) overlying the sound speed perturbations $\delta c/c$ (b) EIT images of polar plumes near the south solar pole at about 1.5 million degrees Celsius in the Fe XII emission line at 195 Å(top) and at somewhat cooler temperature at about 1 million degrees in the Fe IX/X emission lines at 171 Å(bottom). (c) Comet Hyakutake, as imaged by the C3 coronagraph of the LASCO instrument on 2 May 1996. This picture shows the comet to the north of the Sun. The bright region near the Sun is a coronal mass ejection. (d) This composite image of August 27, 1996, consists of a daily map obtained by UVCS in Oxygen VI (1032 Å), while the inner region is a full-disk image of EIT in the Fe XV line at 284 Å.

The "fast" solar wind which originates from coronal holes is completely different. UVCS, the UV coronal spectrometer on board SOHO, has observed similar velocities for heavy ions and protons (Antonucci *et al.* 1997), indicating bulk motions, perhaps in the form of MHD waves, that can drive the wind outflow. Figure 1d shows a composite UV image of EIT (inner region) and UVCS (outer region) tracing magnetic structures from the surface out to several solar radii.

Closer to the surface, SOHO's twin EUV spectrometers, SUMER and CDS, have produced maps of total emission, velocities, plasma densities, and temperatures for a wide variety of solar structures and phenomena, such as prominences, coronal holes, the network, jets, plumes, bright points, and active region loops. Preliminary spectral analysis confirms the impression from time-lapse series of images – ceaseless activity on all scales. Equally important are the synoptic programs of SUMER and CDS, which have resulted in spectral atlases for various objects with an unprecedented accuracy. Over 2000 emission lines have been observed by SUMER, while CDS through its unique combination of high angular spectral and temporal resolution has recorded, for the first time, reliable spectra for several small-scale features, such as jets, loops and prominences. Figure 2a shows a slit image and spectrum of high velocity explosions, observed by SUMER in the Si IV line, with a simultaneous EIT image to provide the location reference.

Finally, in addition to its helioseismology data, MDI provides about 15 full-disk magnetograms and white light images daily (Fig. 2b). While the resolution of these images is less than what is achieved from the ground, the continuity, and absence of atmospheric interference leads to sequences that are extremely well suited to observe the evolution of the surface magnetic field over the course of the solar cycle.

The in-situ particle detectors (ERNE and COSTEP) regularly detect high energy protons, electrons, and alpha particles, in the aftermath of filament eruptions and CME's, while CELIAS, SOHO's solar wind analyzer, has doubled the number of elements and isotopes recorded in the solar wind (Fig. 2c). During a Venus-solar wind line-of-sight passage CELIAS also observed, for the first time from such a large distance, pick-up ion's from Venus' ionosphere, finding a much smaller diffusion cone than anticipated. SWAN, the Solar wind anisotropy detector, has recorded full-sky Lyman alpha maps at a resolution and signal to noise ratio of at least a factor 10 better than previous observations (Fig. 2d). From these maps the latitude distribution of the solar wind is determined.

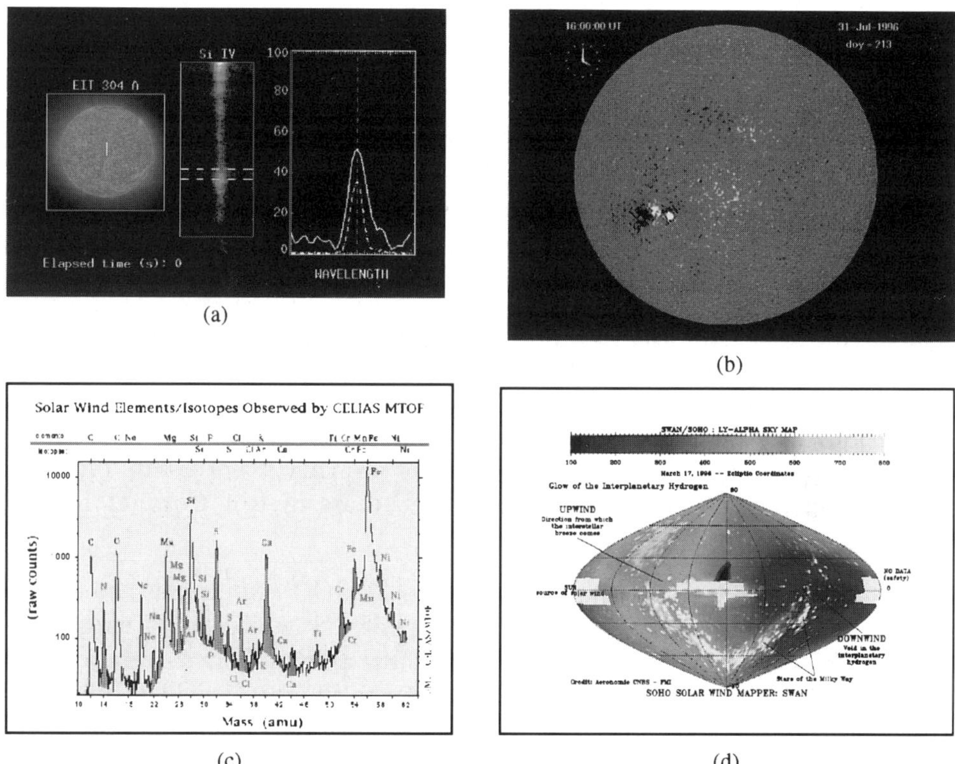

(a) (b) (c) (d)

Figure 2. (a) A SUMER slit spectrum in Si IV at about 1100 Å(center image). The increase in line intensity and width at the location of an explosive event is indicated on the right, while the EIT image on the left gives a location reference, (b) MDI full disk magnetogram. The pixel size is 2.0 arcsecond, and the noise level about 20 Gauss. (c) Element and isotope spectrum as obtained with the MTOF sensor of CELIAS. The elements and isotopes for which SOHO MTOF has given the first in situ spacecraft solar wind observations are in red. Those in shaded green are not observed routinely by conventional solar wind experiments. (d) SWAN full sky Lyman-α intensity map. The colour scale indicates the level of Lyman α intensity scattered by hydrogen atoms in the interplanetary medium. The swan shaped feature near the equator is the blocked out part of the sky near the Sun, while the small region displaced by 180 degrees in longitude represents a blocked out Solar reflection from the spacecraft.

SOHO Science is more than the sum of the efforts from each experiment. From the very beginning the SOHO experiments have devoted much of their observing time to Joint Observing Programs (JOPs), tackling specific physical problems, for example the temperature and density structure in coronal holes, through coordinated observations between SOHO experiments, with other spacecraft, such as Yohkoh and Ulysses, and with numerous ground-based observatories world-wide. In a typical SOHO week three to five different JOPs may be run, each one several times. The schedules, targets, and sequences of the SOHO campaigns have, from the start, effectively been communicated through the World-Wide Web, via the "SOHO Monthly Calendar." See *http://sohowww.nascom.nasa.gov/operations/* for an overview of all SOHO operations.

4. Areas for Joint SOHO - Yohkoh Data Analysis

Although the analysis of the data from the joint observations between Yohkoh and SOHO has only just begun, it is clear that there is an enormous potential for addressing several basic problems of Solar Physics.

Among these are the analysis of the detailed structure of soft X-ray emitting coronal loops, that are observed in broad-band by Yohkoh/SXT, and in four narrow-band filters by SOHO/EIT with the same high spatial resolution. Detailed spectral observations of cooler material in these same loops have been obtained by the CDS and SUMER experiments on board SOHO (SOHO Campaign # 9, "Filling Factors of Hot Coronal Loops", coordinated by Jim Klimchuck).

The innermost C1 coronagraph of the LASCO experiment on SOHO has an occulting disk of 1.1 solar radii. The C1 white light images and the Yohkoh-SXT soft X-ray limb images show an excellent correlation, and allow, for the first time, the inspection of the magnetic field topology all the way from the solar limb out to 30 solar radii (also using the other two LASCO coronographs). The scientific potential of these data is obvious.

Finally, there is the basic physical issue of the connection between magnetic field changes, filament eruptions, flares, and CME's. These phenomena can be observed simultaneously, with great spatial and spectral accuracy, in coordinated observations between MDI, CDS, SUMER, EIT, LASCO, UVCS, for SOHO, and SXT, HXT, and BCS for Yohkoh. A first analysis of this kind, made at the First SOHO - Yohkoh workshop, has led to a detailed event description and timeline for the filament eruption, X-ray flare, and CME of 25 and 26 September 1996
(see: *http://orpheus.nascom.nasa.gov/ zarro/workshop/zarro/sep25.html*).

5. A Series of Joint SOHO - Yohkoh Data Analysis Workshops

The philosophy for the series of SOHO - Yohkoh workshops, originally proposed by Hudson, Zarro, and the author, is similar to that of the very successful series of CDAWs (Coordinated Data Analysis Workshops) instigated by Canfield's group in Hawaii, and organized as a Yohkoh team effort.

We envisage small ($\leq$ 20 people) week-long workshops on highly focussed topics, held bi-annually – around every equinox – with approximately half discussions and half working sessions around workstations.

The requirements for such a workshop are a sufficient number (5-10) of powerful workstations, easy access to SOHO and Yohkoh data archives, sufficient workspaces, and commitment and detailed preparation by participants. For optimal efficiency there will be no posters, brochures, official endorsements, or conference proceedings, but we do require – a new wrinkle – a Web summary of workshop activities, both for post workshop information exchange between the participants, and for disseminating the results within the community (see the URL above for the Web summary of the first workshop). Participants are encouraged to submit their resulting papers to refereed journals, with an appropriate acknowledgement of the workshop.

In the first workshop, held at GSFC in early March 1997, 6 major CME/flare/filament eruptions were studied in separate working groups consisting of SOHO and Yohkoh scientists and ground based observers, and the workshop concept was proven to be valid.

Potential scientific topics for future workshops are:

- Active Region Development
- Emerging/Disappearing Flux and Coronal Effects
- Eruptions and CME's (a repeat of the first workshop)
- Bright Points
- Plumes
- Jets
- Filaments and Prominences
- Coronal Loop Structure Defined by Yohkoh and SOHO
- Yohkoh-SOHO Intercalibrations and Temperature Determinations

Workshop topics are normally selected after internal deliberations within the Yohkoh and SOHO communities, and after discussions at the workshops. Clearly each workshop needs the support of the experiment teams involved.

6. Access to SOHO Data

SOHO, like Yohkoh, is a PI mission. In return for the tremendous effort spent in building and operating their experiments, the PI teams have, for a limited time, proprietary data rights. However, all data will be released in the public domain after one year. Easy and user-friendly Web access will be provided through three European (in Paris, Rutherford, and Torino), and one US archive at GSFC, that is operational now. See the URL provided in the first section for Web access to the Goddard SOHO archive.

There are many coordinated observations between SOHO, Yohkoh, and ground based observatories, usually with the implicit agreement of mutual access to the data. Informal proposals for collaborations (with no financial support) for data analysis and new observing programs are welcomed by most SOHO PI teams, and usually imply access to data before the end of the proprietary period. Participation in one of the many SOHO data analysis workshops – such as the SOHO - Yohkoh workshop, the Whole Sun Month workshop, the ISTP workshops, etc. – provide another means of becoming involved in SOHO data analysis.

Last, but not least, the SOHO Guest Investigator Program, provides an "official" way of becoming associated with one or more of the SOHO PI teams. For the Announcement of Opportunity and the selection list for the first round, see

http://sohowww.nascom.nasa.gov/operations/guest_investigators/.

Selected US investigators will receive NASA funding for their investigation, while successful proposers from other countries have to apply to their national agencies for financial support.

7. Conclusion

In conclusion I want to emphasize that the Yohkoh - SOHO science collaborations represent a truly international effort, linking the Solar Physics communities of Japan, Europe, and North America.

I acknowledge the courtesy of the MDI, SUMER, EIT, UVCS, LASCO, CELIAS, and SWAN consortia for allowing me to reproduce the images presented in this paper.

References

Antonucci, E., Noci, G., Kohl, J.L., Tondello, G., Huber, M.C.E., Giordano, S., Benna, C., Ciaravella, A., Fineschi, S., Gardner, L.D., Martin, R., Michels, J., Naletto, G., Nicolosi, P., Panasyuk, A., Raymond, C.J., Romoli, M., Spadaro, D., Strachan, L., van Ballegooijen, A.: 1997, in *Advances in the Physics of Sunspots*, J.C. Del Toro Iniesta et al. (eds.), ASP Conf. Series, in press.

Duvall Jr., T.L., Kosovichev, A.G., Scherrer, P.H., Bogart, R.S., Bush, R.L., DeForest, C., Hoeksema, J.T., Schou, J., Saba, J.L.R., Tarbell, T.D., Title, A.M., Wolfson, C.J., and Milford, P.N.: 1997, *Solar Physics*, 170, 63-73.

Fleck, B., Domingo, V., and Poland, A.I. (eds.): 1995, *Solar Physics*, 162, 1.

Kosovichev, A.G., Duvall, T.L.: 1997, in *Solar Convection and Oscillations and their Relationship*, eds. J. Christensen-Dalsgaard and F. Pijpers, Proc. of SCORe'96 Workshop, Aarhus (Denmark), Kluwer Acad. Publ, in press.

Sheeley, N.R., Wang, Y.-M., Hawley, S.H., Brueckner, G.E., Dere, K.P., Howard, R.A., Koomen, M.J., Korendyke, C.M., Michels, D.J., Paswaters, S.E., Socker, D.G., St.Cyr, O.C., Wang, D., Lamy, P.L., Llebaria, A., Schwenn, R., Simnett, G.M., Plunkett, S., Biesecker, D.A.: 1997, *Astrophys. J.*, in press.

THE CORONAL DIAGNOSTIC SPECTROMETER ON SOHO

K. J. H. PHILLIPS AND R. A. HARRISON
Space Science Dept., Rutherford Appleton Laboratory, Chilton, Didcot, Oxon. OX11 0QX, U.K.

1. Introduction

The solar extreme ultraviolet (EUV) spectrum, in the range 150–1200 Å, contains numerous emission lines whose intensities and profiles give us much valuable information for probing the upper chromosphere, transition region, and corona. The emissivities of these lines cover the temperature range 10^4 K to more than 10^6 K. Temperatures, densities, element abundances, flows, and turbulent velocities form the information that can be derived from analysis of line intensities and profiles.

Few spacecraft instruments operating in this spectral region, despite its diagnostic importance, have been flown, and consequently there are not many published high-quality solar EUV observations. Among these instruments are the Harvard Extreme Ultraviolet Spectrometer (Reeves, Vernazza, & Withbroe 1976) and the Naval Research Laboratory's XUV spectroheliograph on *Skylab* (Sandlin et al. 1976) and the CHASE coronal helium abundance experiment on *Spacelab 2* (Patchett et al. 1981).

The Coronal Diagnostic Spectrometer (CDS) on the *Solar and Heliospheric Observatory* (SOHO) was designed to provide observations with high temporal, spatial and spectral resolution of the solar atmosphere over long durations and covering a wide range of EUV wavelengths. The quality of the images and spectra it has obtained is considerably improved over that from previous EUV instrumentation.

SOHO was launched on December 2, 1995, and after a cruise phase, was inserted into a 'halo' orbit around the L1 Lagrangian point, located between the Sun and the earth, some 1.5 million km from the earth. The spacecraft does not, therefore, suffer from eclipses by the earth and so in principle continuous observations over long time intervals are possible. The

T. Watanabe et al. (eds.), Observational Plasma Astrophysics: Five Years of Yohkoh and Beyond, 227–236.

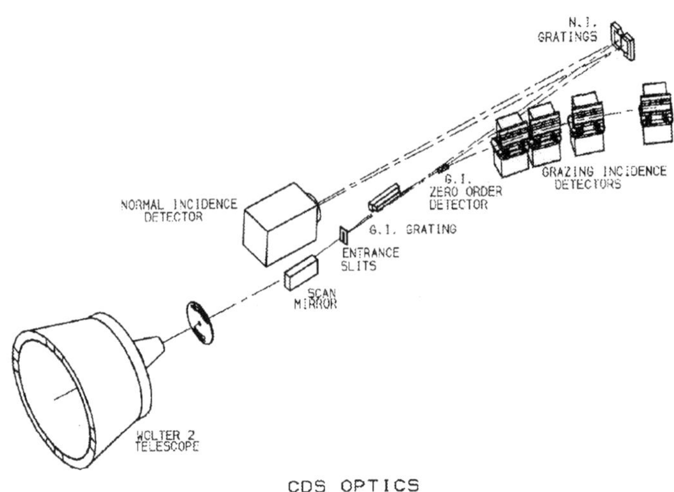

Figure 1. The optical elements of the Coronal Diagnostic Spectrometer on SOHO.

commissioning of SOHO was completed after arriving at the L1 point, and scientific operations begun in March 1996.

Following a brief introduction to the instrument, we will give a summary of some of the chief observations obtained to date.

2. The CDS Instrument

The Coronal Diagnostic Spectrometer is a twin spectrometer operating in the spectral range 151–785 Å. It is described in detail by Harrison et al. (1995). Fig. 1 is a schematic drawing of the instrument. It can be pointed to any direction on the solar disk or off the disk to observe the inner corona. A Wolter Type 2 telescope feeds simultaneously the Normal Incidence Spectrometer (NIS) and the Grazing Incidence Spectrometer (GIS) via a common slit. Either spectrometer (but not both) operates at any time. Two portions of the light from the telescope beam are selected for each spectrometer by a light stop, located just behind the telescope mirror.

The total field of view of the telescope is 4 arcminutes, but the whole solar disc can be imaged by repointing the instrument through angles of $\pm 0.75°$. The instrument is supported on six legs, the front four of which are pivoted at each end and the rear two consisting of linear actuators.

The NIS is a stigmatic spectrometer, making use of two toroidal gratings to image long, thin slits on a two-dimensional detector which is an intensified CCD. Motion of a flat scanning mirror in increments of 2 arcseconds allows images of portions of the Sun to be obtained in chosen spectral lines. The number of spectral lines can be traded with the desired time resolution of the images. The two spectral ranges (appropriate to each of the toroidal gratings) are 308–381 Å and 513–633 Å. The slit sizes are 2×240, 4×240, and 90×240 arcsec.

The GIS is an astigmatic spectrometer, using a spherical grating and a pin-hole entrance slit. The spectrum is sensed by four microchannel plate detectors arranged around the Rowland circle. Images can be formed by moving the scan mirror in one direction and moving the slit in the perpendicular direction. The four wavelength ranges are 151–221 Å, 256–338 Å, 393–493 Å, and 656–785 Å. The prime slits available are 2×2, 4×4, and 8×50 arcsec.

3. Early Results: Spectra

3.1. NIS SPECTRA

Fig. 2 shows images of NIS spectra for a quiet solar region on 1996 April 9. The images are from the two toroidal gratings, and are labelled NIS1 and NIS2. Structure along the length of the slit is of solar origin, though some features, notably the isolated bright points or small streaks, are interactions of cosmic rays with the detector system. (Such 'hits' are expected with SOHO being located well outside the earth's magnetosphere.)

The NIS2 (513–633 Å) region is dominated by the chromospheric He I 584 Å resonance line. Other bright lines include those emitted by transition region ions (O III, $\lambda\lambda$ 526, 600 Å, O IV, 555 Å, O V 630 Å) and coronal ions (Mg X, $\lambda\lambda$ 610, 625 Å, Si XII 521 Å). The bright He II 304 Å line occurs in this spectral range in second order. Thus, a very large temperature range is represented by these lines, from $\sim 10^4$ to about 2×10^6 K, that is, from the upper chromosphere through the transition region to the corona. As many of these lines are bright and unblended, they lend themselves to studies of plasma flows in the various regions of the solar atmosphere.

The lines in the NIS1 (308–381 Å) region are by and large less intense and are emitted by ions formed in the corona. Many are $3s - 3p$ transitions in Fe ions, from Fe XI to Fe XVI, with peak ionization fraction temperatures ranging from 1 to 3×10^6 K (Arnaud & Raymond 1992). Many of these lines are sensitive to electron density (e.g. the line pairs Fe XIII 359 Å/348 Å), and so are being used to study the structures of coronal features.

The full widths at half intensity of these lines are about 0.6 Å for the longer-wavelength lines, and about 0.3–0.4 Å for the shorter-wavelength lines. Although various factors contribute to the line broadening, the pro-

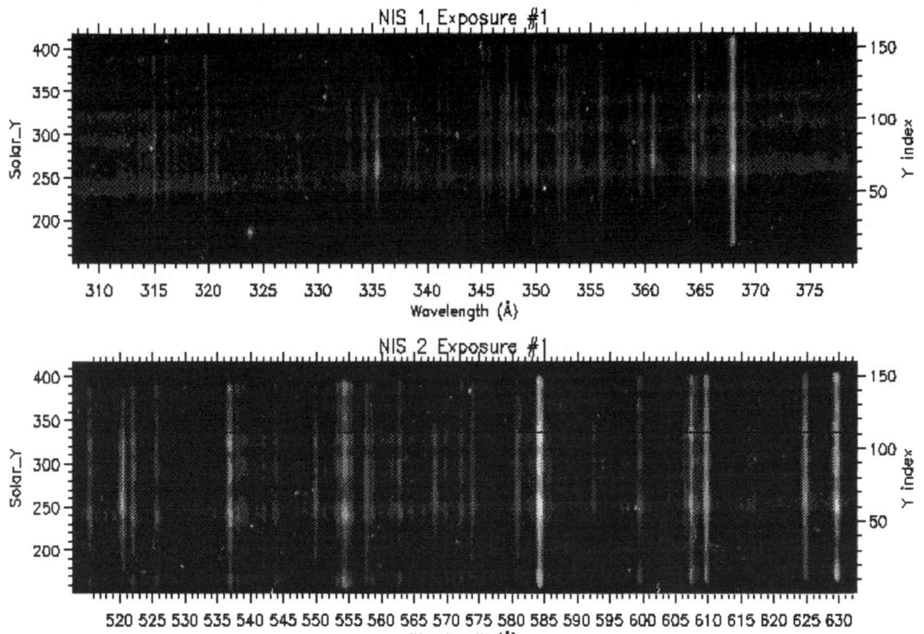

Figure 2. An NIS spectrum of a quiet solar region, as imaged by the two NIS detectors. A 2×240 arcsecond slit was used.

files are dominated by the slit width, i.e. are mostly instrumental in origin, but nonetheless it has been possible to observe large-scale flows (see §4).

3.2. GIS SPECTRA

A large number of GIS spectra have already been obtained in the first year of SOHO's lifetime, with a variety of targets ranging from quiet-Sun to active regions. Some crude estimate of the temperature of the emitting region can be made by simple inspection of the GIS1 (first GIS band) spectrum (see Fig. 3). This range contains prominent resonance ($3p - 3d$) lines of Fe ions from Fe IX to Fe XIV, with wavelengths 171 Å to 211 Å. When an active region is observed, higher ionization stages still are represented, with $3s-3p$ transitions occurring. Thus, the lines due to Fe XV at 284 Å and Fe XVI at 335 Å are sometimes very strong, depending on the temperature of the region. The large oscillator strengths of some of these resonance lines means that the emission is not always optically thin, and programmes underway at present are using this fact to establish mean electron densities and optical path-lengths for comparison with atomic ratio methods; this has

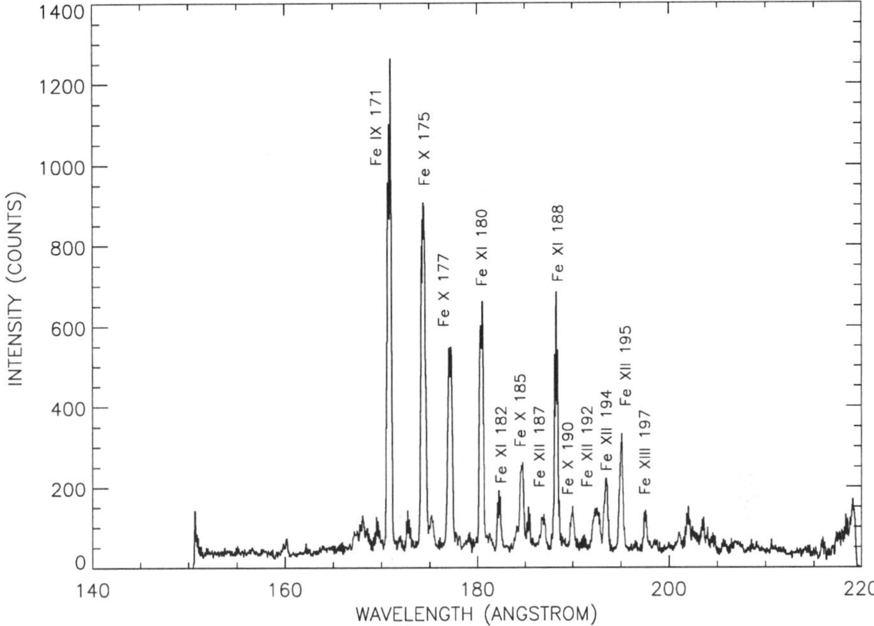

Figure 3. GIS spectrum (GIS1, λλ 151–221 Å) of a quiet solar region.

been done with some effect in the X-ray region with Fe XVII lines (Phillips et al. 1996).

Much useful diagnostic information is available through the intensity ratios of lines in all four GIS bands. Thus, the ratio of the Mg VIII 315 Å and 430 Å lines is sensitive to electron density. There are a number of lines in the GIS3 and GIS4 bands which contain lines due to ions present at chromospheric temperatures, such as O II, O III, Ne I, as well as lines emitted in the transition region, with temperatures of $\sim 10^5$ K, such as C III, C IV, and Ne III. The diagnostic capabilities are many of these lines as well as the atomic data needed for interpretation are reviewed by Mason & Monsignori Fossi (1994) and in the special edition of *Atomic Data and Nuclear Data Tables*, vol. 57, nos. 1/2 (J. Lang, editor).

4. Active Regions

Although solar activity has been at a very low level over the lifetime of the SOHO mission so far, a number of extensive active regions have been observed with the CDS experiment, and in particular image data obtained with the NIS. Simultaneous images in four lines, ranging from chromospheric to coronal temperatures, are shown in Fig. 4 for an active region in March 1996. The rasters were formed by moving the 2×240 arcsecond slit over the 244×240 arcsecond area, i.e. approximately 200 000 km square area on the Sun. In the coronal line emission (Mg IX and Mg X), emitted at temperatures of $\sim 10^6$ K, we see the loop structure that is familiar from, e.g., the soft X-ray images from the SXT instrument on *Yohkoh*. The loops apparently extend to several tens of thousands of km in height. Some of this loop structure is visible in the transition-region Mg VI (emitted at about 400 000 K), but in the chromospheric He I line emission (emitted at about 200 000 K) most of the structure is at small heights, with the brightest patches of emission corresponding to the footpoints of the coronal loops. It is clear from the Mg IX and Mg X images that the coronal loops are considerably inclined to the radial direction, with east–west-oriented loops leaning towards both the north and south.

5. Prominences

Fig. 5 shows images in various lines made with the NIS of a limb region, including a prominence which shows clearly in emission off the limb in the relatively cool, chromospheric He I 584 Å line. The whole prominence is about 2 arcminutes (almost 100 000 km) long, and extends to a height of about 50 000 km. There is little evidence of the prominence in other lines, though perhaps the northern end is visible in the O V line at 630 Å. The coronal Mg IX and Si XII line images show a bright structure, with a temperature of approximately 2×10^6 K, not far from the southern end of the prominence.

Other images of prominences show not only a structure in emission in the chromospheric He I line but also absorption features lying across the brightening at the limb in the Mg IX and Si XII lines. This occurred for example with a prominence at the west limb on 1996 March 5. In the O III 599 Å line, one end of the prominence appears as both an emission and absorption feature. It would not seem that prominences are often seen in absorption at the limb in *Yohkoh* SXT images, a fact which could be of interest in determining the properties of the source function of prominence material.

6. Quiet Sun structures

The chromospheric network structure that was a familiar characteristic of the *Skylab* EUV images has been extensively observed with the CDS instrument (Fig. 6). It shows up clearly as sharply defined emission structures in chromospheric and transition region lines, but is barely visible in coronal lines where the dominant features are more ill-defined structures presum-

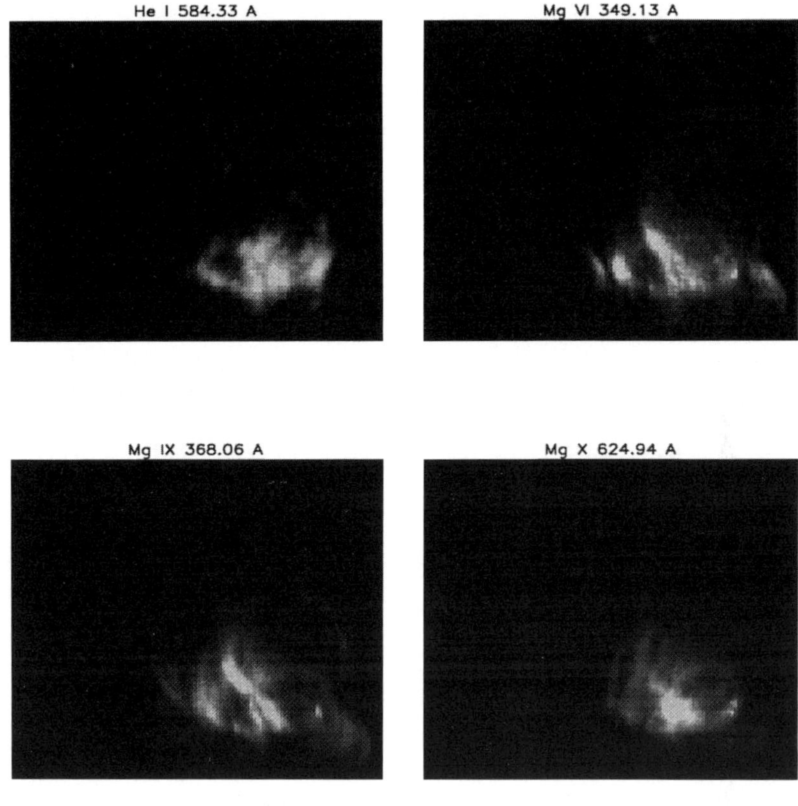

Figure 4. Active region images (size 244 × 240 arcseconds) made with NIS in emission lines indicated. The He I line is chromospheric, Mg VI transition region, and Mg IX and Mg X are coronal in origin. The region was located approximately 10 arcminutes from Sun centre in a south-west direction.

ably lying at greater heights, often associated with bipolar regions or having feet that show up as bright points in the lower-temperature line images. Movies of such quiet-Sun regions show brightness changes occurring all the time, with some network features brightening or fading. As was deduced from the *Skylab* images, the network structures are features associated with the magnetic field at the edges of photospheric supergranules.

Of some interest is the association of some network features that become particularly bright over a period of some tens of minutes with magnetic

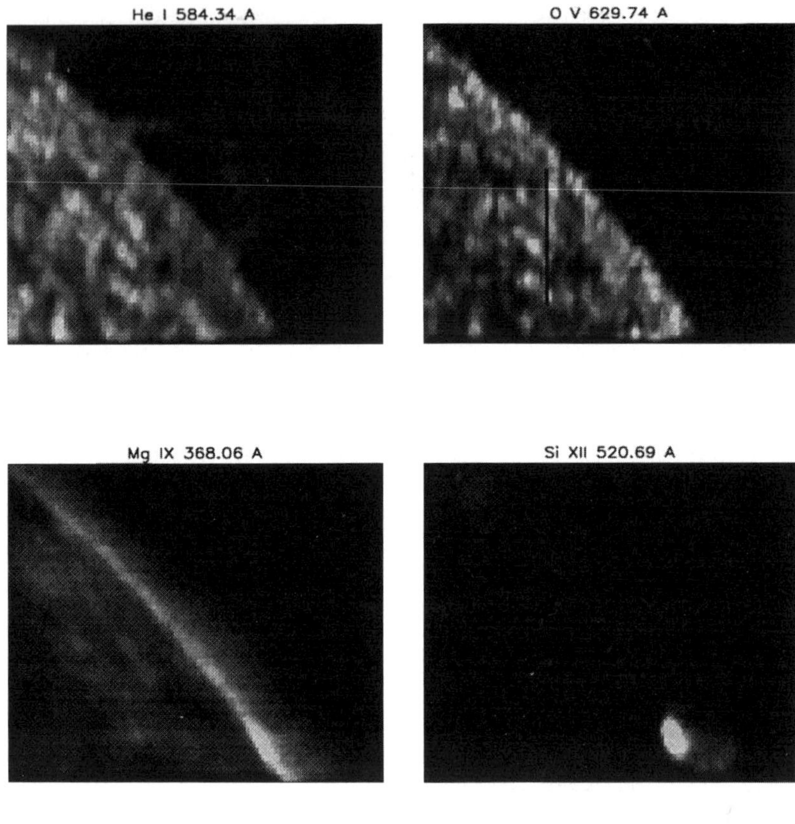

Figure 5. NIS images (size 244×240 arcseconds) taken at the limb in the lines indicated (see caption to Fig. 4 for further details). A prominence shows clearly as an emission structure off the limb in the He I 584 Å line.

field. Such brightenings have been called 'blinkers' by Harrison (1997) in a recent description. How such features are related to the constantly changing magnetic field that has been observed with ground-based magnetograms (Martin 1988; see also Moses et al. 1994) and can be examined with the MDI instrument on SOHO is the subject of investigations now being carried out.

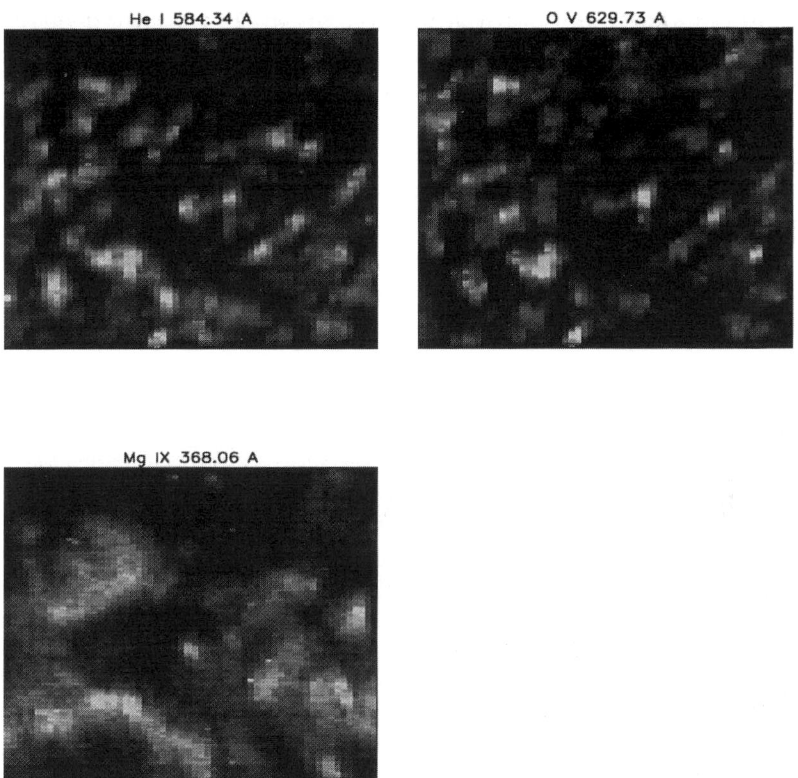

Figure 6. NIS images (size 244 $\times$ 240 arcseconds) of a quiet-Sun region on the disk in three lines.

7. Concluding Remarks

The CDS instrument on SOHO has operated very successfully over the first year of the spacecraft's operations and has obtained high-quality images and spectra from a large variety of solar features. Images of the quiet Sun, prominences, bright points and active regions are being examined intensively at present to find out information on their structure at temperatures between 10^4 and several 10^6 K. Spectra too from both the NIS and GIS instruments show a wealth of possible diagnostic information which will help us understand the temperatures and densities of various structures visible in the EUV, as well as flows, explosive events (from Doppler signatures), element abundances, and (with other SOHO instruments) associations with e.g. the photospheric magnetic field. Of particular interest from a number of points of view is the comparison of image data from the CDS and other SOHO instruments with the X-ray images that continue to be produced with the SXT on *Yohkoh*. Some interesting problems are being thrown up with comparison between images from the coronal lines from CDS and the EIT instruments, e.g., and those from SXT. As our investigators analyze the large amounts of data already accumulated, we expect some exciting results and possible changes to our understanding of many solar phenomena.

We thank Mr George Hanoun and Dr J. Lang for help with this paper.

References

Arnaud, M., & Raymond, J. 1992, ApJ, 398, 394
Harrison, R. A. et al. 1995, Solar Phys., 162, 233
Harrison, R. A. et al. 1997, Solar Phys., in press
Lang, J. (ed.) 1994, At. Data Nucl. Data Tab., 57, nos. 1/2
Martin, S. F. 1988, Solar Phys., 117, 243
Mason, H. E., & Monsignori Fossi, B. C. 1995, Astr. Ap. Rev., 6, 123
Moses, D. et al. 1994, ApJ, 430, 913
Patchett, B. E., Norman, K., Gabriel, A. H., & Culhane, J. L. 1981, Space Sci. Rev., 29, 431
Phillips, K. J. H., Greer, C. J., Bhatia, A. K., & Keenan, F. P. 1996, ApJL, 469, L57
Reeves, E. M., Vernazza, J. E., & Withbroe, G. L. 1976, Phil. Trans. Roy. Soc. A281, 319
Sandlin, G. D., Brueckner, G. E., Scherrer, V. E., & Tousey, R. 1976, ApJ, 205, L47

NOAA 7978: THE LAST BEST OLD-CYCLE REGION?

H. S. HUDSON
Solar Physics Research Corp., Tucson, AZ, USA[1]
B. J. LABONTE
Institute of Astronomy, U. Hawaii, Honolulu, HI, USA
A. C. STERLING
Computational Physics, Inc., Fairfax, VA, USA[1]
AND
T. WATANABE
National Astronomical Observatory, Mitaka, Japan

Abstract. Sunspots in NOAA active region 7978 first appeared on July 6, 1996, while the region was at E15. The spots reached an area of 400 millionths within four days and the active complex remained interesting for at least five solar rotations, generating an X class flare, a long-duration event, and coronal mass ejections well-observed by both *Yohkoh* and SOHO. During these five rotations, only a few short-lived competing active regions appeared. The occurrence of a good center of activity during such quiet solar-minimum conditions means that whole-Sun instruments, such as those of the *Yohkoh* BCS spectrometers or the GOES photometers, responded primarily to a single localized source. Although the active region had large horizontal and vertical dimensions, the most intense X-ray emission came from low coronal altitudes. We discuss the overall configuration and evolution of this region, which appears destined to be the focus of many specialized studies. The discussion includes a description of the effect of AR7978 on the total solar irradiance, and of a preliminary search for "relaxation oscillator" evidence for coronal energy build-up and release.

[1]Current address ISAS, 3-1-1 Yoshinodai, Sagamihara, Kanagawa 229, Japan.

T. Watanabe et al. (eds.), Observational Plasma Astrophysics: Five Years of Yohkoh and Beyond, 237–244.

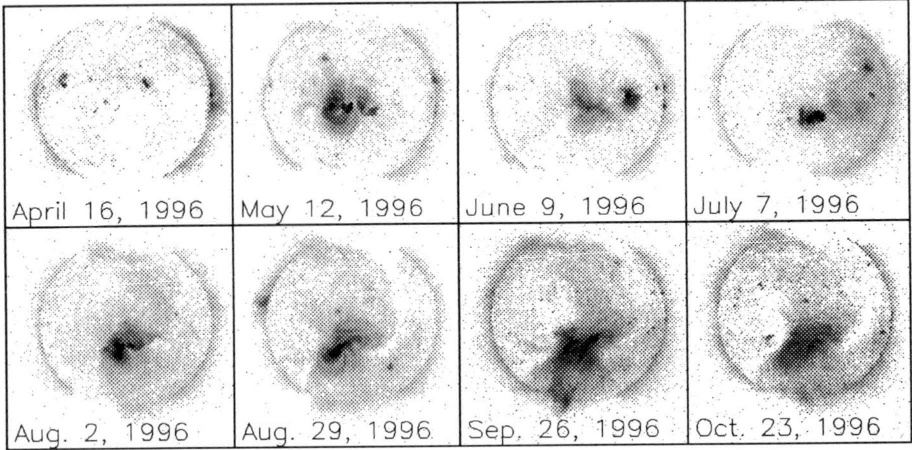

Figure 1. *Yohkoh* images of the 1996 activity complex over eight rotations. The "last best" region, NOAA AR7978, initially formed on July 7, 1996. Note the rapidly-formed connections to high latitudes, and the CME aftermath visible in the September image. North is up, east to the left; the images are from the AlMg filter of SXT with logarithmic compression.

1. Introduction

Solar activity in 1996 was dominated by an activity complex near a Carrington longitude of 250 at about S10. As described in the accompanying paper by Harvey & Hudson (1997), this complex showed eruptions of new flux as early as April, and the dispersing magnetic flux could be traced through the end of the year. The high point of the region development was the birth on the disk of NOAA region 7978, its rapid growth, and the occurrence of an X-class flare on 9 July 1996. Figure 1 shows *Yohkoh* SXT images of this complex from a time span of eight rotations. The main objective of this paper is to introduce the data on this active complex, and to propose some research directions that take advantage of it.

The concentration of so much activity into a limited region of the solar surface, which otherwise showed typically passive solar-minimum behavior, provides us with several opportunities. First, the emerging flux acts as a "test particle" probing the magnetic reactions (such as reconnection and coronal-hole formation) as outlined by Harvey & Hudson (these proceedings). Second, we can assume that most of the activity during this time period came from this complex. Such an assumption makes for a more straightforward interpretation of whole-Sun data such as those of the *Yohkoh* BCS (Sterling, 1997; Sterling *et al.*, 1997) or the GOES whole-Sun X-ray photometers. This paper surveys some of the properties of this activity complex, including a brief description of a search for "relaxation oscilla-

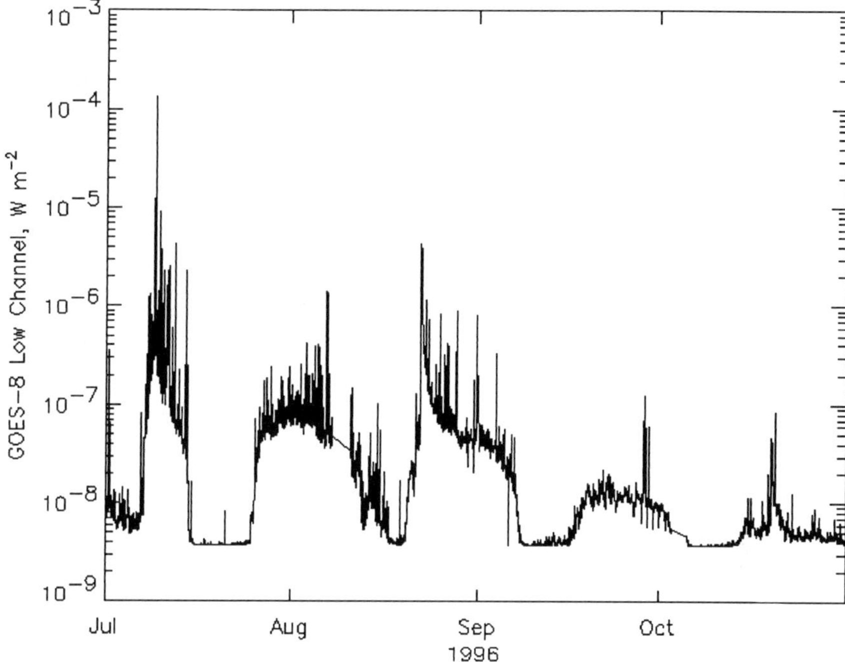

Figure 2. Time-series data in the low channel (1-8Å) of GOES-8, for five months following the eruption of AR7978. Note the sharp occultation signatures.

tor" behavior, which (if detectable) would support the concept of coronal build-up of flare energy.

2. Morphology

The active regions associated with this complex tended to be large in horizontal extent, but with a strong concentration at low altitudes. In Figure 1 one can see that the coronal effects of AR7978 extend rather quickly to high latitudes. The frame from September shows the aftermath of a coronal mass ejection extending over the S polar region.

Figure 2 illustrates the small vertical extent of the active region by showing a difference image comparing two images taken at a 3-hour time interval. This difference image shows that the coronal sources, partially occulted by rotation, lie in a thin strip elongated tangent to the limb. The dimensions of this difference source perpendicular and parallel to the limb are 3×10^5 km and 2×10^4 km (FWHM), respectively. The latter corresponds to a scale height of some 8×10^3 km, much less than that expected (4–7×10^4 km) from hydrostatic equilibrium at the observed

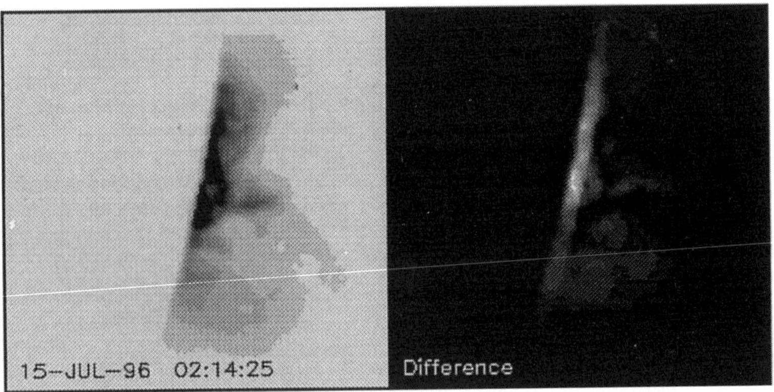

Figure 3. Soft X-ray image of AR7978 at its west limb passage (left; image time July 15, 1996, 02:24 UT) and a difference image (right; comparison with the image of July 14, 23:29 UT) over a 3-hour span. This shows the concentration of X-ray intensity into a thin layer. The color scale is reversed.

temperatures. We thus conclude that the active-region structure consists of low-lying magnetic loops with substantial gas overpressure. This evidence from the difference image is supported by the time profiles of soft X-rays as observed by SXT and GOES (Sterling *et al.*, 1997); we note that this pattern persisted at successive occultations of the active complex. Hara (1997) has inferred a general tendency for low-altitude concentration of solar active regions from a statistical analysis of the SXT image histograms.

3. The Total Irradiance Variations

The sun was virtually spotless when AR7978 erupted, and within four days the new spot group had reached a total area exceeding 400 millionths of the hemisphere. This is large enough to produce a detectable decrease in the total solar irradiance (Willson *et al.*, 1981; Hudson, 1986), and as Figure 3 shows, a well-correlated dip (about 0.04%) did take place. Here we have used the sunspot areas reported in the NOAA *Solar-Geophysical Data* for the comparison with total-irradiance observations from the VIRGO experiment on SOHO (Wehrli *et al.*, 1997). The tabulated sunspot data do not have the sampling or precision of the total irradiance data but are easily available.

The onset of the July 1996 dip was caused by the physical growth of the spots, rather than by foreshortening resulting from solar rotation. Do these differing causes produce different correlations? This question remains open from the ACRIM data analysis, and the current data appear to be much more suitable for this type of study because of the existence of SOHO white-light images from the MDI instrument. These can be used to determine sunspot and facular area variations with good control – for the first time

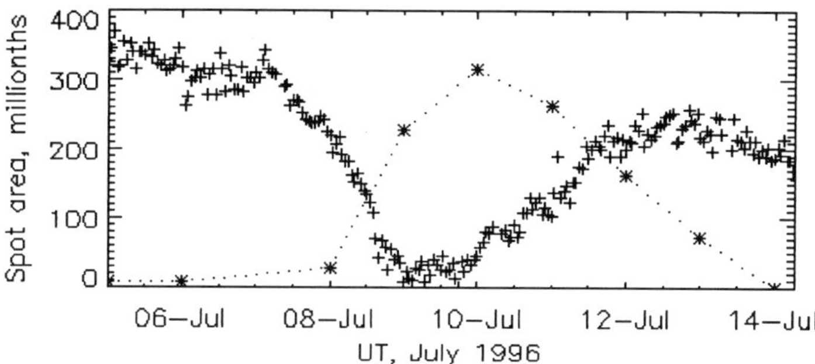

Figure 4. A total-irradiance "dip" associated with the AR7978 sunspots in July, 1996, as recorded by VIRGO (*e.g.* Fröhlich *et al.*, 1997). Frequent points, the total irradiance (arbitrary units; the depth of the dip is about 400 parts per million); infrequent points, the projected sunspot area from the NOAA/SGD archive.

– of the pernicious effects of seeing, scattered light, and day/night cycles present in ground-based data.

4. Relaxation Oscillations

In the most popular view, flare energy is stored in coronal fields before its release. This is the basis of the classical large-scale magnetic reconnection models of solar flares (see respectively Tsuneta, 1996, and Hudson and Khan, 1996, for arguments pro and con). There is some difference of opinion (*e.g.* Melrose, 1995) as to whether the energy is stored remotely or locally, or even if a substantial part of flare energy enters the energy-release region from below the photosphere. In the latter case, because of the low Alfvén speed in the photosphere, the effect might only be noticed in the gradual phase of a flare or LDE.

In any case, the concept of coronal energy storage suggests a search for relaxation-oscillator behavior as at least a sufficient condition. In such a relationship, the energy of a flare event correlates with the interval before or after its occurrence, as explained below. The correlation-after relationship is well-known in astrophysics from X-ray "burster" sources (Lewin *et al.*, 1970); in that case the oscillation arises from a build-up of gravitational energy, followed by an unstable infall into a compact star. In the solar MHD case, the build-up would consist of non-potential energy in the tangled structure of the coronal magnetic loops, and the release would come from an unstable process involving reconnection. Because there would be many ways in which to dilute this effect, its absence would hardly be proof that coronal energy build-up does not occur. Thus the oscillation-relaxation

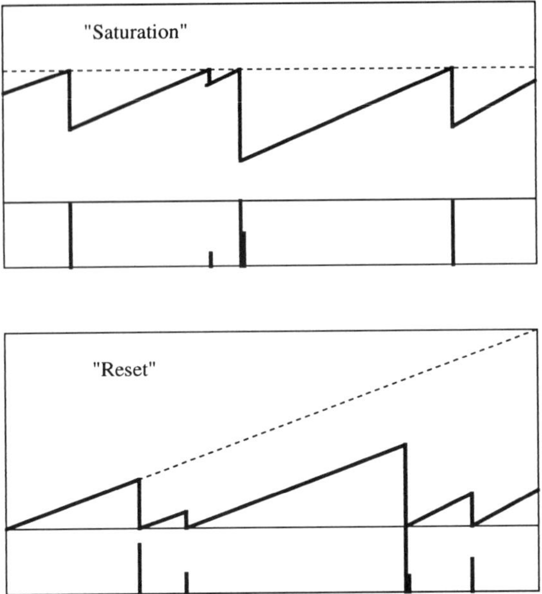

Figure 5. Sketches of two forms of relaxation oscillation. The heavy lines represent coronal energy content *vs.* time in the two models, the vertical lines the flare energy releases. bottom, the "reset to zero" case; top, the "saturation" case. In the "reset" case, the system energy is assumed to increase steadily; in the "saturation" case, it builds back to a trigger level. In both cases we assume single systems so that there is no confusion from independent relaxations.

pattern would be a sufficient but not a necessary condition.

We can distinguish two cases, sketched in Figure 4, both of which envisage a steady rate of energy build-up. In the "reset" case, the instability hypothetically would reduce the stored energy to zero. The energy released in a given event would therefore correlate with the time interval *before* the flare. In the idealized second case, "saturation", the system would approach a limiting stored energy, resulting in an instability when that limit was reached. In this case the correlation would be with the time *after* the flare. The X-ray burster correlation is of the saturation type. This also appears to be the the pattern expected from a "sandpile" (self-organized critical state) flare model (Lu and Hamilton, 1991).

We have studied the GOES time history for several one-day periods within the lifetimes of AR7978 and AR7981 (the successor region on the second rotation), searching for both "reset" and "saturation" effects. Figure 5 shows an example from one day's GOES photometric data in this kind of analysis. No correlation shows up either way, but the analysis could be improved in several ways. In particular, one might suspect that different regions within an activity complex would show more physical linkage,

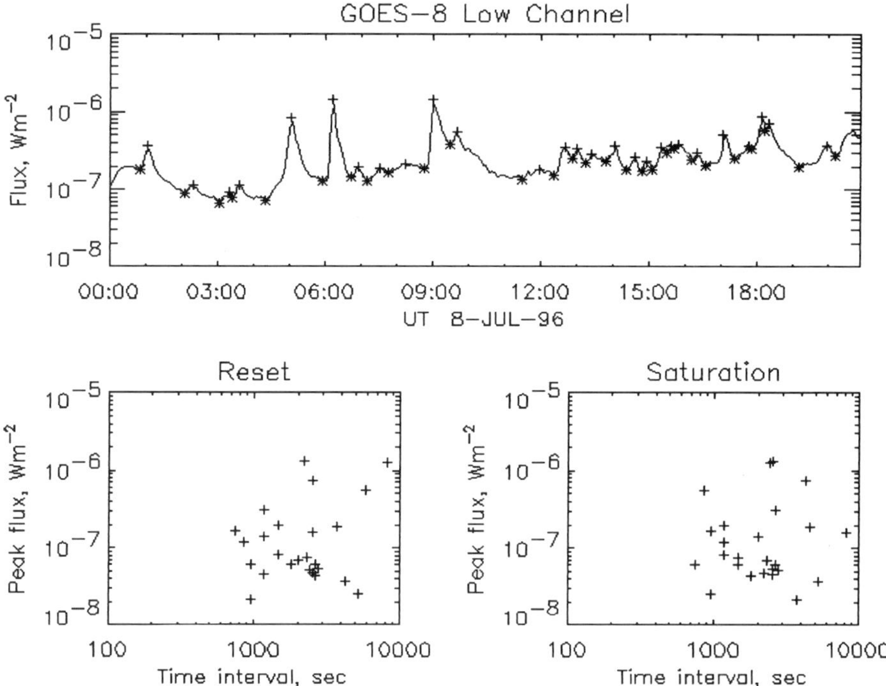

Figure 6. Typical results of a naive search for a relaxation-oscillator signature in the GOES photometry for July 8, 1996. The upper panel shows the GOES data with the locations of interactively selected peaks (*) and valleys (+). The lower panels show the correlations for the two cases ("reset" and "saturation") described in the text. Neither shows a significant correlation.

i.e. that widely separated regions might have independent flare build-up and release. If so, the screening of flares and microflares by location – using the SXT data – might improve the correlation to the point of detectability. Such improvements are beyond the scope of the work described here.

5. Conclusions

The activity complex focusing on NOAA active region 7978 was quite remarkable. It had the effect of making the Sun look lopsided in the *Yohkoh* SXT standard movie. The absence of competing activity made it easier to detect the sometimes subtle X-ray effects accompanying the launch of CMEs, a subject not discussed in this paper (see Hudson & Webb, 1997). The reconnection of the newly emerged flux and the previously existing coronal structure, with effects on the coronal hole population, is the subject of the accompanying paper by Harvey & Hudson (1997). We have mentioned several other investigations this kind of solar activity makes possible, and hope that there will be future work. These include the study of active-region effects on the solar total irradiance, with possible clues about the development of magnetism in the solar interior. Other areas of interest include morphology, connectivity, and the pattern of flare occurrence; all of these benefit from the solar-minimum conditions resulting in essentially one unique active complex, a solar activity "test particle."

Acknowledgements

NASA supported this work under contract NAS 8-37334 (Hudson). Sterling received support from the NRL/ONR basic research program. The *Yohkoh* satellite is a project of the Institute of Space and Astronautical Sciences of Japan. We thank the VIRGO team for providing data in advance of publication.

References

Fröhlich, C., and 21 co-authors, 1997, *Solar Physics* **170**, 27.
Hara, H., 1997, these proceedings.
Harvey, K., and Hudson, H. S., 1997, these proceedings.
Hudson, H. S., 1988, *Annu. Revs. Astron. Astrophys.* **26**, 473.
Hudson, H. S., and Khan, J. I., 1996, in R. D. Bentley and J. T. Mariska (eds.), *Magnetic Reconnection in the Solar Atmosphere*, ASP Conf. Proc. 111, 135.
Hudson, H. S., and Webb, D. A., 1997, to be published.
Lewin, W., Van Paradijs, J., and Van den Heuvel, E., 1995, *X-ray Binaries* (Cambridge).
Lu, E. T., and Hamilton, R. J., 1991, *Astrophys. J.* **380**, 89L.
Melrose, D. A., 1995, *Astrophys. J.* **451**, 391.
Sterling, A. C., 1997, *Astrophys. J.* **478**, 807.
Sterling, A. C., Hudson, H. S., and Watanabe, T., 1997, *Astrophys. J.* **479**, L149.
Tsuneta, S., 1996, in R. D. Bentley and J. T. Mariska (eds.), *Magnetic Reconnection in the Solar Atmosphere*, ASP Conf. Proc. 111, 409.
Wehrli, C., Appourchaux, T., Crommelynck, D., Finsterle, W., Fröhlich, C., and Pap, J., 1997, in "Sounding Solar and Stellar Interiors," F. Schmieder & J. Provost (Eds.), Proc. IAU Symposium 181, Nice 1997, in press.
Willson, R. C., Gulkis, S., Janssen, M., Hudson, H. S., and Chapman, G. A., 1981, *Science* **211**, 700.

S XV SPECTRAL PROPERTIES OF AN ACTIVE REGION FROM THE *YOHKOH* BRAGG CRYSTAL SPECTROMETER

ALPHONSE C. STERLING[1,2]
Computational Physics Inc.
2750 Prosperity Ave., Suite 600, Fairfax, VA, USA

Extended Abstract

Yohkoh's Bragg Crystal Spectrometer (BCS; Culhane et al. 1991, Lang et al. 1992) package contains four channels, covering the H- and He-like stages of iron (Fe XXVI and Fe XXV, respectively), He-like calcium (Ca XIX), and He-like sulfur (S XV). For typical solar phenomena and standard spectral accumulation times, these channels detect emissions from plasmas with electron temperatures in the ranges of about 15—37 MK for Fe XXVI (Pike et al. 1996), 12—24 MK for Fe XXV (e.g., Sterling, Doschek & Pike 1994; Feldman et al. 1995), 7—20 MK for Ca XIX (e.g., Mariska, Sakao, Bentley 1996, Sterling et al. 1997b), and 3—17 for S XV (e.g., Harra-Murnion et al. 1996, 1997). While only flares contain significant amounts of plasmas in the temperature ranges of the first three channels, the S XV channel covers temperatures of active region plasmas also. Only with this channel is it possible to routinely study non-flaring active region plasmas with the BCS.

Watanabe et al. (1992, 1995) have used the S XV channel to study the non-flaring Sun during times near solar maximum, when several active regions were on the solar disk. These studies yield useful information about the average properties of these plasmas, but because the BCS is a full-Sun instrument, they do not tell us about the spectral properties of individual active regions. Such information can only be garnered from observations at times when there is only a single active region on the solar disk. As the solar cycle declined toward the mid-1990's it became more common for isolated regions to appear on the Sun. Here we summarize our findings in S XV for

[1]Also, E. O. Hulburt Center for Space Research, Naval Research Laboratory, Washington, D. C.
[2]Current address: ISAS, Yoshinodai 3-1-1, Sagamihara, Kanagawa 229, Japan

T. Watanabe et al. (eds.), Observational Plasma Astrophysics: Five Years of Yohkoh and Beyond, 245–247.

a region, NOAA active region AR 7953, which was alone on the Sun over the period 1996 March 22–24.

We analyzed some 150 spectra over the three-day period. Integration times ranged from about 15 to 3000 seconds, with the weakest spectra having the longest integration times. Intensity light curves in soft X-rays from *GOES* spacecrafts, as well as in S XV from BCS, show many transient intensity brightenings (e.g., Shimizu 1995) over the time period. Most of these are at the *GOES* "A" and "low-B" levels, although there was also a C1 flare on March 23, and a "high-B" flaring event on March 22. From the spectra, we determine S XV fluxes, electron temperatures, excess line broadening information (described in terms of a "non-thermal velocity" parameter), and emission measures of the active region plasmas.

We examined spectra for times when flare-like activity was nearly absent, as well as during times when flare-like activity was present. Even for the non-flaring active region the flux level varies with time. It is relatively high throughout March 22 and until about 4 UT on the 23rd, and then reduced at later times. Over the higher flux-level time, the non-flaring active region plasma has an average temperature of 6.1 ± 0.8 MK, and an average non-thermal velocity of 43.5 ± 11.8 km s^{-1}. During the later, relatively reduced flux-level times, the average temperature is 5.4 ± 0.9 MK. For these quieter times, the non-thermal velocities are harder to determine reliably. This is because the line widths in BCS are affected by spatial extent of the source in the north-south direction, and the effective spatial extent of the active region is larger when the flux is lower and integration times longer. During the times of the flare-like activity, the temperatures and non-thermal velocities are increased over those of the non-flaring times, with temperatures typical of those seen in other flares in S XV.

We have investigated the correlation in time between the various spectral properties of the active region. Outside the times of flares there is a strong correlation between flux and electron temperature, between the electron temperature and the emission measure, and between flux and emission measure. The correlation between either the flux or the electron temperature with the non-thermal velocity is weak. During times of the flare-like features alone, or over the entire three-day time period (including flaring and non-flaring times) there are strong correlations in time between all these parameters, i.e., between the flux and electron temperature, flux and non-thermal velocity, electron temperature and non-thermal velocity, electron temperature and emission measure, and between flux and emission measure.

Our non-flaring non-thermal velocity values of about 45 km s^{-1} are similar to those found by Saba & Strong (1991), who observed an active region in a line formed at lower temperature (Mg XI). Also, this value is not too much different from the non-thermal velocities seen late in the decay phases of flares in Fe XXV and Ca XIX (e.g., Doschek et al. 1980; Fludra et al. 1989). And thus the late-flare excess line broadenings may be due to heated background active region plasmas, and only indirectly associated with the actual flare-producing mechanism. This would distinguish the late-phase flare broadenings from those of the rise and peak phases of the flare.

Our strong correlations between electron temperature, emission measure, and flux indicates that more involved processes are occurring in transient brightenings than a simple heating leading directly to a brightening of a portion of the active region. There must be an associated increase in the density, the volume, or both in order to drive up the emission measure also. Thus our observations restrict the nature of the heating occurring in the active region corona. A strong correlation between electron temperature, emission measure, and flux is also well known to hold in flares.

A full report of this study appears in Sterling (1997), and related work appears in Sterling, Hudson & Watanabe (1997a).

References

Culhane, J.L. et al. 1991, Solar Phys., 136, 89.
Doschek, G. A. 1990, ApJS, 73, 117.
Feldman, U., Doschek, G. A., Mariska, J. T., & Brown, C. M. 1995, ApJ, 450, 441.
Fludra, A., Lemen, J. R., Jakimiec, J., Bentley, R. D., & Sylwester, J. 1989, ApJ, 344, 991.
Harra-Murnion, L., et al. 1996, Astron. & Ap., 308, 670.
Harra-Murnion, L., Akita, K., & Watanabe, T. 1997, ApJ, 479, 464.
Lang, J. et al. 1992, PASJ, 44, L55.
Mariska, J. T., Sakao, T., & Bentley, R. D. 1996, ApJ, 459, 815.
Pike, C. D. et al. 1996, ApJ, 464, 487.
Saba, J. L. R., & Strong, K. T. 1991, ApJ, 375, 789.
Shimizu, T. 1995, PASJ, 47, 251.
Sterling, A. C., Doschek, G. A., & Pike, C. D. 1994, ApJ, 435, 898.
Sterling, A. C. 1997, ApJ, 478, 807.
Sterling, A. C., Hudson, H. S., & Watanabe, T. 1997a, ApJ, 452, L149.
Sterling, A. C., Hudson, H. S., Lemen, J. R., & Zarro, D. A. 1997b, ApJS, 110, 115.
Watanabe, T., et al. 1992, PASJ, 44, L141.
Watanabe, T., et al. 1995, Solar Phys., 157, 185.

THE RELATIONSHIP BETWEEN UV AND X-RAY BRIGHTENINGS

L.K. HARRA-MURNION AND S.A. MATTHEWS
Mullard Space Science Laboratory, University College London
Holmbury St. Mary, Surrey, RH5 6NT, UK

Extended Abstract

The relationship between structures in the transition region and corona is not well understood, and this knowledge is vital for a complete understanding of the heating of active region plasma. There have been many previous studies of active regions in the UV and X-ray using the XUV spectroheliograph onboard Skylab. Cheng et al.(1980) discuss the relationship between X-ray emission at 2–3$\times10^6$ K and UV emission at 6$\times10^5$ K of active region loops. They found that while hot coronal loops are low-lying and closely packed, the cool loops are fewer in number and large with smaller aspect ratios. Dere (1982) found that the transition region emission exists in loops or in segments of loops whereas the coronal emission is always found in loops. Transition region loops and coronal loops were found not to exist coaxially. Feldman & Laming (1994) conclude in their study that the cold and hot loops sometimes appear to occupy the same location, and also frequently the cool and hot loops emerge in different locations. They do not observe the cool plasma being part of a large structure which extends into the corona.

The Coronal Diagnostic Spectrometer on SoHO enables us to observe different temperature structures ranging from $10^4 \rightarrow > 10^6$K simultaneously. The first results of active regions observed by CDS are described by Harrison et al.(1997) and Brekke et al.(1997). The latter paper illustrates how a small difference in line formation temperature (Mg IX and Mg X) results in different loop structures.

In this extended abstract, observations from the Coronal Diagnostic Spectrometer (CDS) on SoHO and the Soft X-ray Telescope (SXT) on *Yohkoh* are described to study the correlation between 2.4$\times10^5$ K and 1-3$\times10^6$ K plasma during a small brightening observed in active region NOAA 7981. The small brightening rose above the quiescent GOES 1–8 Å level of A7 to A9 over a period of $\approx$ one hour. We investigate the morphological

T. Watanabe et al. (eds.), Observational Plasma Astrophysics: Five Years of Yohkoh and Beyond, 249–250.

relationship between the transition region plasma at 2.4×10^5 K to the hot coronal plasma > 1MK.

Our observations show clear loop structures visible in O V(2.2×10^5 K), rather than loop leg or footpoint emission, which corresponds to loops visible in the higher temperature coronal lines (e.g. Fe XIV at 1.8×10^6 K and Fe XVI at 2.2×10^6 K). We note the existence of co-spatial emission in SXT-temperature plasmas (3 MK), as well as in regions where Fe XIV and O V lines are formed. The cool O V loop brightens before the SXR loop, excluding the possibility of rapid cooling of the coronal loop as an explanation for its existence. Also apparent is the absence of cooler transition region emission in a separate soft X-ray loop structure which brightens repeatedly over the space of a day. In the Fe XVI image, we observe the presence of diffuse emission coincident with the SXT brightening. Corresponding emission in Fe XIV is absent. This implies the presence of a very definite temperature cut-off, with plasma of differing temperatures being confined to separate structures. This suggests that the plasma is being continually heated, and hence unable to cool down to temperatures lower than 1.8×10^6 K (Fe XIV). Both of these results are consistent with the findings of Feldman & Laming(1994) in which they find that hot and cold loops often occupy the same spatial location (we find a loop structure which has cospatial O V and SXT emission), whilst in other cases hot and cold loops can be seen to emerge from different locations (in another SXT loop structure there is no evidence of transition region emission). For this active region the observations seem to provide no evidence to support the idea of continuous loop structures extending through the upper solar atmosphere. There also does not appear to be a connection between higher temperature coronal plasmas and the lower temperature transition region structures. A more detailed, high time resolution study of a similar region at the solar limb, would provide us with the opportunity to investigate further the possibility that UV brightenings contribute to active region heating and to obtain accurate estimations of the altitudes of different temperature loops.

References

Brekke, P., Kjeldseth-Moe, O., Brynildsen, N., Maltby, P., Haugan, S.V.H., Harrison, R.A., Thompson, W.T., and Pike, C.D. (1997), *Solar Phys.*, 179,163.

Cheng, C-C., Smith, J.B., and Tandberg-Hanssen, E. (1980), *Solar. Phys.*, **67**, pp. 259–265

Dere, K.P. (1980) *Solar. Phys.*, **75**, pp. 189–203

Feldman, U., and Laming, J.M. (1994), *Astrophys. J.*, **434**, pp. 370–377

Harrison, R.A. *et al.* (1997), *Solar Phys.* in press

JOINT OBSERVATIONS OF AN ACTIVE REGION WITH SOHO AND YOHKOH

R.D.BENTLEY
Mullard Space Science Laboratory
Holmbury St. Mary, Dorking, Surrey RH5 6NT, UK

1. Introduction

The Solar and Heliospheric Observatory (SoHO) was launched in December 1995 and because fully operational in April 1996. As well as making studies of helio-seismology, and in-situ measurements of the solar wind, it carries a compliment of instruments designed to observe the quiet Sun. The spacecraft is in a halo orbit around the Sun-Earth L1 Lagrangian point, and makes continuous observations of the Sun.

SoHO has undertaken a number of programmes of collaborative observations with ground-based observatories and other spacecraft. Here we present observations made between August 23 and 30, 1996. These involved the CDS (Harrison *et al.*, 1995), SUMER (Wilhelm *et al.*, 1995) and EIT (Delaboudinière *et al.*, 1995) instruments on SoHO, the SXT instrument (Tsuneta *et al.*, 1991) on the Japanese Yohkoh mission, and the Swedish telescope on La Palma. A detailed analysis has not yet been performed on the data and thus this report is therefore very much a *"show and tell"* of what was seen.

The observations were made as targets of opportunity since this was the first time an equatorial coronal hole had been observed by SoHO and the active region was perhaps at its most extensive over the several rotations that it had been observed (Slater, 1997). They were made around other programmes that were being executed by SoHO, but an attempt was made to optimize them with those of the other observatories.

2. Observations

The SoHO mission makes continuous observations of the Sun. Images made by the Extreme Ultraviolet Imaging Telescope (EIT) provide the context

T. Watanabe et al. (eds.), Observational Plasma Astrophysics: Five Years of Yohkoh and Beyond, 251–255.

within which other observations can placed. At the time of the observations discussed here, restrictions in the bandwidth allocated to the EIT instrument limited the usefulness of these images (although this difficulty has now been resolved). Consequently it was necessary to make maps with the Coronal Diagnostic Spectrometer (CDS) instrument to provide the context for the higher cadence observations by the same instrument. These programmes take a long time to execute (40 minutes), but since the regions under study were evolving slowly this was not a problem.

During the week these observations were made, the SoHO mission was also following programme of "Global Sun" measurements. These committed the SoHO instruments for large parts of the day, but the residual time was concentrated to coincide with best time of day for observations by the Swedish Telescope on La Palma (0800-1300 UT). The CDS and SUMER planners worked with the SoHO Science Operations Leader, who was also associated with the team in La Palma, to select target that the ground-based observers would also study the next day.

As the region rotated around the disk and became more extended, studies would concentrate on particular parts of the active region: the spot, the active region filament the equatorial coronal hole, and the extended filament that stretched out ahead and behind of the active region. When observing the sun spot, the Michelson Doppler Imager (MDI) was used to provide pointing coordinates. The spot was particularly well developed and the CDS was able to perform a number of detailed studies of it. Luckily the active region, although extensive, was not flaring. If it had been, it would have been impossible to make observations with SUMER, and very difficult with CDS. Even so, some hotter parts of the region were avoided by the SUMER instrument.

In addition to the full disk images made several times each orbit, the Yohkoh Soft X-ray Telescope (SXT) made observations that were centered on the active region. Around three orbits correspond to the optimal observing interval, although Yohkoh is unable to make observations during orbital night.

3. Discussion

A comparison between the images taken by the Yohkoh SXT and SoHO EIT instruments (see Figure 1) highlights the differences in the nature of the two sets of observations. The EIT images are formed using *normal incidence* optics that employ multi-layers to select the desired (narrow) waveband. The SXT uses *grazing incidence* optics and filters; it makes broad-band images with the filters allowing some measure of temperature to be made. The SXT instrument always sees hotter plasma than the EIT and this

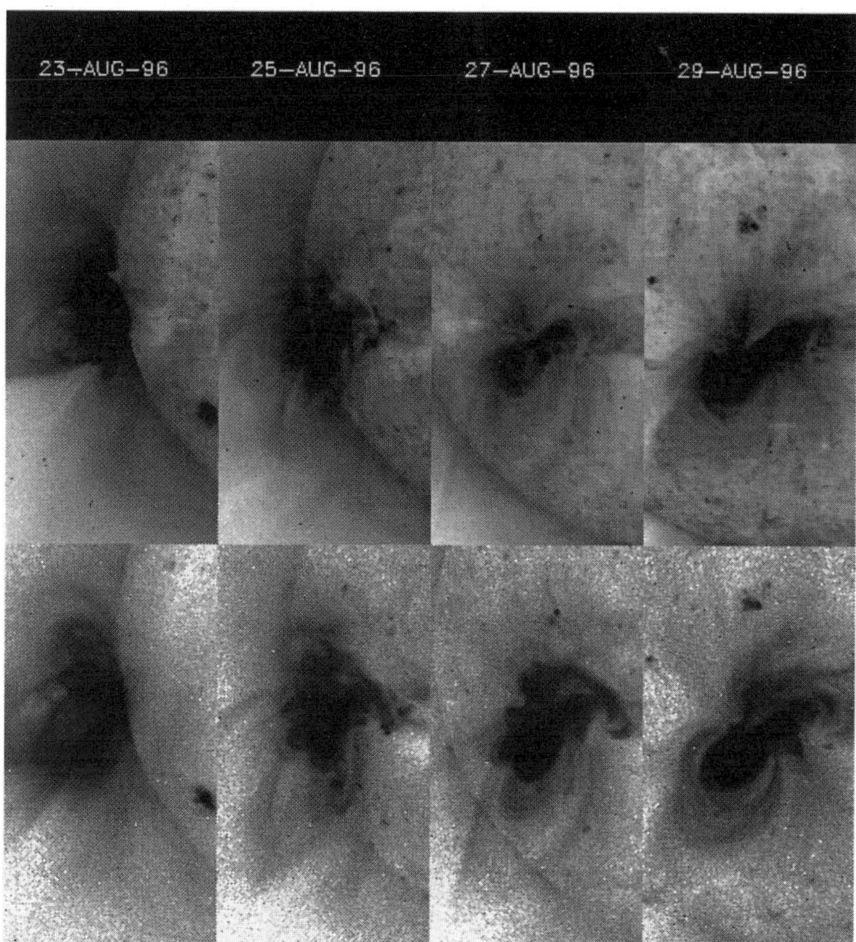

Figure 1. The active region is shown rotating over the limb in SoHO EIT 284 Å channel (top) and the Yohkoh SXT AlMg filter (bottom). The panels show the apperance every two days: 23 August 1996, 25 August 1996, 27 August 1996 and 29 August 1996 (left to right). Note the "V-shape" in the image taken by the EIT on 23 August 1996 (top left) which marks the boundaries of open field lines of the equatorial coronal hole as it extends into the corona; the hole can be seen a lighter region (in this inverted colour table) in the same image extending westwards and northwards from where the "V" points to the solar limb.

sometimes gives the SXT images the appearance of being less distinct that the EIT images. EIT sees the lower parts of loops very well in some lines, and its hotter lines start to show similar things to the SXT.

Observation of the equatorial coronal hole and active region as they cross the limb show a marked difference in appearance between the EIT

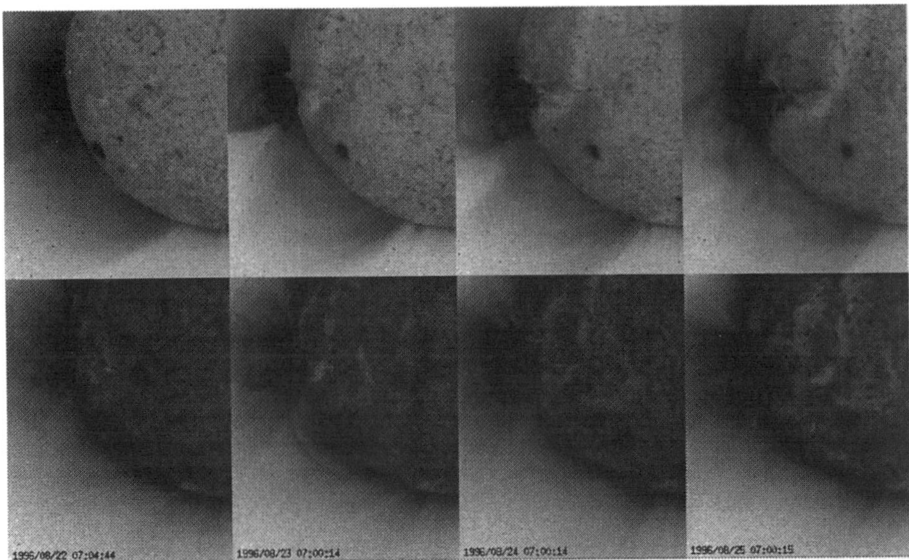

Figure 2. Another view of the equatorial coronal hole rotating over the limb is given by these images from the SoHO EIT. The *top panel* shows observations made in the 284 Å channel, and the *bottom panel* those in the 195 Å channel. The images were taken on successive days: 22 August 1996, 23 August 1996, 24 August 1996, 25 August 1996. The coronal hole can be seen most clearly above the limb on August 23 and 24 in the 284 Å channel (top). In the 195 Å channel, loops obscure the boundaries of the hole. In this reverse colour table, the coronal hole can be seen as a lighter feature extending westwards and northwards from just below the active region loops.

and SXT instruments, and between channels within the EIT instrument. In Figure 1, the image formed in the EIT represents cooler plasma than that made by the SXT and many small bright features are seen in the EIT image that are absent from the SXT image. Also, loops seen in the two instruments are not co-spatial in the images of August 27 and 29. In the 284 Å line of EIT (peak temperature 2.0 MK) the open nature of the field in the coronal hole is clearly visible with a large "V-shape" visible on the limb (see Figures 1 and 2). However, in SXT and the 195 Å EIT channel (peak temperature 1.6 MK) this structure is not early so clear as material at greater altitudes obscures the structure. Because of the complex nature of the band-pass in the EIT channels, it is not possible to make a clear estimate of the temperature in the same way that filter rations can be used in the SXT. Also, the nature of the normal incidence optics tends to compress that range of intensities and the instrument is less prone to saturation than the SXT is.

The CDS is able to produce maps that consists of scans of selected wavelength ranges at each point in a raster. This allows the temperature, density and velocity structure of features to be determined. Some features are more visible at one wavelength compared to others. Filaments can be seen in CDS lines, particularly when they become active. The equatorial coronal hole was more distinct, although also more ragged in Mg X (624.94 Å) than in Si XII (520.67 Å) since the Mg X picks out the lower layers of the atmosphere where the coronal hole was rooted.

4. Conclusions

Combining observations from SoHO and other sources provides and interesting insight into what is happening in the solar atmosphere. It is intended that these observations will be the subject of further work and other ground-base observations are being sought in the context. This process would be simpler if the "Whole Sun Catalogue" (Sánchez Duarte *et al.*, 1997) was in place. This is designed to contain observations from all site around the world and should mean that complementary observations are always easy to locate.

References

Delaboudinière, J.-P., Artzner, G.E., Brunaud, J., Gabriel, A.H., Hochedez, J.F., Millier, F., Song, X.Y., Au, B., Dere, K.P., Howard, R., Krepli, R., Michels, D.J., Moses, J.D., Defise, J.M., Jamar, C., Rochus, P., Chauvineau, J.P., Marioge, J.P., Catura, R.C., Lemen, J.R., Shing, L., Stern, R.A., Gurman, J.B., Neupert, W.M., Maucherat, A., Clette, F., Cugnon, P., and Van Dessel, E.L. (1995), EIT: Extreme-Ultraviolet Imaging Telescope for the SOHO Mission, *Sol. Phys.* **162**, 291–312

Harrison, R.A., Sawyer, E.C., Carter, M.K., Cruise, A.M., Cutler, R.M., Fludra, A., Hayes, R.W., Kent, B.J., Lang, J., Parker, D.J., Payne, J., Pike, C.D., Peskett, S.C., Richards, A.G., Culhane, J.L., Norman, K., Breeveld, A.A., Breeveld, E.R., Al Janabi, K.F., McCalden, A.J., Parkinson, J.H., Self, D.G., Thomas, P.D., Poland, A.I., Thomas, R.J., Thompson, W.T., Kjeldseth-Moe, O., Brekke, P., Karud, J., Maltby, P., Achenbach, B., Bräuninger, H., Kühne, M., Hollandt, J., Siegmund, O.H.W., Huber, M.C.E., Gabriel, A.H., Mason, H.E., and Bromage, B.J.I. (1991), CDS: The Coronal Diagnostic Spectrometer for SOHO, *Sol. Phys.* **162**, 233–290

Sánchez Duarte, L., Fleck, B., and Bentley, R.D. (1997), The Whole Sun Catalogue, Proceedings of the ASPE meeting held in Teneriffe in October 1996, in press.

Scherrer, P.H., Bogart, R.S., Bush, R.I., Hoeksema, J.T., Kosovichev, A.G., Schou, J., Rosenberg, W., Springer, L., Tarbell, T.D., Title, A., Wolfson, C.J., Zayer, I., and the MDI Enginerring Team (1995), The Solar Oscillations Investigation - Michelson Doppler Imager, *Sol. Phys.* **162**, 129–188

Slater, G.L., ???? (1997), this volume.

Tsuneta, S., Acton, L., Bruner, M., Lemen, J., Brown, W., Caravalho, R., Catura, R., Freeland, S., Jurcevich, B., Morrison, M, Ogawara, Y., Hirayaman, T., and Owens, J. (1991), *Sol. Phys.* **136**, 37–68

Welheim, K., Curdt, W., Marsch, E., Schühle, U., Lemaire, P., Gabriel, A.H., Vial, J.-C., Grewing, M., Huber, M.C.E., Jordan, S.D., Poland, A.I., Thomas, R.J., Kühne, M., Timothy, J.G., Hassler, D.M., and Siegmund, O.H.W (1995), SUMER - Solar Ultraviolet Measurements of Emitted Radiation, *Sol. Phys.* **162**, 189–231

IV. High-Energy Particle Acceleration in Cosmic Plasmas

CORONAL HARD X-RAY SOURCES IN SOLAR FLARES OBSERVED WITH YOHKOH/HXT

S. MASUDA
STEL, Nagoya University
3-13 Honohara, Toyokawa, Aich 442, Japan

AND

T. KOSUGI, T. SAKAO AND J. SATO
National Astronomical Observatory
2-21-1 Osawa, Mitaka, Tokyo 181, Japan

Abstract.

In solar flares the *Yohkoh* Hard X-ray Telescope (HXT) has revealed, in addition to the most prominent double-footpoint sources, several types of coronal hard X-ray sources. (1) A compact source that appears only during the impulsive phase. It is located above the soft X-ray loop and has a relatively hard spectrum. This strongly suggests that magnetic reconnection takes place above (outside of) the bright soft X-ray loop. It is maybe that the reconnection outflow impinges upon the underlying loop, forms a shock, and energized electrons resulting in a hard X-ray source above the loop. (2) A gradual source that begins in the impulsive phase but becomes dominant later in the gradual phase with a characteristic time scale of 5–10 min. It is located near the soft X-ray loop apex and is characterized by a very soft spectrum, which is interpreted as originating from an isothermal plasma with the electron temperatures of 30–40 MK. The plasma seems to be directly heated *in situ* to those super-hot temperatures. (3) Another gradual source seen in the so-called super-hot thermal flares (Type A flares). This source is quite similar to the gradual source mentioned above, but appears with quite weak, if any, impulsive signatures. (4) A coronal source seen in Long Duration Event (LDE) flares. LDE flares usually show a clear cusp-shaped loop structure in soft X-rays, in which a soft-spectrum, relatively large-sized (1–2 arcmin), long-lasting (typically 30 min) hard X-ray source appears. It is located below the soft X-ray cusp but slightly above the densest portion of the bright soft X-ray loop.

T. Watanabe et al. (eds.), Observational Plasma Astrophysics: Five Years of Yohkoh and Beyond, 259–267.

1. INTRODUCTION

Hard X-rays from solar flares, especially those above $\sim$ 30 keV and seen in the impulsive phase, are mostly emitted from double footpoint sources. The two sources are located low in the corona near the chromosphere and vary almost synchronously with each other. This indicates that the double-footpoint structure is fundamental, in the sense that they represent the magnetically-conjugate footpoints of a flaring loop along which high-energy electrons precipitate down and radiate thick target Bremsstrahlung emission (Sakao 1994; Sakao *et al.* 1994). This is one of the most important findings from the Hard X-ray Telescope (HXT; Kosugi *et al.* 1991) onboard *Yohkoh* (Ogawara *et al.* 1991).

Where are these electrons energized? The *Yohkoh* HXT also provides a hint to solve this question by revealing several types of hard X-ray sources at higher altitudes in the solar corona. In this paper, we focus on these coronal hard X-ray sources. Four types of coronal hard X-ray sources are reviewed individually in Sections 2–5, and a summary is given in Section 6.

2. ABOVE-THE-LOOPTOP IMPULSIVE SOURCES

The presence of an above-the-looptop impulsive hard X-ray source was first unveiled with the *Yohkoh* HXT (Masuda 1994*a*; Masuda *et al.* 1994, 1995). In this section, we first summarize characteristics of this source, taking as an example one of the canonical events, the 13 January 1992 flare, and then discuss their implications.

- The above-the-looptop source appears together with the double footpoint sources in the impulsive phase.
- It is located well above (by more than $\sim$ 10 arcsec or twice the HXT resolution) the corresponding soft X-ray loop apex. Figure 1 compares a hard X-ray image (contours) with a soft X-ray image (grey scale) taken with the *Yohkoh* Soft X-ray Telescope (SXT; Tsuneta *et al.* 1991). The solid line indicates the west solar limb. The hard X-ray source is clearly seen above the corresponding soft X-ray loop.
- The temporal behavior of this source is almost the same as that of the footpoint sources (within the time resolution of about 10 seconds).
- The hard X-ray spectrum of this source is as hard as that of the footpoint sources; the power-law spectral index is $\sim$ 4 (on the basis of a single power-law spectrum interpretation) or the electron temperature is $1 - 2 \times 10^8$ K (on the basis of a thermal emission interpretation from an isothermal plasma).
- This source is less intense than the footpoint sources by a factor of five. [Note, however, that the above-the-looptop source of the 13 January

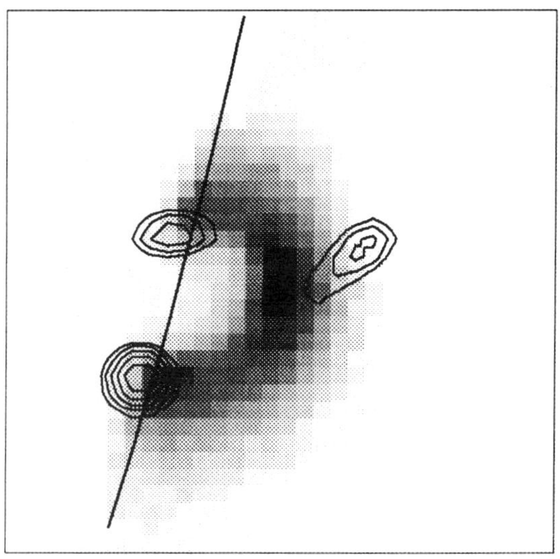

Figure 1. Hard X-ray (contours) and soft X-ray (grey scale) images of the 13 January 1992 flare that occurred near the west solar limb. The hard X-ray image was synthesized in the M2-band (33–53 keV) with photon counts accumulated over 17:26:52 – 17:27:39 UT (the rise phase). Contour levels are 70.7, 50.0, 35.4, 25.0, and 17.7 % of the peak intensity. The soft X-ray image was taken with the Be filter at 17:27:01 UT. The field of view covers 78.4 × 78.4 arcsec^2 (5.7 × 10^4 km^2. Solar north is to the top, west to the right. The solar limb is shown by a solid line. The accuracy of coalignment is estimated to be within ~ 1 arcsec.

1992 flare is the most intense (in the relative sense with respect to the footpoint sources) among the flares which have this type of source. See, *e.g.*, Wang *et al.* (1995).]

- This source occupies a small portion of the high-temperature (>20MK) region seen in soft X-rays. This high-temperature region evolves into a two-ridge structure, where the temperature peaks at the two ridges, and the above-the-looptop hard X-ray source is located in between the highest-temperature ridges (Tsuneta *et al.* 1997).
- This source is located at slightly higher altitudes at higher X-ray energies (not shown here; see Figure 1 of Masuda *et al.* 1994).
- This source is compact in size (~ 7000km × 7000 km in the M2-band), and does not show any significant change during the impulsive phase.

These observations as a whole are best interpreted in the scheme of magnetic reconnection. The current sheet where anti-parallel magnetic fields merge is located even higher above the photosphere than the above-the-looptop hard X-ray source. The flare energy is primarily released there and particles are energized. One of the reconnection outflows, the downward-directed one, impinges upon the underlying loop and forms a shock. This process further energizes electrons in the region just above the soft X-ray loop, resulting in the above-the-looptop hard X-ray source (See, *e.g.*, Somov and Kosugi 1997).

There is additional circumstantial evidence that support the above scenario. Shibata *et al.* (1995) reported a hot-plasma ejection seen very faintly in soft X-ray images during the impulsive phase of the 13 January 1992 flare (as well as in some other events), which is suggestive of the formation of a current sheet and further of the presence of upward reconnection outflow. Aschwanden *et al.* (1996*a*) found delays of low-energy hard X-rays with respect to higher-energy X-rays in the very sensitive CGRO/BATSE data for the 13 January 1992 flare (as well as for many other events; see Aschwanden *et al.* 1996*b*,*c*), and discussed the time-of-flight distance of electrons to the double footpoint sources. Their conclusion was that the most plausible site for electron acceleration is inside the above-the-looptop hard X-ray source.

The number of flares, however, in which the presence of an above-the-looptop source is confirmed is still small, say less than half-a-dozen. Furthermore, the events so far detected are not intense enough for a detailed time-series analysis. Thus there is still much left to be studied.

3. LOOPTOP SOURCES DURING THE GRADUAL PHASE

A different type of coronal hard X-ray source appears late in the impulsive phase and eventually becomes dominant towards the gradual phase. Figure 2 shows as an example hard and soft X-ray images of the 4 October 1992 flare taken during the gradual phase. In the M1-band image (right panel), a footpoint source (low-altitude source) is as intense as the looptop coronal source even in this phase. On the other hand, no footpoint source is seen in the L-band image (left panel). Characteristics of the looptop source in the gradual phase can be summarized as follows (Masuda 1994*a*,*b*):

- The brightest portion of this source is located at the apex of the soft X-ray flaring loop, with its centroid being at a slightly higher altitude than the brightest region in soft X-rays.
- This source is larger and more diffuse than the above-the-looptop impulsive source. The shape sometimes seems to trace the soft X-ray loop.
- This source dominates in the X-ray energy range < 30 keV throughout the gradual phase. In the HXT L-band (14–23 keV), other sources such as footpoint sources are rarely seen (*cf.* Figure 2).
- The spectrum of this source is quite soft. It can be well explained as originating from an isothermal plasma of ~ 30 MK.
- This type of source is seen in each of the ten events selected in an unbiased manner and analyzed by Masuda (1994*a*).

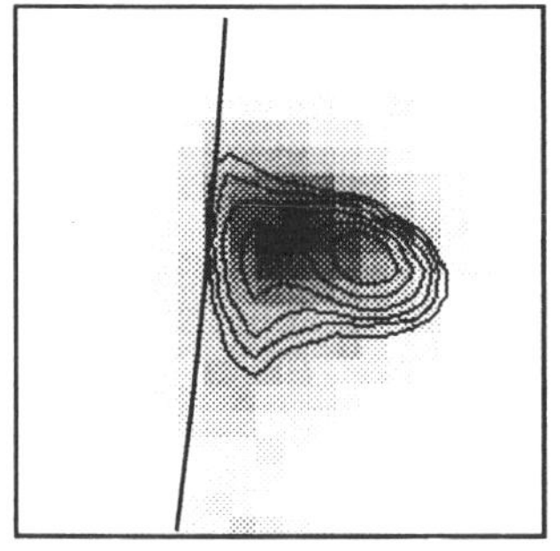
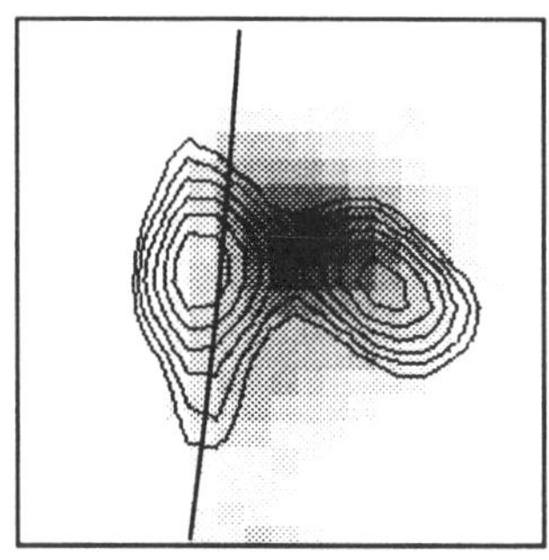

Figure 2. Hard and soft X-ray images of the 4 October 1992 flare during the gradual phase. The flare occurred near the west solar limb. The left and right panels show HXT L-band (22:20:35–22:20:43 UT) and M1-band (22:20:15–22:20:43 UT) images (contours), respectively, superposed upon a grey-scale SXT Be-filter image from 22:20:33 UT. Contour levels are 70.7, 50.0, 35.4, 25.0, 17.7, and 12.5 % of the peak intensity. The field of view covers 49 $\times$ 49 arcsec^2. Solar north is to the top, west to the right. The solar limb is shown by a solid line.

The timing of appearance and the hard X-ray spectrum of this source are the same as those of Lin *et al.*'s (1981) "new component", which was detected with a balloon-borne hard X-ray spectrometer with a high energy resolution and a very low background noise. Thus there is no doubt that the looptop gradual source is the spatially-resolved Lin *et al.*'s source.

The looptop source is a super-hot, thermal ($T \sim 30$ MK) plasma. Since it is the brightest at the apex and since it does not extend to the footpoints (although it appears to trace the loop), it is unlikely that chromospheric evaporation is the cause of this source. Rather the plasma seems to be directly heated *in situ* at the location of the source. If so, the looptop source in the gradual phase may also reflect the energy-release process as the above-the-looptop impulsive source does, though the rate of energy-release is much smaller than the latter.

4. CORONAL SOURCES IN SUPER-HOT FLARES

Sometimes, but only rarely, we observe hard X-ray flares without any distinct impulsive-phase signatures. This type of flare was first found by the *Hinotori* satellite and named Type A flares, or "super-hot thermal flares" (see, *e.g.*, Tanaka 1987). The *Yohkoh* HXT observed such a flare on 6 February 1992 for the first time with sufficient spatial and temporal resolution (Kosugi *et al.* 1994). The following gives a summary.

- The most outstanding is its soft spectrum in the hard X-ray energy range. The peak counts in the HXT L-, M1-, and M2 bands are 807,

144, and 11 counts/s/subcollimator, respectively, which is unusually soft for an impulsive-phase spectrum, but is well explained by thermal emission from a plasma with $T \sim 30$ MK.

- Hard X-rays in the HXT higher energy bands is quite weak in comparison with the flare's intense soft X-rays; the GOES X-ray class of this flare, M7.6, is more intense than those of the events discussed in the previous sections, *e.g.*, M2.0 for the 13 January 1992 flare and M2.4 for the 4 October 1992 flare, while the hard X-ray counts in the M2-band is much less than the latter two.
- The hard X-ray source first appears near the apex of a soft X-ray flaring loop and then gradually expands downward. At the same time, a bright soft X-ray source first appears at the northern footpoint of the loop and gradually expanded upwards (Kosugi *et al.* 1994).
- The source shows a gradual time behavior without spikes (except for weak spikes that are seen in the very initial phase and originate from a different flare loop).

Most of these characteristics are similar to those of the looptop gradual source discussed in the previous section. We have no reason to classify this source as different from the looptop gradual source. We note, however, that the third item above could be easily obtained because this type of flare allows us to see the looptop gradual source from the very beginning since it is not contaminated by the intense footpoint emission of the impulsive phase. More interesting is the fact that the Type A source does not appear in the same loop as the impulsive source; as mentioned in the fourth item above, one flaring loop corresponds to the type A source and a different loop corresponds to the impulsive source. This suggests that the hard X-ray emission in a loop, or more generally the particle acceleration there, is strongly controlled by the plasma conditions in that loop. A further study is needed in order to verify this.

5. CORONAL SOURCES IN LONG DURATION EVENTS

Long Duration Events (LDEs) are flares whose soft X-ray duration exceeds, say, one hour. The *Yohkoh* SXT has revealed that LDEs are large-sized, cusp-shaped loop brightenings in soft X-rays. The first clear example was reported for the 21 February 1992 flare by Tsuneta *et al.* (1992) (see also, *e.g.*, Tsuneta 1996). The hard X-ray intensity of that flare (as well as of most of the LDEs so far observed by *Yohkoh*) was weak, so that only a few images taken in the HXT L-band (14–23 keV) are available in hard X-rays. Hard X-ray images in the higher energy bands have been synthesized for a limited number of LDEs. Hence, the following is just a preliminary summary, derived from a limited number of LDEs. A detailed study is in progress (Sato, in preparation).

- Hard X-rays in the HXT L-band are emitted mainly from an extended source whose size corresponds to the arcade of loops seen in soft X-rays.
- The brightest core of the hard X-ray source is located below the soft X-ray cusp, but at a slightly higher altitude than the brightest portion of the soft X-ray loop.
- Usually the hard X-ray spectrum is quite soft. We see almost nothing in the higher energy range > 30 keV.

The hard X-ray source in LDEs resembles the looptop source in the gradual phase, as summarized above. There are, however, some differences. The LDE hard X-ray source lasts longer (typically 30 min or more) and is larger. Further study is needed to see whether these differences are due to the large number of flaring loops, seen as an arcade of loops.

6. CONCLUSION

The *Yohkoh* HXT has revealed a compact source in the impulsive phase of flares, which is located well above the soft X-ray flaring loop. It shows a relatively hard spectrum (as hard as the footpoint sources). Also it varies in a similar fashion as the footpoint sources (so far as the effective time resolution of ~ 10 s is concerned). As a whole, these observations strongly suggest that the magnetic reconnection takes place above (outside of) the bright soft X-ray loop. It is possible that outflow from the reconnection results in a shock just above the soft X-ray loop, leading to the formation of the above-the-looptop source, as described above.

A different type of coronal source predominates during the gradual phase. It is characterized by a very soft spectrum, which is well interpreted as originating from an isothermal plasma of $T \sim 30 - 40$ MK. Sometimes, especially in the HXT L-band (14–23 keV), this component is seen even during the impulsive phase; in such cases the source appears first as a loop-shaped source, similar to the soft X-ray loop, and evolves into a compact source at the apex of the soft X-ray loop. Since this component is brightest at the apex and its extension along the loop does not reach the footpoints, it is difficult to interpret it as resulting from chromospheric evaporation. Rather, we speculate that the plasma is directly heated *in situ* at the location of the hard X-ray source.

Two other types of coronal sources, namely the looptop source in superhot thermal flares and the coronal source in LDEs, exhibit similar characteristics to the coronal source in the gradual phase. Each of these source is of thermal origin from a $T \sim 30 - 40$ MK plasma, located near the soft X-ray loop apex, and vary only gradually.

In conclusion, a majority of flares show some type of coronal hard X-ray source. These coronal sources may give us clues to the site(s) and mechanism(s) of the primary energy-release of flares. Further studies of these sources, analyzing already existing data in more detail as well as using the *Yohkoh* HXT in the coming solar maximum around 2000, are highly desirable.

References

Aschwanden, M. J., Hudson, H., Kosugi, T., and Schwartz, R. A. (1996*a*) Electron Time-of-Flight Measurements during the Masuda Flare, 1992 January 13, *Astrophys. J.*, **464**, pp. 985–998

Aschwanden, M. J., Wills, M. J., Hudson, H. S., Kosugi, T., and Schwartz, R. A. (1996*b*) Electron Time-of-Flight Distances and Flare Loop Geometries Compared from CGRO and Yohkoh Observations, *Astrophys. J.*, **468**, pp. 398–417

Aschwanden, M. J., Kosugi, T., Hudson, H. S., Wills, M. J., and Schwartz, R. A. (1996*c*) The Scaling Law between Electron Time-of-Flight Distances and Loop Lengths in Solar Flares, *Astrophys. J.*, **470**, pp. 1198–1217

Kosugi, T., Makishima, K., Murakami, T., Sakao, T., Dotani, T., Inda, M., Kai, K., Masuda, S., Nakajima, H., Ogawara, Y., Sawa, M., and Shibasaki, K. (1991) The Hard X-ray Telescope (HXT) for the Solar-A Mission, *Solar Phys.*, **136**, pp. 17–36

Kosugi, T., Sakao, T., Masuda, S., Hara, H., Shimizu, T., and Hudson, H. S. (1994) Hard and Soft X-ray Observations of a Super-hot Thermal Flare of 6 February, 1992, in S. Enome and T. Hirayama (eds.), "New Look at the Sun with Emphasis on Advanced Observations of Coronal Dynamics and Flares", pp. 127–129

Lin, R. P., Schwartz, R. A., Pelling, R. M., and Hurley, K. C. (1981) A New Component of Hard X-rays in Solar Flares, *Astrophys. J. Lett.*, **251**, pp. L109–L114

Masuda, S. (1994*a*) Hard X-ray Sources and the Primary Energy Release Site in Solar Flares, Ph.D. Thesis, Univ. of Tokyo.

Masuda, S. (1994*b*) Vertical Structure of Hard X-ray Sources in Solar Flares, in S. Enome and T. Hirayama (eds.), "New Look at the Sun with Emphasis on Advanced Observations of Coronal Dynamics and Flares", pp. 209–212

Masuda, S., Kosugi T., Hara H., Tsuneta S., and Ogawara Y. (1994) A Loop-Top Hard X-ray Source in a Compact Solar Flare as Evidence for Magnetic Reconnection, *Nature*, **371**, pp. 495–497

Masuda, S., Kosugi T., Hara H., Sakao T., Shibata K., and Tsuneta S.

(1995) Hard X-ray Sources and the Primary Energy-Release Site in Solar Flares, *Publ. Astron. Soc. Japan*, **47**, pp. 677–689

Ogawara, Y., Takano, T., Kato, T., Kosugi, T., Tsuneta, S., Watanabe, T., Kondo, I., and Uchida, Y. (1991) The Solar-A Mission: An Overview, *Solar Phys.*, **136**, pp. 1–16

Sakao, T. (1994) Characteristics of Solar Flare Hard X-ray Sources as Revealed with the Hard X-ray Telescope aboard the Yohkoh Satellite, Ph.D. thesis, Univ. of Tokyo.

Sakao, T., Kosugi, T., Masuda, S., Yaji, K., Inda-Koide, M., and Makishima, K. (1994) Hard X-ray Imaging Observations of Footpoint Sources in Impulsive Flares, in S. Enome and T. Hirayama (eds.), "New Look at the Sun with Emphasis on Advanced Observations of Coronal Dynamics and Flares", pp. 169–172

Shibata, K., Masuda, S., Hara, H., Yokoyama, T., Tsuneta, S., Kosugi, T., and Ogawara, Y. (1995) Hot-Plasma Ejections Associated with Compact-Loop Solar Flares, *Astrophys. J.*, **451**, L83-L85

Somov, B. V. and Kosugi, T. (1997) Collisionless Reconnection and High-Energy Particle Acceleration in Solar Flares, *Astrophys. J.*, in press

Tanaka, K. (1987) Impact of X-Ray Observations from the Hinotori Satellite on Solar Flare Research, *Publ. Astron. Soc. Japan*, **39**, pp. 1–45

Tsuneta, S. (1996) Structure and Dynamics of Magnetic Reconnection in a Solar Flare, *Astrophys. J.*, **456**, pp. 840–849

Tsuneta, S., Acton, L., Bruner, M., Lemen, J., Brown, W., Caravalho, R., Catura, R., Freeland, S., Jurcevich, B., Morrison, M., Ogawara, Y., Hirayama, T., Owens, J. (1991) The Soft X-ray Telescope for the Solar-A Mission, *Solar Phys.*, **136**, pp. 37–67

Tsuneta, S., Hara, H., Shimizu, T., Acton, L.W., Strong, K.T., Hudson, H.S., and Ogawara, Y. (1992) Observation of a Solar Flare at the Limb with the Yohkoh Soft X-ray Telescope, *Publ. Astron. Soc. Japan*, **44**, pp. L63–L69.

Tsuneta, S., Masuda, S., Kosugi, T., and Sato, J. (1997) Hot and Super-Hot Plasmas above an Impulsive-Flare Loop, *Astrophys. J.*, **478**, pp. 787–798

Wang, H., Gary, D. E., Zirin, H. Kosugi, T., Schwartz, R. A., and Linford, G., (1995) The Microwave and Hα sources of the 1992 January 13 Flare, *Astrophys. J. Lett.*, **444**, pp. L115–L118

HARD X-RAY EMISSION FROM A MIRROR TRAP AT THE TOP OF RECONNECTING LOOPS

P.C.H.MARTENS
ESA/SSD, at NASA/GSFC, Greenbelt, MD 20771, USA
AND
L.FLETCHER
ESA/SSD, ESTeC, 2200 AG Noordwijk, The Netherlands

1. Introduction

We present a model for the generation of hard X-ray emission at the top of flaring loops. We find that non-thermal particles in a Syrovatskii current-sheet geometry can be trapped near the current sheet by the convergent magnetic field. There they produce looptop hard X-ray emission in observable quantities, for reasonable loop parameters. We suggest that the model may apply to the hard X-ray looptop sources observed by Masuda *et al.* (1994) with HXT instrument onboard Yohkoh.

This contribution is an extended abstract of an article to be published in full at a later date (Fletcher and Martens, 1997)

2. The Model

We consider a 'standard' two-ribbon flare geometry, having a current sheet and magnetic bottle at the top of a post-flare loop. The current sheet/magnetic field geometry is from Syrovatskii (1971). Particles accelerated in the current sheet in the centre of the geometry are transported along the field lines away from the acceleration region, but a proportion of them are trapped by the magnetic bottle and generate HXR bremsstrahlung emission in the coronal portion of the loop.

T. Watanabe et al. (eds.), Observational Plasma Astrophysics: Five Years of Yohkoh and Beyond, 269–271.

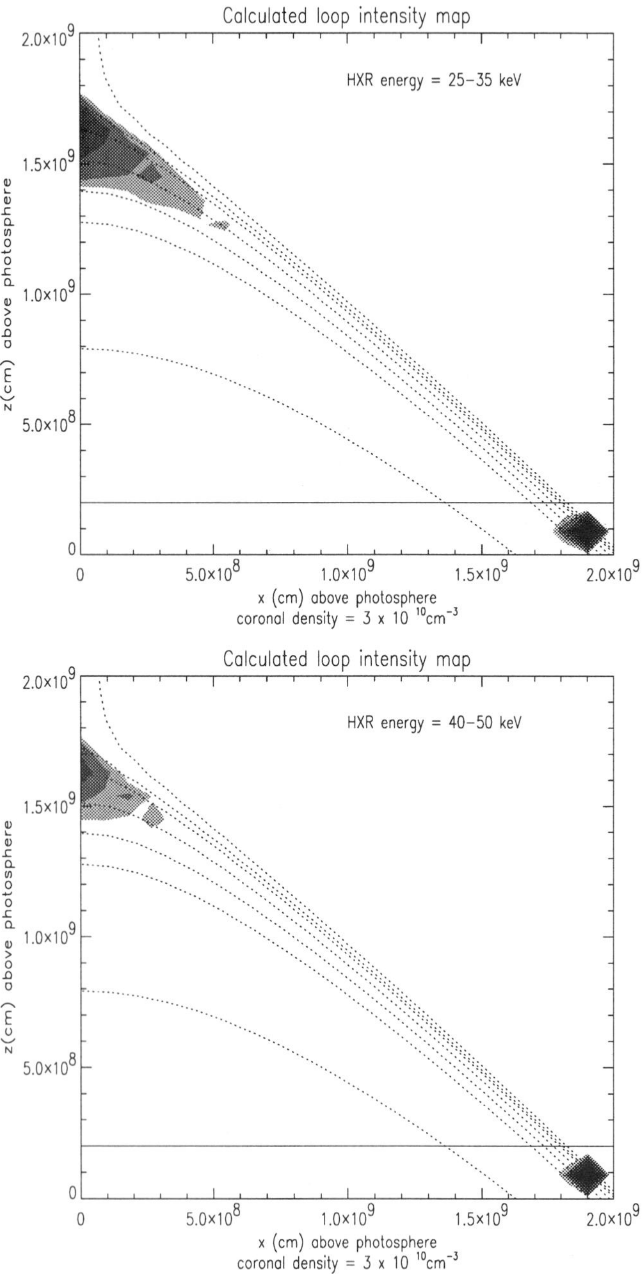

Figure 1. Maps of hard X-ray emission generated by electrons trapped near the current sheet. The maps are made in the 25-35 and 40-50 keV photon energy bands, and the contours are at 12.5%, 25%, 34%, 50% and 70% of the maximum counts in the map.

We do not model acceleration: a power-law injected electron spectrum is assumed, having also an angular distribution uniform in pitch angle. Modelling of the particle transport in the collisional, magnetised medium of the corona/chromosphere is done with a stochastic particle simulation.

3. Results

In figure 1 we show selected results from the simulations. The plot is of one of the lower quadrants of the structure - this should be mirrored about the z axis. A coronal density of $3 \times 10^{10}\text{cm}^{-3}$ is used, and the structure considered is 2.4×10^{9}cm long along the separatrix leading from the end of the current sheet to the bottom of the calculation box. The chromospheric density (below 2×10^{8}cm) is exponential, with a maximum density of $9.26 \times 10^{15}\text{cm}^{-3}$, matched to the coronal density. The current sheet is orientated along the z-axis and extends from $1.7 - 2.0 \times 10^{9}$cm. A source at the top of the structure, whose intensity decreases with increasing energy, can be clearly seen.

4. Conclusions

We have shown with the aid of numerical simulations of particle transport in a current-sheet geometry that it is possible to generate HXR bremsstrahlung at the top of flaring structures. Confinement is by the converging magnetic field present around current sheet in a reconnection region. This provides a possible explanation for the looptop hard X-ray sources reported by Masuda *et al.* (1994), which they suggest coincide with the reconnection region at the top of postflare loops.

References

Fletcher, L. and Martens, P.C.H, *Astronomy and Astrophysics*, in preparation.
Masuda, S., Kosugi, T., Hara, H., Tsuneta, S., and Ogawara, Y.: 1994, *Nature*, 371, 495.
Syrovatskii, S.I.:1971, *Soviet Physics, J.E.T.P.*, 33, 933.

ENERGY RELEASE AND PARTICLE ACCELERATION IN SOLAR FLARES WITH RESPECT TO FLARING MAGNETIC LOOPS

T. SAKAO AND T. KOSUGI
National Astronomical Observatory
Mitaka, Tokyo 181, Japan

AND

S. MASUDA
Solar-Terrestrial Environment Laboratory, Nagoya University
Toyokawa, Aichi 442, Japan

Abstract. We study the spatial evolution of hard X-ray double footpoint sources in 14 intense impulsive flares observed with Hard X-ray Telescope (HXT) aboard *Yohkoh.* In 7 out of the 14 events, the separation between the double footpoints increased (at the 3 σ confidence level) during the impulsive phase, with a typical separation velocity, $v_{\rm sep} \sim 50$ km/s (corrected for projection effects on the solar surface). The remaining 7 flares, however, do not show significant increase in separation; some of them even showed decreasing footpoint separation ($v_{\rm sep} < 0$). The magnetic field configuration inferred from the former class of events (*i.e.*, with $v_{\rm sep} > 0$) suggests a "loop-with-a-cusp" picture where magnetic reconnection takes place at a magnetic X- (or Y-) point above a closed loop. The latter class of events (with $v_{\rm sep} \leq 0$) may imply an "emerging-flux-type" field configuration.

We also find a significant difference between these two classes of flares. Events with separating footpoints show a hard X-ray source in the L-band of HXT (14 – 23 keV) located in between the footpoint sources. This source most likely originates from a super-hot plasma at the top of the flaring loop, producing spectral enhancement below 20 – 30 keV. In contrast to this, events with non-separating footpoints show double sources in the L-band, without the loop-top source.

We conjecture that (1) two kinds of magnetic field configuration, one is "loop-with-a-cusp" and the other is "emerging-flux-type", are responsible

T. Watanabe et al. (eds.), Observational Plasma Astrophysics: Five Years of Yohkoh and Beyond, 273–284.

for solar flare energy release via magnetic reconnection, and that (2) superhot plasma is produced at the loop top during the impulsive phase only in flares with the loop-with-a-cusp field configuration. Little such plasma is created in flares with the emerging-flux-type configuration, resulting in a (spatially averaged) hard X-ray spectrum which breaks down towards higher energies at around 20 – 30 keV, so that (3) this spectrum may reflect purely non-thermal emission from the footpoints of flaring magnetic loop(s).

1. Introduction

Magnetic fields in the solar corona play a vital role in flares. The source of flare energy originates from non-potential magnetic fields (Sweet 1969), and energized plasma particles propagate along or are trapped by the field lines. To investigate magnetic field configurations responsible for flares is, therefore, one of the key means to understand energy release, particle acceleration, and plasma heating processes which are operating in solar flares.

It has become widely accepted that the "loop-with-a-cusp" magnetic field configuration is responsible for coronal energy release. This configuration implies that magnetic energy is released via X- (or Y-) type reconnection taking place at the tip of a cusp-shaped field above a closed loop. A number of observations as well as theoretical consideration have thus far suggested the correctness of this picture; for instance, in Hα two-ribbon events and post-flare loops (Hirayama 1974, Kopp and Pneuman 1976, Švestka *et al.* 1987, Schmieder *et al.* 1987, Forbes and Malherbe 1991), in soft X-ray arcade formation (Tsuneta *et al.* 1992a), and in long duration events (LDEs; Tsuneta *et al.* 1992b, Tsuneta 1996). In 1994, by combining images of limb flares observed with hard and soft X-ray telescopes aboard *Yohkoh* (HXT and SXT, respectively; Kosugi *et al.* 1991, Tsuneta *et al.* 1991), Masuda made a remarkable observation that this picture is also applicable to energy release in impulsive flares where a large amount of energy is released with much smaller temporal and spatial scales than the events mentioned above (Masuda 1994, Masuda *et al.* 1994, 1995).

Besides the loop-with-a-cusp magnetic field configuration, there has been a growing number of observational evidence for the "emerging-flux (or loop-loop interaction) -type" picture for energy release in microflares (Shimizu *et al.* 1994) and in flares (Hanaoka 1996, Nishio *et al.* 1997). The relationship between the two pictures, however, has so far remained unclear.

In the meanwhile, observations with HXT have revealed various fundamental characteristics of hard X-ray emission in solar flares. Sakao (1994) has shown that in impulsive flares the bulk of hard X-rays above 30 keV is emitted from the footpoints of a flaring loop, manifesting itself most frequently as double sources. The hard X-rays from the footpoints are emitted by electrons accelerated near the top of the loop, streaming down along the loop towards both ends. This result provides us with a new method for investigating magnetic field configurations in which energy is released. As reconnection proceeds, field lines outside the diffusion region reconnect successively, one after another. This would result in systematic "activation" of field lines. Since electrons which emit hard X-rays at the footpoints propagate along a magnetic field line that is possibly connected to the region where they are accelerated, the spatial evolution of hard X-ray double footpoint sources gives us direct information on the evolution of hard X-ray flaring loop(s), through which we can infer the field configuration. In this article, we present spatial evolution of hard X-ray footpoints in 14 flares observed with HXT. We focus on temporal behavior of separation between the footpoints and discuss possible implications of the observations.

2. Observation

2.1. EVENT SELECTION AND METHOD OF ANALYSIS

To investigate the spatial evolution of hard X-ray footpoints, we have selected events for study according to the following criteria: (1) impulsive flares observed with HXT between October 1991 (the beginning of HXT observation) and December 1994, (2) peak count rate in the M2-band of HXT (33 – 53 keV) exceeding 30 counts/s/subcollimator, and (3) a double-source structure in the M2-band at the time of peak M2-band hard X-ray emission. A total of 17 events met the criteria. Among these, three (2 February 1992, 11:33 UT; 11 August 1992, 13:47 UT; and 11 February 1993, 18:31 UT) were excluded from the study either because the duration of the impulsive phase was too short ($\sim$ 4 s; 2 February 1992 event), and therefore not suited for studying the evolution of the footpoints, or because the apparent double source corresponds to footpoints of different flaring loops seen in SXT (11 August 1992 and 11 February 1993 events). Table 1 summarizes the remaining 14 events studied.

For quantitative evaluation of the double footpoint location, we use the Gaussian fitting procedure described in Sakao (1994). This procedure gives parameters (position as well as flux) of two (circular) Gaussian sources which reproduce observed count data set within statistical uncertainty. Hence, it is independent of image reconstruction algorithms used for HXT. The fitting was performed to M2-band data, with a constant data accumulation interval (typically 1 s) for each fitting, as long as the double-source structure is clearly discernible in the M2-band during the impulsive phase and the accumulated X-ray counts provide a stable fitting result. Before or after the peak of M2-band emission, some events showed double sources at a well-separated location from those seen at the peak. We excluded such sources from the analysis as they may represent a different flaring loop from the one around the M2-band peak. The fitting provides the center position $(x_i(t), y_i(t))$ of the i-th Gaussian source ($i = 1, 2$) at each time t. The time-dependent separation, $s(t)$, of the footpoints is then given by

$$s(t) = \sqrt{(x_1(t) - x_2(t))^2 + (y_1(t) - y_2(t))^2} \ . \tag{1}$$

Note that $s(t)$ is free from spacecraft jitter. In the present article, we concentrate on the separation $s(t)$ rather than the location of each individual footpoint. The spatial evolution of the footpoints themselves will be the subject for a future paper.

2.2. SEPARATING / NON-SEPARATING FOOTPOINTS

Figure 1 shows an example of the events which show separating footpoints (15 November 1991 flare). Hard X-ray flux from each of the double sources, also derived from the fitting, is shown in Figure 1(a). The double sources emit hard X-rays simultaneously within a fraction of a second, which implies that hard X-rays are emitted at the footpoints by electrons accelerated near the loop top. The time profile of $s(t)$ is presented in Figure 1(b) at every 0.5 s. The solid line in the figure indicates a running average of $s(t)$ with 4 data points (*i.e.*, smoothed over a 2-second window). The separation does not increase monotonically, but rather shows repeated episodes of increase and decrease, while the overall separation increasing. Such behavior is generally seen in events with increasing footpoint separation. The average separation velocity, $v_{\rm sep}$, during the time interval analyzed (22:37:32 UT – 22:38:08 UT) is 58.0 ± 3.7 km/s (corrected for projection effects on the solar surface).

Table 1 summarizes measured $v_{\rm sep}$ for the 14 events studied, where $v_{\rm sep} > 0$ indicates that the footpoints are separating. Among these events, seven (events 1, 3, 4, 11 – 14) show significant footpoint separation above the 3 σ level while the rest show either only marginal increase in the separation or even decreasing separation (*e.g.*, events 2, 6). The typical value of $v_{\rm sep}$ for the seven events with significantly-increasing footpoint separation is about 50 km/s. It is worthwhile mentioning that events with non-separating footpoints, especially those with $v_{\rm sep} < 0$, tend to have shorter impulsive phase duration than those with separating footpoints. The typical duration, τ, *i.e.*, the period of time over which the (analyzed) double sources are seen, is shorter than 30 s for the $v_{\rm sep} < 0$ events and larger for the others. Little,

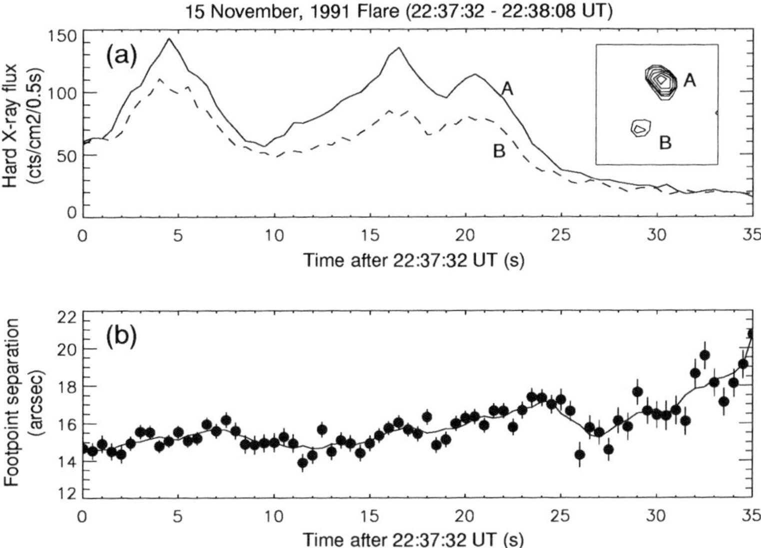

Figure 1. Separating footpoints observed in the 15 November 1991 flare. (a) Time profiles of M2-band hard X-ray fluxes from the double footpoint sources shown in the inset. (b) Footpoint separation $s(t)$ (corrected for projection effects) at every 0.5 s. The solid line shows a running average of $s(t)$ with a 2-second window. Error bar for each data point represents the 1 σ statistical level.

if any, correlation is seen between the distance of the footpoints and their separating/non-separating behavior.

For the events with separating footpoints, the observations suggest that the loop-with-a-cusp magnetic field configuration, together with magnetic reconnection at the X-point, is responsible for the impulsive phase energy release. The typical value of $v_{\rm sep} \sim 50$ km/s is significantly larger than that for Hα two ribbon events (typically 10 km/s, at most less than 20 km/s) and arcade formation events (a few km/s). This implies that the reconnection proceeds at a higher rate, hence higher velocity for footpoint separation, in energetic flares than in not-very-energetic events.

On the other hand, there are a certain number of events which emit intense hard X-rays without any apparent separation of the footpoints. Such events may have a different field configuration from the loop-with-a-cusp picture. We will discuss this issue in Section 3.

2.3. FOOTPOINT SEPARATION AND SPECTRAL CHARACTERISTICS

We observe that the behavior of the footpoints, whether they are separating or not, is deeply related to the shape of the spatially integrated hard X-

TABLE 1. Double-footpoint separation velocity.

No.	Flare (UT)[1]		Hα Loc.	GOES/Hα	M2[2]	τ (s)[3]	v_{sep} (km/s)[4]	Note
1	91/11/02	06:45:45	S14W62	M9.1/1B	251	58.0	126.8 ± 8.4	
2	91/11/10	06:48:34	S14E56	M2.2/2B	151	14.0	–35.8 ± 30.7	(a)
3	91/11/15	22:37:50	S14W19	X1.5/3B	528	35.5	58.0 ± 3.7	
4	91/12/03	16:36:17	N17E72	X2.2/2B	502	44.0	67.7 ± 22.4	
5	91/12/04	17:43:24	N17E58	M4.1/SN	36	82.0	–7.3 ± 12.9	
6	91/12/16	04:56:07	N04W45	M2.7/SF	154	15.0	–71.4 ± 25.5	(a)
7	92/02/07	11:54:29	S21W53	M3.7/2B	35	42.0	17.8 ± 16.7	
8	92/02/15	21:29:40	S16W13	M5.5/1B	33	32.0	7.8 ± 22.4	
9	92/09/10	22:52:26	N12E41	M3.2/2B	138	22.0	–0.1 ± 16.6	(b)
10	93/03/11	15:15:12	S04W34	M1.1/SF	82	25.0	–15.2 ± 27.0	(a)
11	93/04/22	14:08:26	N11E04	M1.5/SB	75	32.0	91.3 ± 13.4	
12	94/01/27	05:09:17	N11W65	M2.7/1B	37	70.0	73.6 ± 13.7	
13	94/06/30	21:21:10	S12E27	M2.5/1B	86	72.0	48.0 ± 4.0	(c)
14	94/08/14	17:36:31	S12W08	M3.9/1N	56	38.0	41.9 ± 11.6	

(a) No SXT observation.
(b) PFI offpointed.
(c) The latter half of the impulsive phase may be missing due to the entrance of the spacecraft into the South Atlantic Anomaly.
[1]Peak time in the M2-band count rate.
[2]M2-band peak count rate in units of counts/s/subcollimator.
[3]Duration of the impulsive phase (the period of time over which the analyzed double sources are seen).
[4]Corrected for projection effects on the Sun.

ray spectrum of the event. Figure 2 shows the relationship between v_{sep} (corrected for projection effects) and $\Delta\gamma \equiv \gamma_{M1/L} - \gamma_{M2/M1}$, where $\gamma_{M2/M1}$ is the photon index obtained from M1- and M2-band count data pair assuming a single power-law incident spectrum, and likewise for $\gamma_{M1/L}$. It is clearly discernible that events with separating footpoints tend to show positive $\Delta\gamma$, which means that the spectrum breaks up toward lower energies at 20 – 30 keV. On the other hand, events with non-separating footpoints tend to show a hard X-ray spectrum which breaks down towards higher energies ($\Delta\gamma < 0$). So the relationship between the footpoint motion and the spectral shape can be summarized as

$$v_{sep} \gtrsim 0 \Longleftrightarrow \Delta\gamma \gtrsim 0 \, . \tag{2}$$

The value of $\gamma_{M2/M1}$ ranges between 2 and 5 without showing systematic differences between the two classes of events. On the contrary, events with separating footpoints tend to have $\gamma_{M1/L} = 3 - 6$ while those without

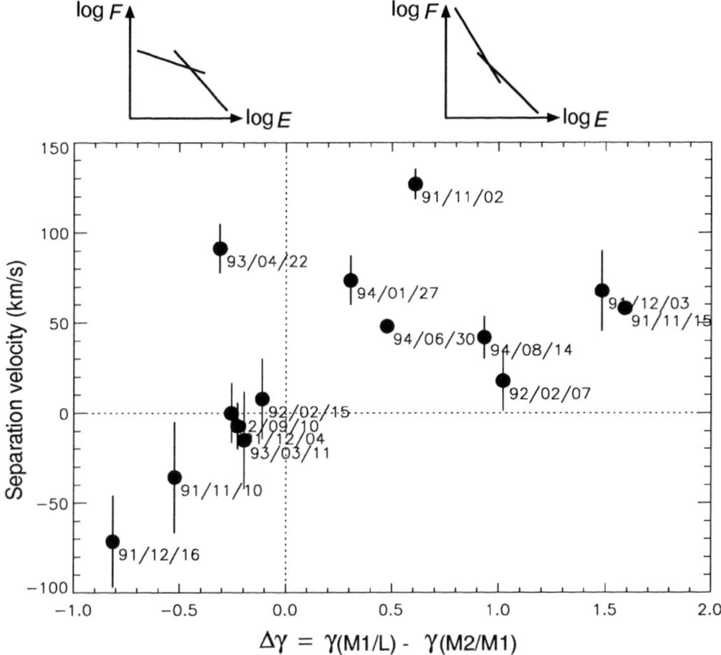

Figure 2. Relationship between separation velocity v_{sep} of the footpoints (corrected for projection effects on the solar surface) and shape of hard X-ray spectrum. Positive $\Delta\gamma$ indicates that spectrum breaks up toward lower energies at around 20 – 30 keV, while negative $\Delta\gamma$ opposite. Note that events with $v_{sep} > 0$ tend to have positive $\Delta\gamma$ while the opposite is the case for events with $v_{sep} < 0$. Error bars in the figure represent the 1 σ statistical level.

separating footpoints show harder spectra; $\gamma_{\mathrm{M1/L}} = 2 - 3.5$. Hence we can say that the observed difference in $\Delta\gamma$ for the two classes of events is mainly caused by the difference in the spectrum below 20 – 30 keV.

The enhancement of lower energy hard X-rays ($\Delta\gamma > 0$), observed in events with increasing footpoint separation, is most likely due to the presence of super-hot plasma, with temperature around 30 – 50 MK, located at the top of the flaring loop. Figure 3 shows HXT images in the L- and M2-bands (contours), for events 2 and 3 in Table 1. In event 3 (15 November 1991 flare), which shows separating footpoints, we see that the L-band image is morphologically different from the M2-band image; the latter shows double footpoint sources at both ends of the flaring loop while the former shows a hard X-ray source in between the double sources of the M2-band. The third source, which has a softer spectrum than the footpoints, is most likely located at the top of the loop. On the other hand, in event 2 (10 November 1991 flare), which does not show separating footpoints, the L-

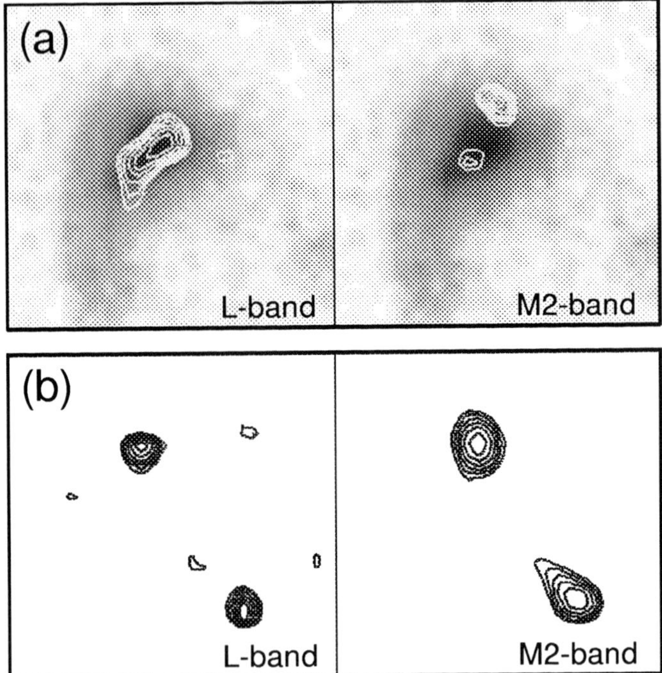

Figure 3. Hard X-ray images seen in the L- and M2-bands of HXT at the peak of emission in the M2-band (contours). (a) The flare on 15 November 1991 (event 3 in Table 1), which shows separating footpoints visible in the M2-band. Soft X-ray image taken with the Be filter of SXT is overlaid on each map (grayscale). (b) The flare on 10 November 1991 (event 2 in Table 1), which lacks separating footpoints. No SXT observation was made for this flare. The contour levels are 70.7, 50.0, 35.4, 25.0, 17.7, 12.5 % of the peak brightness in each map. The field of view of each map covers $79'' \times 79''$. Solar north is to the top, east is to the left.

and M2-band images look similar to each other. They both show double sources with no additional source present in the L-band image. This morphological characteristic is also found in the rest of the 14 events. Hence, from spectral and morphological consideration, we argue that, during the impulsive phase, flares with increasing footpoint separation tend to produce super-hot plasmas at the loop top, while in flares without separating footpoints, little or no super-hot plasma is produced there.

It is interesting to note that events which do not show separating footpoints tend to lack corresponding SXT observations (events 2, 6, and 10; event 9 is observed with offpointed SXT PFI). This may suggest that, prior to the flares, the corresponding active regions had not been bright enough in soft X-rays to be selected as the observing region of SXT. We will investigate this in more detail in a separate study.

3. Discussion and Conclusions

3.1. MAGNETIC FIELD CONFIGURATION INFERRED FROM FOOTPOINT MOTION

3.1.1. *Loop-with-a-cusp picture*

Among our 14 impulsive flares, half showed significant (above the 3 σ confidence level) increasing separation of hard X-ray footpoints. The separating footpoints suggest a rising energy release (particle acceleration) site during the course of the flare. The most likely scenario for this observation is that magnetic reconnection and subsequent energy release take place at an X-point located at the tip of a cusp-shaped magnetic field above a closed loop (loop-with-a-cusp picture), such as the one proposed for Hα two-ribbon events, evolution of post-flare loops, soft X-ray arcade formation, and long duration events. By identifying a hard X-ray source above the closed loop visible in soft X-rays, Masuda *et al.* (1994, 1995) argued that this picture is also applicable to at least some impulsive flares. Our observation has confirmed this by studying spatial evolution of the hard X-ray footpoints.

For the events with the loop-with-a-cusp configuration, the typical velocity of footpoint separation during the impulsive phase is about 50 km/s, which is larger than the velocity of separating Hα two-ribbons ($\sim$ 10 km/s; *e.g.*, Švestka 1976) and of footpoint separation in soft X-ray arcade formation events and LDEs (a few km/s and $\sim$ 20 km/s, respectively; Tsuneta *et al.* 1992a, 1992b). This would imply that in intense impulsive flares, magnetic reconnection indeed proceeds at a higher rate than that for less energetic events.

3.1.2. *Emerging-flux-type picture*

Although at least half of the events analyzed suggest the loop-with-a-cusp picture with energy release at the X-point, we still observe some events which do not significantly show separating footpoints (some even show decreasing footpoint separation) despite their intense hard X-ray emissions. A possible explanation to this is that, while the overall field has the loop-with-a-cusp configuration with separating footpoints of field lines, the apparent hard X-ray footpoint motion in the M2-band is mimicked by super-hot plasmas produced at the footpoints and/or hard X-rays are emitted at progressively higher altitudes near the footpoints due to upward motion of high density plasmas ($\sim 10^{11}$ cm^{-3}) caused by chromospheric evaporation (*e.g.*, Antonucci *et al.* 1982, Culhane *et al.* 1994). However, both of these cases have flaws. In the former case, if thermal emission at the footpoints from plasma with a temperature of 30 – 50 MK were to be observed in the M2-band, it would produce hard X-ray counts in the L-band higher than what was observed. In the latter case, typical velocities of upflowing plasmas (a few $\times\ 10^2$ km/s) would be too small to explain the observed decrease in the footpoint separation, especially when we observe the loop from the top.

There are some interesting observations which may have a close relationship with non-separating footpoint flares. Canfield *et al.* (1996) have observed converging Hα footpoint motion in an Hα surge / X-ray jet event. They proposed that an emerging flux which interacts with the overlying magnetic field could explain the observation. Recently, microwave images from Nobeyama Radio Observatory, combined with soft and hard X-ray images from *Yohkoh*, suggest that the interaction of two magnetic loops (one with a typical scale size of 1.5×10^4 km and the other $(2-6) \times 10^4$ km) which share a common footpoint is responsible for flare initiation (Nishio *et al.* 1997; see also Hanaoka 1996). Nishio *et al.* have shown that hard X-rays are preferentially emitted from the smaller loop while the larger loop is visible only in microwaves and soft X-rays. If hard X-rays are to be emitted (for some reason) from the smaller loop, then we may expect that newly reconnecting field lines have smaller footpoint separation than the previous ones. Based on these observations, we speculate that events with non-separating footpoints are caused by interaction of magnetic loops, such as interaction between an emerging flux and the overlying coronal field, although some other possibilities might exist.

3.2. THERMAL/NON-THERMAL ASPECTS OF FLARES WITH RESPECT TO MAGNETIC FIELD CONFIGURATION

We have found that in flares which suggest loop-with-a-cusp magnetic field configurations ($v_{\rm sep} > 0$), the hard X-ray spectrum tends to break up at around 20 – 30 keV toward lower energies ($\Delta\gamma > 0$). Also this class of flares shows an L-band source located in between the double sources. The L-band source, which may produce the spectral enhancement below 20 – 30 keV, most likely originates from a super-hot plasma at the loop top (Lin *et al.* 1981). On the other hand, in flares without separating footpoints (which may be caused by the emerging-flux-type field configuration), little such plasma is observed. The absence of the superhot plasma may result in a spectrum which breaks down toward higher energies above 20 – 30 keV ($\Delta\gamma < 0$). This spectrum would reflect "bare" non-thermal emission from the footpoints by accelerated electrons.

It is interesting to mention that all of the 10 limb flares studied by Masuda (1994) which led him to the loop-with-a-cusp picture have HXT spectra with $\Delta\gamma > 0$. In contrast to this, the 14 events studied by Nishio *et al.* (1997), from which they derived the emerging-flux-type picture, are dominated by spectra with $\Delta\gamma < 0$ (at least 9 out of 14).

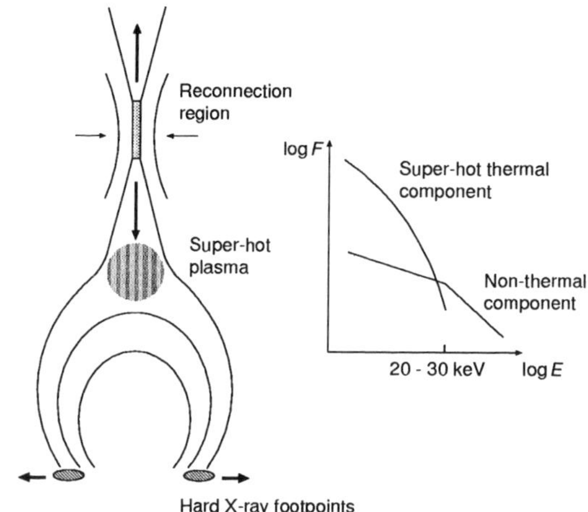

Emerging-flux-type Configuration

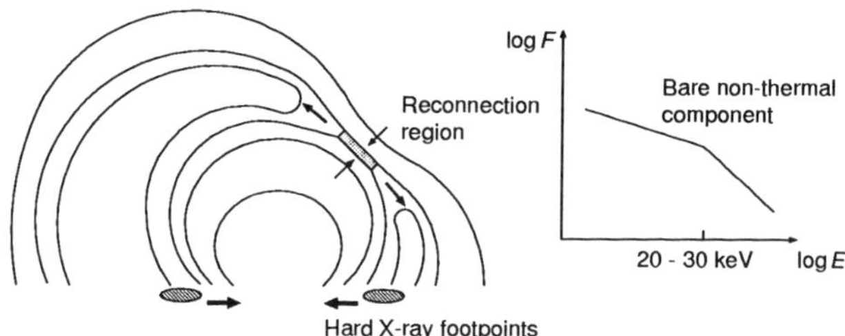

Figure 4. Relationship between magnetic field configuration in two dimensions and thermal/non-thermal hard X-ray production in solar flares. Flares with the loop-with-a-cusp configuration produce super-hot plasmas at the loop top while little such plasma is produced in flares with the emerging-flux-type configuration.

Based on the above observations, we conjecture the following (see Figure 4): (1) Two different magnetic field configurations, *i.e.*, loop-with-a-cusp and emerging-flux-type, are responsible for solar flares with energy release via magnetic reconnection. (2) Electrons are accelerated in both two configurations, while super-hot plasma at the loop top is created if (and only if) the magnetic field has the loop-with-a-cusp configuration. (3) Hard X-ray spectra that break down toward higher energies at around 20 – 30 keV, which is observed in flares with the emerging-flux-type configuration, reflect bare non-thermal emission from the footpoints. Meanwhile, spectra of flares with the loop-with-a-cusp configuration are composed of thermal emission from super-hot plasma and non-thermal emission. (4) Particle (electron) acceleration takes place at the reconnection site, *i.e.*, irrespective of the configuration of the surrounding magnetic fields. On the other hand, creation (and confinement) of super-hot plasmas during the impulsive phase is affected by the macroscopic field configuration (the field configuration around the reconnection site).

Acknowledgements We would like to express our sincere thanks to ISAS, NASA, SERC, and the *Yohkoh* Team for their continuous and valuable support of the mission.

References

Antonucci, E., Gabriel, A.H., Acton, L.W., Culhane, J.L., Doyle, J.G., Leibacher, J.W., Machado, M.E., Orwig, L.E., and Tapley, C.G. (1982) *Solar Phys.*, **78**, 107.
Canfield, R.C., Reardon, K.P., Leka, K.D., Shibata, K., Yokoyama, T., and Shimojo, M. (1996) *Astrophys. J.*, **464**, 1016.
Culhane, J.L., Phillips, A.T., Inda-Koide, M., Kosugi, T., Fludra, A., Kurokawa, H., Makishima, K., Pike, C.D., Sakao, T., Sakurai, T., Doschek, G.A., and Bentley, R.D. (1994) *Solar Phys.*, **153**, 307.
Forbes, T.G. and Malherbe, J.M. (1991) *Solar Phys.*, **135**, 361.
Hanaoka, Y. (1996) *Solar Phys.*, **165**, 275.
Hirayama, T. (1974) *Solar Phys.*, **34**, 323.
Kopp, R.A. and Pneuman, G.W. (1976) *Solar Phys.*, **50**, 85.
Kosugi, T., Makishima, K., Murakami, T., Sakao, T., Dotani, T., Inda, M., Kai, K., Masuda, S., Nakajima, H., Ogawara, Y., Sawa, M., and Shibasaki, K. (1991) *Solar Phys.*, **136**, 17.
Lin, R.P., Schwartz, R.A., Pelling, R.M., and Hurley, K.C. (1981) *Astrophys. J. Lett.*, **251**, L109.
Masuda, S. (1994) Ph.D. thesis (University of Tokyo).
Masuda, S., Kosugi, T., Hara, H., Tsuneta, S., and Ogawara, Y. (1994) *Nature*, **371**, 495.
Masuda, S., Kosugi, T., Hara, H., Sakao, T., Shibata, K., and Tsuneta, S. (1995) *Publ. Astron. Soc. Japan*, **47**, 677.
Nishio, M., Yaji, K., Kosugi, T., Nakajima, H., and Sakurai, T. (1997) *Astrophys. J.*, in press.
Sakao, T. (1994) Ph.D. thesis (University of Tokyo).
Schmieder, B., Forbes, T.G., Malherbe, J.M., and Machado, M.E. (1987) *Astrophys. J.*, **317**, 956.
Shimizu, T., Tsuneta, S., Acton, L.W., Lemen, J.R., Ogawara, Y., and Uchida, Y. (1994) *Astrophys. J.*, **422**, 906.
Švestka, Z.F. (1976) Solar Flares (Reidel) pp. 45.
Švestka, Z.F., Fontenla, J.M., Machado, M.E., Martin S.F., Neidig, D.F., and Poletto, G. (1987) *Solar Phys.*, **108**, 237.
Sweet, P. (1969) *Ann. Rev. Astron. Astrophys.*, **7**, 149.
Tsuneta, S., Acton, L., Bruner, M., Lemen, J., Brown, W., Caravalho, R., Catura, R., Freeland, S., Jurcevich, B., Morrison, M., Ogawara, Y., Hirayama, T., and Owens, J. (1991) *Solar Phys.*, **136**, 37.
Tsuneta, S., Takahashi, T., Acton, L.W., Bruner, M.E., Harvey, K.L., and Ogawara, Y. (1992a) *Publ. Astron. Soc. Japan*, **44**, L211.
Tsuneta, S., Hara, H., Shimizu, T., Acton, L.W., Strong, K.T., Hudson, H.S., and Ogawara, Y. (1992b) *ibid.*, L63.
Tsuneta, S. (1996) in *Magnetodymanic Phenomena in the Solar Atmosphere – Prototype of Stellar Magnetic Activity*, eds. Uchida, Y., Kosugi, T., and Hudson, H. (Kluwer), 161.

WHAT DID YOHKOH AND COMPTON CHANGE IN OUR PERCEPTION OF PARTICLE ACCELERATION IN SOLAR FLARES ?

MARKUS J. ASCHWANDEN
University of Maryland
Astronomy Department
College Park, MD 20742, USA

Abstract. We review some key hard X-ray (HXR) observations obtained from the *Yohkoh* and the *Compton Gamma Ray Observatory (CGRO)* spacecraft that recast our understanding of particle acceleration in solar flares:

1. Above-the-loop-top hard X-ray sources.
2. Conjugacy and simultaneity of double footpoint hard X-ray sources.
3. Electron time-of-flight measurements and scaling law.
4. Dichotomy of Hard X-ray time structures.
5. Electron trapping time measurements.
6. Bi-directionality of electron beams.

These key observations provide promising new constraints for particle acceleration models.

1. Prolog

The combination of HXR imaging observations from the *Yohkoh Hard X-Ray Telescope (HXT)* with high time resolution (64 ms) observations from the most sensitive *Burst and Transient Source Experiment (BATSE)* onboard *CGRO* provide a powerful instrumental framework to probe acceleration and propagation of nonthermal particles in solar flares. Statistics obtained from hundreds of solar flares observed with *CGRO*, and with *Yohkoh*, whose 5-year anniversary was celebrated during this workshop, have led to significant new discoveries in solar flare physics (see also summaries by Kosugi 1996; Tsuneta 1996; Emslie 1996). The following review entails a selection of key observations that are currently believed to represent the most sensible diagnostics of particle acceleration processes operating in solar flares.

T. Watanabe et al. (eds.), Observational Plasma Astrophysics: Five Years of Yohkoh and Beyond, 285–294.

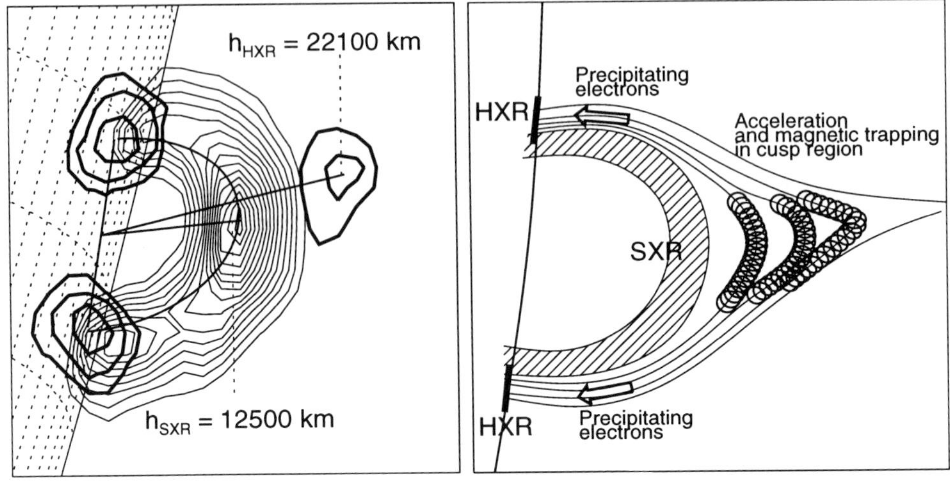

Figure 1. Observations and theoretical model of above-the-loop-top HXR source: *Yohkoh/HXT* 23-33 keV image (thick contours) and Be119 *SXT* image (thin contours) of the 92-Jan-13, 1728 UT flare (Masuda 1994; left panel). Theoretical model envisioning (stochastic or shock) acceleration and magnetic trapping (in the reconnection outflow) occurring in the cusp region beneath the magnetic reconnection point (right panel).

2. Above-the-Loop-Top Hard X-Ray Sources

Before *Yohkoh*, nonthermal HXR emission was dominantly observed from the location of flare loop footpoints (e.g. Duijveman, Hoyng & Machado 1982; Takakura, Tanaka, & Hiei 1984), interpreted as collisional bremsstrahlung produced by electrons precipitating from the flare loop towards the chromosphere (thick-target model of Brown 1971). While thermal ($\lesssim$ 25 keV) HXR emission is often also observable from most parts of the flare loop, including the top portion, Masuda (1994) identified with *Yohkoh/HXT* for the first time HXR emission originating from an *above-the-loop-top* site. The physical nature of this "cuspal" HXR emission is still debated: Masuda (1994) proposed a thermal interpretation, which was criticized by Hudson & Ryan (1995) due to inplausible time scales, while recent work with pixon imaging corroborate that the cuspal HXR source exhibits the same nonthermal spectrum as the footpoints (Alexander & Metcalf 1997). Also the correlated time evolution of footpoint and above-the-loop-top sources support a topological connection. Because the electron density in the cusp region is too low ($n_e \lesssim 10^{10}$ cm^{-3}) to produce detectable HXR bremsstrahlung, the nonthermal HXR yield is probably enhanced by magnetic trapping in the cusp region (Martens 1996). The greatest significance of Masuda's discovery lies in the prospect that the cuspal HXR source may trace out directly the location of the primary energy release and acceleration region.

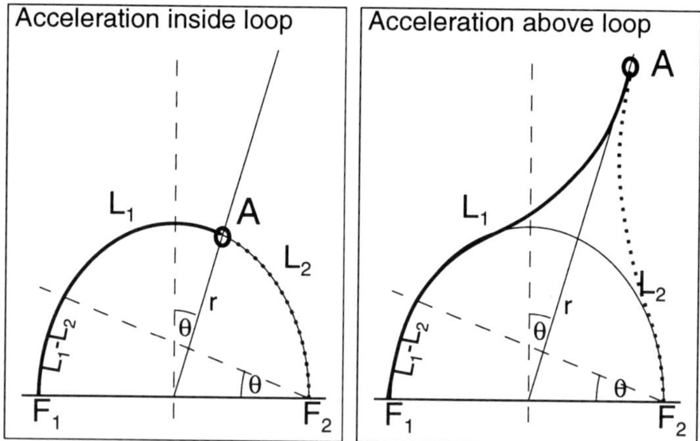

Figure 2. Geometry of electron path trajectories (with lengths L_1, L_2) from a common acceleration site (A) to two conjugate footpoints (F_1, F_2). The inclination angle ϑ of an asymmetric location A is related to the path difference by $(L_2 - L_1) = 2r\vartheta$. The two cases illustrate the geometry for an acceleration source inside a semi-circular flare loop (left panel) and for an acceleration site above the flare loop (right panel).

3. Conjugacy and Simultaneity of Double Footpoint Sources

Sakao (1994) measured the simultaneity of HXR emission from two conjugate footpoints by cross-correlating the HXR fluxes from both footpoints (F_1, F_2). The cross-correlation time delays were found to be near-simultaneous in 6 cases out of 7 investigated flares, with uncertainties of $\pm 0.1 - 0.5$ s. Using the geometry indicated in Fig.2 we obtain the following relation between the asymmetry angle ϑ and the travel times $t_1 = L_1/v$ and $t_2 = L_2/v$ for particles propagating with speed v from a common acceleration site A to two conjugate footpoints F_1 and F_2:

$$\vartheta = \frac{v}{2r} \frac{(t_1 - t_2)}{q_\alpha q_H} \tag{1}$$

with r the semi-circular loop radius, $q_\alpha \approx 0.64$ is a correction factor for the average pitch angle α, and $q_H \approx 0.85$ a correction factor for helical twist of the magnetic field line along the trajectory (Aschwanden et al. 1996a). The path difference $(L_1 - L_2)$ does not depend on the height of the acceleration source, or on the assumption whether acceleration takes place inside or outside of the loop. In Table 1 we list the upper limits of the travel time differences $(t_2 - t_1)$ measured by Sakao (1994), the loop radii r of the corresponding flares, and calculate the upper limit for the asymmetry angles ϑ for three different speeds: (a) for electrons that produce 33-53 keV HXR emission in the *HXT/M2* channel ($E_{kin} \approx 2\varepsilon_{HXR} = 66$ keV, $v/c = 0.46$), (b) for 0.1-1 MeV protons ($v/c = 0.015 - 0.046$), and (c) for conduction fronts ($v \approx 1000$ km/s). The results show that the simultaneity limits do not require a highly symmetric acceleration site for electrons (in contrast to widely-held statements), but would require an unplausibly high symmetry for protons or conduction fronts, which can thus be ruled out as energy carriers to produce HXR emission at the footpoints.

TABLE 1. Upper limits on asymmetry angle of acceleration site for Sakao's flares

Flare Date	Simultaneity $(t_1 - t_2)$ [s]	Loop radius r[Mm]	Electron asymmetry ϑ	Proton asymmetry ϑ	Conduction fr asymmetry ϑ
91/11/15	± 0.1	5.6	$\pm 38^0$	$\pm 1 - 4^0$	$\pm 0.3^0$
91/12/03	± 0.2	4.7	$> 90^0$	$\pm 3 - 9^0$	$\pm 0.7^0$
92/02/07	± 1.0	6.4	$> 90^0$	$\pm 10 - 33^0$	$\pm 2.4^0$
91/12/16	± 0.2	9.7	$\pm 44^0$	$\pm 1 - 4^0$	$\pm 0.3^0$
92/09/10	± 0.3	8.0	$\pm 80^0$	$\pm 2 - 8^0$	$\pm 0.6^0$
91/11/10	± 0.5	15.7	$\pm 67^0$	$\pm 2 - 7^0$	$\pm 0.5^0$
91/11/02	-1.5	11.8	$> 90^0$	$\pm 8 - 27^0$	$\pm 2.0^0$

4. Electron Time-of-Flight Measurements and Scaling Law

A small, but systematic, time delay has been discovered in sub-second HXR pulses between different energies (Aschwanden, Schwartz, & Alt 1995a). The low-energy (25-50 keV) HXR pulses were found to lag high-energy (50-100 keV) pulses by an amount of typically $\approx 10 - 100$ ms. Thanks to the high sensitivity of the large-area detectors of *BATSE/CGRO*, these small time delays could be determined with uncertainties of $\approx 5 - 20$ ms in large flares with high count rates ($10^4 - 10^5$ cts/s), using burst-trigger data with a time resolution of 64 ms (Aschwanden & Schwartz 1995). These systematic time delays were found to be consistent with the functional dependence $\tau(E)$ expected for electron time-of-flight differences,

$$\tau^{TOF}(E, E_0) = \frac{L}{c}[\frac{1}{\beta(E_0)} - \frac{1}{\beta(E)}] \ . \qquad (2)$$

The energy conversion factor $q_E(\varepsilon, E) = E/\varepsilon$ between electron kinetic energies E and HXR photon energies ε is $q_E \approx 1.5 - 2.0$ for typical HXR spectra with a power-law slope of $\gamma = 4 - 6$ in the framework of the thick-target model (Aschwanden & Schwartz 1996). Using these conversion factors, the measured HXR delays $\tau(\varepsilon)$ can directly be scaled into electron time-of-flight (TOF) differences $\tau^{TOF}(E)$, leading to a determination of the time-of-flight distance L (Aschwanden et al. 1996a,b,c).

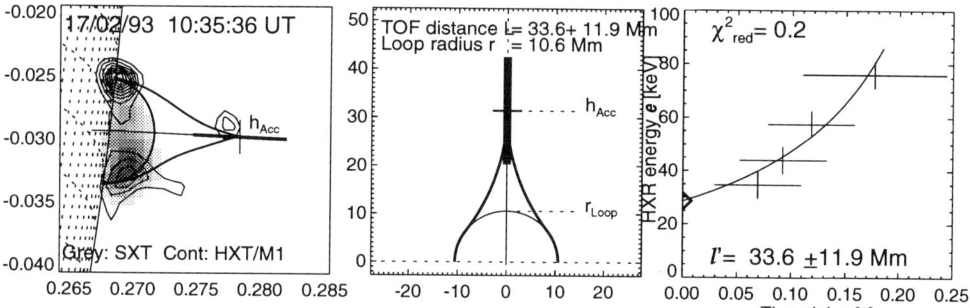

Figure 3. Electron time-of-flight measurement of the 1993-Feb-17, 1035 UT, flare. HXR delays measured from *BATSE/CGRO* are found to increase systematically up to 180 ms between 30 keV and 80 keV HXRs (right panel), yielding an electron TOF distance of $L = 33.6 \pm 11.9$ Mm. The footpoint separation in the *Yohkoh/HXT/M1* (23-33 keV) image (left panel) yields a flare loop radius of $r = 10.6$ Mm. The ratio of the TOF distance L to the loop half length s is $L/s = 2.02$ [Aschwanden et al. 1996c]

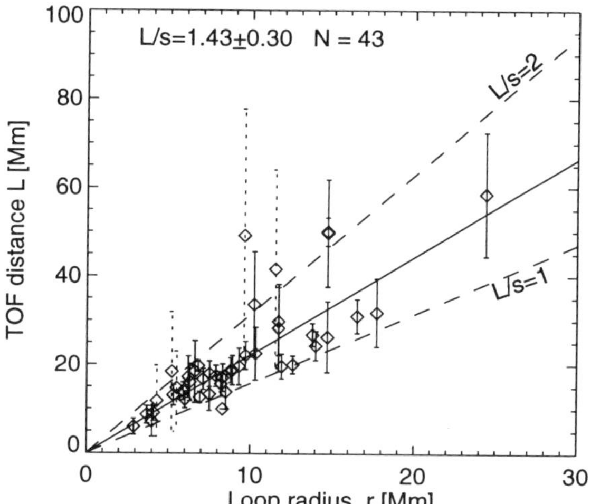

Figure 4. Scaling law between electron time-of-flight distances L (measured with *Compton*) and flare loop half lengths $s = r * \pi/2$ (measured with *Yohkoh*), from a set of 43 commonly observed flares [Aschwanden et al. 1996c].

These electron TOF distances L scale with the loop size. From a set of 43 flares with loop radii varying from $r = 3000$ to 25,000 km, a scaling law of $L/s = 1.43 \pm 0.30$ was found between the TOF distance L and the loop half length s (Aschwanden et al. 1996c). Because the average timing of electrons streaming to two conjugate footpoints would yield $L/s = 1.0$ for every (asymmetric) location inside a flare loop (with equal footpoint brightness), the observed scaling law seems to indicate an acceleration site *above the loop-top*, consistent with Masuda's (1994) finding.

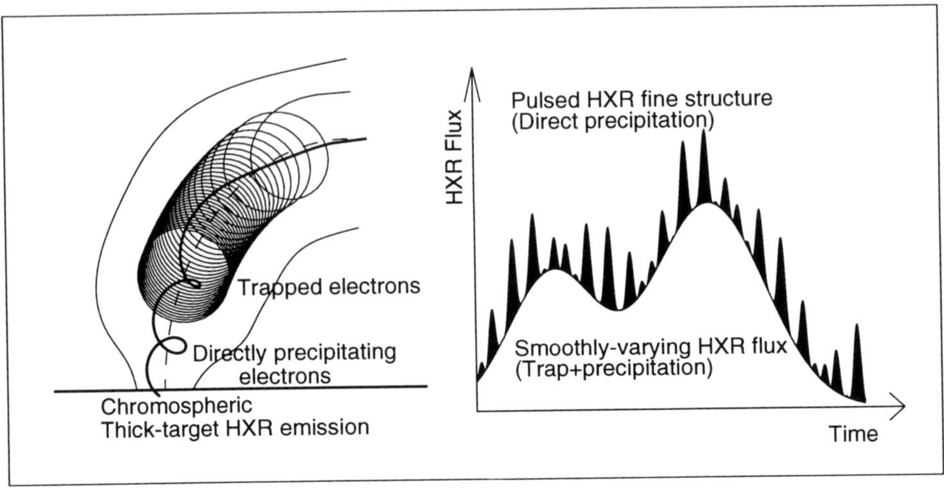

Figure 5. Relation between electron propagation and observed HXR time structures: Electrons with small pitch angles (injected near the flare loop top) precipitate directly and produce rapidly-varying HXR pulses with time-of-flight delays, while electrons with large pitch angles become trapped and produce a smoothly-varying HXR flux when they eventually precipitate, with a timing that corresponds to trapping time scales.

5. Dichotomy of Hard X-Ray Time structures

The energy-dependent timing of HXR emission can now clearly be understood in terms of three physical processes: (1) electron time-of-flight (TOF) propagation time differences, (2) electron trapping time differences, and (3) thermal heating and cooling time scales. The TOF timing has been verified in sub-second HXR pulses, while trapping was found to dominate the timing of the smoothly-varying (nonthermal) HXR flux. The thermal effects play only a role for energies $\lesssim 40$ keV. The dichotomy in the opposite timing of the fast-fluctuating and smoothly-varying (nonthermal) HXR flux has a simple explanation: Assuming that fast HXR time structures are attributed to the acceleration or injection function, only the directly-precipitating electrons can preserve this temporal modulation pattern in chromospheric HXR emission, while temporarily trapped electrons lose their memory about the time phase of their injection during the many bounces of their mirror motion in the flare loop. When they eventually precipitate, they produce a HXR time profile that represents a "smeared-out" (and thus slowly-varying) function of the injection profile. Consequently, the trapping delays are largely manifested in the slowly-varying HXR flux, while the fast sub-second pulses are dominated by time-of-flight differences of directly-precipitating electrons.

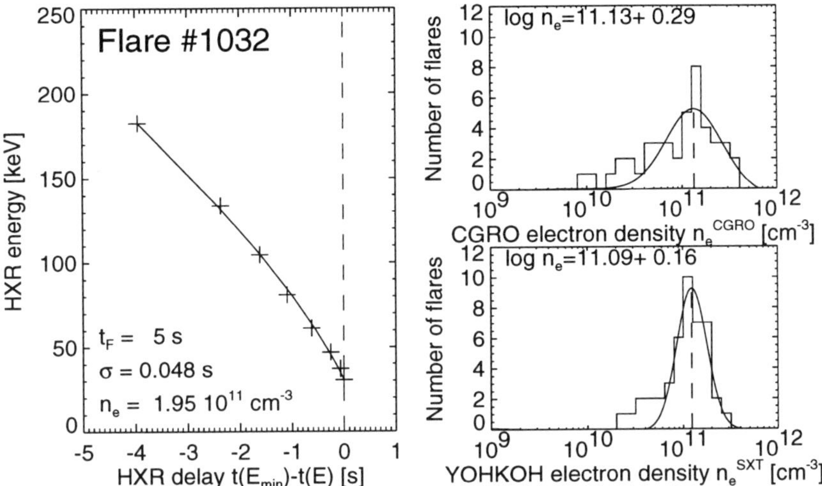

Figure 6. Time delay measurement of the smoothly-varying HXR flux during the 91-Nov-09, 2105:44 UT, flare (left panel). A trap model in terms of the electron collisional deflection time $\tau^{Trap}(E) \propto E^{3/2}$ (thin line) is fitted to the datapoints (crosses), yielding a trap density of $n_e = 1.95 \cdot 10^{11}$ cm^{-3}. Statistics of the so determined trap densities with *Compton* in 44 flares show a distribution with a mean of $n_e^{Trap} \approx 10^{11}$ cm^{-3} (right top panel). Independently, we determined electron densities from soft X-ray emission measures in the corresponding 44 flare loops with *Yohkoh/SXT*, yielding a similar distribution with $n_e^{SXR} \approx 10^{11}$ cm^{-3} (right bottom panel). [Aschwanden et al. 1997]

6. Electron Trapping Time Measurements

Since the collisional energy loss time is comparable to or shorter than the flare loop transit time for low-energy ($\lesssim 30$ keV) electrons, they thermalize in the coronal part of the flare loop, contributing to the well-known *thermal* HXR emission observed over the entire loop volume at $\lesssim 30$ keV. For higher-energy electrons, the main energy loss occurs in chromospheric layers, either when electrons precipitate directly or after some intermediate trapping until they become pitch-angle scattered into the loss-cone. There are two limits of trapping times: the *weak-diffusion limit*, which is dominated by collisional deflection times, and the *strong-diffusion limit*, which occurs on faster time scales due to wave-resonant pitch-angle scattering. In a systematic time delay study on 44 flares observed with *COMPTON* and *Yohkoh* it was found that the time delays of the smoothly-varying HXR flux in the 20-200 keV energy range is consistent with trapping times in terms of collisional deflection time differences (Aschwanden et al. 1997),

$$\tau^{Trap}(E, E_0) = 0.95 \cdot 10^8 \, (\frac{20}{\ln \Lambda}) \, [\frac{(E_0^{3/2} - E^{3/2})}{n_e}] \tag{3}$$

with E the electron energies (related to the HXR photon energies by the conversion factor $q_E = E/\varepsilon \approx 1.5 - 2$) in units of keV, n_e the electron density (cm^{-3}), and $\ln \Lambda \approx 20$ the Coulomb logarithm. An example of such a measurement with the fit of Eq.3 is shown in Fig.6 (left). The so-determined trap densities were found within a range of $n_e^{Trap} \approx 10^{10} - 10^{12}$ cm^{-3} (Fig.6 top right), consistent with independent density measurement from SXR emission measures in the flare loops simultaneously observed with *Yohkoh/SXT* (Fig.6 bottom right). This result provides evidence that electron trapping is governed in the *weak-diffusion limit*, while wave-resonant pitch-angle scattering in the *strong-diffusion limit* seems to be less relevant, in the energy range of 20-200 keV.

7. Bi-Directional Electron Beams

Simultaneous detection of normal-drifting radio type III bursts (with a frequency drift rate $d\nu/dt < 0$) and oppositely-drifting reverse-slope (RS) bursts (with $d\nu/dt > 0$) allow us to pinpoint the plasma frequency ν_p of the common startfrequency, and thus to probe the electron density n_e in the acceleration site. During some flares, a clear separatrix between the two burst types has been observed, demarcating the height of the acceleration region (see insert in Fig.7). The electron density of the acceleration region (inferred from plasma frequency at the separatrix) is found to be typically $n_e^{acc} \approx 10^9 - 10^{10}$ cm^{-3}, indicating a low-density site outside the SXR-bright flare loops. The start of many type III and type RS bursts was found to exhibit a detailed one-to-one correlation with sub-second HXR pulses (observed with *SMM/HXRBS* or *BATSE/COMPTON*) within a coincidence of $\lesssim 100$ ms, implying a topological connection between the coronal acceleration site and the chromospheric HXR emission site (e.g. Aschwanden, Benz, & Schwartz 1993; Aschwanden et al. 1995b). These observational signatures in upward and downward direction constitute strong evidence that particle acceleration occurs in some bi-directional symmetry with respect to the (vertical) magnetic field. The symmetry of the accelerating fields does, however, not convey an equal number of accelerated particles to the chromosphere and into interplanetary space, since many magnetic field lines above flare regions seem to be closed on a larger scale, based on the frequent observation of metric type U-bursts (see cartoon in Fig.7).

Radio signatures (type III, U, RS-bursts) of electron beams have been detected within the context of *Yohkoh/SXT* imagery recently, generally outlining large-scale closed magnetic field structures above flare loops (Pick et al. 1994; Aurass, Klein, & Martens 1995) or collimated magnetic corridors pointing away from the flare region, sometimes detectable as SXR jets (Kundu et al. 1995; Raulin et al. 1996).

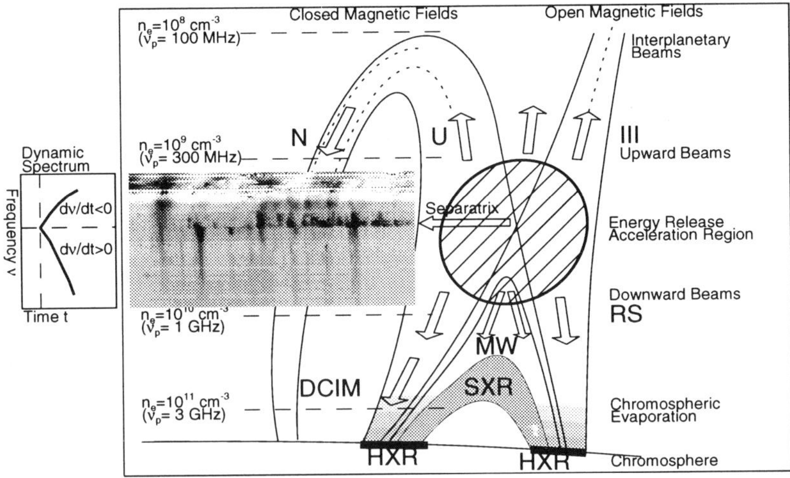

Figure 7. A synthesized standard flare scenario: The acceleration region is located in the reconnection region above the SXR-bright flare loop, accelerating electron beams in the upward direction (type III, U, N bursts) and in the downward direction (type RS, DCIM bursts). Downward moving electron beams precipitate to the chromosphere (producing HXR emission and driving chromospheric evaporation), or remain transiently trapped (producing microwave [MW] emission). SXR loops become subsequently filled up, with increasing footpoint separation as the X-point rises. The insert shows a dynamic radio spectrum (ETH Zurich) of the 92-Sept-06, 1154 UT, flare, showing a separatrix between type III and type RS bursts at $\approx$ 600 MHz, probably associated with the acceleration region.

8. Observational Constraints for Particle Acceleration Models

HXR observations from *COMPTON* and *Yohkoh* convey new constraints for theoretical (electric DC-field, stochastic, shock) models of particle acceleration in solar flares:

1. **The acceleration site** is most likely located *above* the SXR-bright flare loop, based on the electron time-of-flight measurements and Masuda's discovery of above-the-loop-top HXR sources. The electron density of the acceleration site is in the regime of $n_e \approx 10^9 - 10^{10}$ cm^{-3}. The accelerator is required to have some spatial symmetry in order to produce bi-directional electron beams.
2. **Acceleration time scales** need to be shorter than the observed subsecond HXR pulses. The TOF delays require that acceleration times are shorter than the free-streaming propagation times to the chromosphere (see also timing models in Aschwanden 1996), being $\approx 100-300$ ms in typical flare loops.
3. **The pitch-angle distribution** of accelerated particles needs to be nearly isotropic to explain the observed large variation of trapping

efficiencies from flare to flare (inferred from the HXR modulation depths). Parallel acceleration (e.g. DC-electric fields) cannot explain high trapping ratios, while perpendicular acceleration would produce larger TOF delays than observed.

Acknowledgements: We thank Takeo Kosugi, Hugh Hudson, Brian Dennis, Richard Schwartz, Daniel Alt, Meredith Wills, and Robert Bynum for collaborations and helpful discussions. We gratefully acknowledge generous access and support by the *CGRO*, *Yohkoh*, and *GSFC* data centers. This work has been supported by NASA/SRT grants NAG-5-3593 and NSGW-5078.

9. References

Alexander,D. and Metcalf,T.R. 1997, ApJ, in press
Aschwanden,M.J. 1996, in "High Energy Solar Physics", (eds. R.Ramaty, N.Mandzhavidze, and X.-M.Hua), AIP Conf. Proc. 374, p.300
Aschwanden,M.J. and Benz,A.O. 1997, ApJ 480, 825
Aschwanden,M.J., Benz,A.O., Dennis,B.R., and Schwartz,R.A. 1995b, ApJ 455, 347
Aschwanden,M.J., Benz,A.O., and Schwartz,R.A. 1993, ApJ 417, 790
Aschwanden,M.J., Bynum,R.M., Kosugi,T., Hudson,H.S., and Schwartz,R.A. 1997, ApJ 487, (October 1 issue), in press
Aschwanden,M.J., Hudson,H., Kosugi,T., and Schwartz,R.A. 1996a, ApJ 464, 985
Aschwanden,M.J., Kosugi,T., Hudson,H.S., Wills,M.J., and Schwartz,R.A. 1996c, ApJ 470, 1198
Aschwanden,M.J. & Schwartz 1995, ApJ 455, 799
Aschwanden,M.J. & Schwartz 1996, ApJ 464, 974
Aschwanden,M.J., Schwartz,R.A., and Alt,D.M. 1995, ApJ 447, 923
Aschwanden,M.J., Wills,M.J., Hudson,H.S., Kosugi,T., and Schwartz,R.A. 1996b, ApJ 468, 398
Aurass,H., Klein,K.L., and Martens,P.C.H. 1995, Solar Phys. 155, 203
Brown,J.C. 1971, Solar Phys. 18, 489
Duijveman,A., Hoyng,P., & Machado,M.E. 1982, Solar Phys. 81, 137
Emslie,G.A. 1996, ASP Conf.Ser. 111, 419
Hudson,H. and Ryan,J. 1995, Ann.Rev.Astron.Astrophys. 33, 239
Kosugi,T. 1996, in "High Energy Solar Physics", (eds. R.Ramaty, N.Mandzhavidze, and X.-M.Hua), AIP Conf.Proc. 374, p.267
Kundu,M.R., Raulin,J.P., Pick,M., and Strong,K.T. 1995, ApJ 444, 922
Martens,P. 1996, this workshop
Masuda,S. 1994, PhD thesis, Univ. Tokyo
Pick,M., Raoult,A., Trottet,G., Vilmer,N., Strong,K., Magalhaes,A. 1994, Proc. Kofu Symp., NRO Rep. 360, 263
Raulin,J.P., Kundu,M.R., Nitta,N., and Raoult,A. 1996, ApJ 472, 874
Sakao,T. 1994, PhD thesis, Univ. Tokyo
Takakura,T., Tanaka,K., & Hiei,E., 1984, Adv.Space Res., 4, 143
Tsuneta,S. 1996, ASP Conf.Ser. 111, 409

MICROWAVE AND HARD X-RAY OBSERVATIONS OF AN X-CLASS LIMB FLARE

HIROSHI NAKAJIMA AND KEN'ICHI FUJIKI
Nobeyama Radio Observatory
Nobeyama, Minamisaku, Nagano 384-13, Japan

THOMAS R. METCALF
Lockheed-Martin Palo Alto Research Laboratory
O/H1-12, Bldg. 252, 3251 Hanover St., Palo Alto, CA 94304, USA

SHARAD R. KANE
Space Sciences Laboratory, University of California
Berkeley, CA 94720, USA

AND

MAKI AKIOKA
Hiraiso Solar Terrestrial Research Center
3601 Isozaki, Hitachnaka, Ibaraki 311-12, Japan

Extended Abstract

We study two intense, extended microwave and hard X-ray flares which occurred slightly behind the west limb on 1992 November 2, using data from the Nobeyama Radioheliograph at 17 GHz and the Hard X-Ray and Soft X-Ray Telescopes onboard the Yohkoh. We get the following observational results:

(1) There are two weak enhancement phases prior to the onset of the impulsive phase, i.e., a weak preflare heating phase accompanied by unpolarized microwave emission and an initial phase accompanied by polarized microwave emission from high energy electrons. The preflare region in the soft X-ray image consists of a streamer-like structure and overlying large closed loops. The most intense part of the enhancements is located in the lower part of the streamer-like structure. The microwave source in the ini-

T. Watanabe et al. (eds.), Observational Plasma Astrophysics: Five Years of Yohkoh and Beyond, 295–296.

tial phase evolves at the base of the streamer-like structure seen in the preflare phase.

(2) The major hard X-ray emission (23-33 keV) in the impulsive phase is confined in two locations above the brightest top of the corresponding soft X-ray loop. The corresponding microwave source covers the whole hard X-ray and soft X-ray sources.

(3) There are two components of electron spectra in the impulsive phase, i.e., one is a quite soft component which contributes to hard X-ray emissions at energies $\leq$ 100 keV and the other is a quite hard component which contributes to microwave emissions.

(4) Fitting of a gyrosynchrotron formula to the observed microwave spectrum in the optically-thin part (17, 35, and 80 GHz) suggests that electrons at energies $\geq$ 500 keV mainly contribute to the microwave emissions at frequencies $\geq$ 17 GHz and that the magnetic field is rather strong, more than 250 Gauss.

The streamer-like structure seen in the preflare phase suggests initial formation of a magnetic neutral sheet.

The hard X-ray source in the impulsive phase is confined in a small region considerably higher than the corresponding soft X-ray loop, suggesting that the softer electron component is accelerated near the magnetic neutral sheet. On the other hand, higher-energy ($\geq$ 500 keV) electrons which contribute to microwave emission are distributed in the larger region than the soft X-ray loop, suggesting acceleration and confinement of higher-energy ($\geq$ 500 keV) electrons inside and outside the soft X-ray loop. As the time profile of the 17 GHz total emission is globally similar to the hard X-ray time profile observed by ULYSSES, acceleration of lower-energy ($\leq$ 100 keV) and higher-energy ($\geq$ 500 keV) electrons should be closely related to each other. One of the most probable acceleration mechanisms to meet these observations is two-step acceleration, in which the site of higher-energy electrons is located near the acceleration site of lower-energy electrons, and lower-energy electrons are further accelerated to higher energies and distributed inside and outside the soft X-ray loop.

A full paper on this subject will be submitted shortly to Solar Physics.

MAGNETIC RECONNECTION AND PARTICLE ACCELERATION IN SOLAR FLARES

T. KOSUGI
National Astronomical Observatory
Osawa 2-21-1, Mitaka, Tokyo 181, Japan

AND

B.V. SOMOV
Astronomical Institute, Moscow State University
Universitetskii Prospekt 13, Moscow 119899, Russia

Abstract. We briefly review the main observational and theoretical results based on *Yohkoh* data on the site and mechanism of magnetic energy transformation into kinetic and thermal energies of 'super-hot' plasma and accelerated particles in solar flares. Open issues are the focus of our attention.

1. Introduction

Hard and soft X-ray observations on board *Yohkoh* (Kosugi et al., 1991; Tsuneta et al., 1991) suggest that magnetic reconnection is responsible for many non-steady phenomena in the corona. In particular, in many solar flares reconnection of magnetic field lines appears to take place. Reconnection seems to be common to impulsive (compact) and gradual (large-scale) flares (Shibata, 1996; Tsuneta *et al.*, 1997). However, in the interpretation of the *Yohkoh* data, 'the basic physics of the reconnection process remains uncertain' (Masuda *et al.*, 1994).

We advocate the point of view that collisionless reconnecting current sheets play the role of the source of flare energy and, at least, the first-step mechanism in a two-step acceleration of electrons and ions to high energies (Somov and Kosugi, 1997). Bearing this concept in mind, we summarize briefly what we know, on the basis of *Yohkoh* observations, and what else we would like to know on reconnection and acceleration in solar flares.

T. Watanabe et al. (eds.), Observational Plasma Astrophysics: Five Years of Yohkoh and Beyond, 297–306.

2. Key Evidence of Reconnection

Three well established properties of X-ray structures observed by *Yohkoh* allow us to assume that solar flare energy release proceeds via magnetic reconnection.

First, Masuda *et al.* (1994, 1995) discovered a hard X-ray source above the soft X-ray loop (Figure 1). The location of this source is consistent with a general picture of reconnection in flares (Hirayama, 1974). Moreover, Shibata *et al.* (1995), Ugai (1996), and Tsuneta *et al.* (1997) suggest that the hard X-ray above-the-loop-top source may be heated by the fast MHD shock associated with super-Alfvénic outflow from the reconnection site.

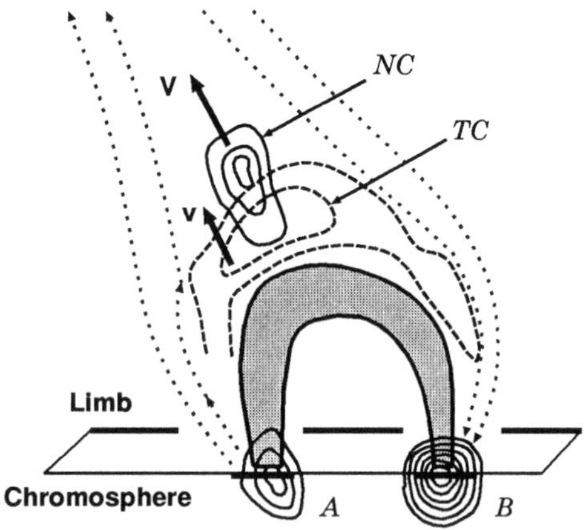

Figure 1. Sketch of non-thermal (NC) and quasi-thermal (TC) components of the coronal hard X-ray emission and their apparent motion. A and B are the chromospheric 'footpoint sources'.

Second, an apparent rising motion of the flaring loop together with the above-the-loop-top source and an increase in separation of the hard X-ray double 'footpoint sources' (Sakao *et al.,* 1997) seem to be consistent with an assumed increasing height of the reconnection site and an increasing size of the reconnected field lines as reconnection proceeds (e.g., Tsuneta, 1995). So, the observed motions of X-ray sources can be understood as a result of the magnetic reconnection process, for example, in a high-temperature turbulent-current sheet (HTTCS) which accelerates particles and fast flows of 'super-hot' plasma with reconnected frozen-in magnetic field lines (Somov and Kosugi, 1997).

Third, a 'loop-with-a-cusp' structure is further evidence of magnetic reconnection in flares observed by *Yohkoh* (Tsuneta *et al.*, 1992; Masuda *et al.*, 1994; Tsuneta, 1996). It would be really hard to create sharp cusp-like structure at the loop top without invoking reconnection.

3. Indirect Evidence of Reconnection

There is growing qualitative evidence showing that flares in the corona are due to reconnection. Each piece of evidence might be considered as one which does not give us the unique interpretation, but taken all the evidence appears to unambiguously point to reconnection.

For example, coronal X-ray *jets*, i.e., transitory soft X-ray enhancements with a strongly collimated apparent motion, presumably result from reconnection between new emerging flux and pre-existing coronal magnetic fields (Shibata *et al.*, 1996; Yokoyama and Shibata, 1996). Blob-like X-ray ejections, often referred to as *plasmoids*, are also typical consequences of reconnection in current sheets (Horiuchi and Sato, 1994; Schumacher and Kliem, 1996; Shibata, 1996; Ugai, 1996).

Observed changes of large-scale coronal structures and reformation of coronal streamers that are associated with coronal mass ejections cannot be understood without invoking the concept of reconnection (Acton, 1996). *Yohkoh*'s wide dynamic range and high-cadence observations revealed unexpected structures and their changes that are very essential in understanding the key role of the actual 3D reconnection in the faint preflare stages as well as during arcade flares observed on the limb (Uchida, 1996; Uchida *et al.*, 1997).

The 'super-hot' ($T \geq 30$ MK) plasma observed by the HXT above the soft X-ray loops in flares (Sakao *et al.*, 1992; Sakao and Kosugi, 1996; Tsuneta *et al.*, 1997) can be produced by the MHD reconnection process of the Petschek type or by collisionless reconnection in the HTTCS. The high-temperature ridges located on both sides of the soft X-ray loop (Tsuneta *et al.*, 1997) are presumably heated by MHD slow shocks of the Petschek type or, alternatively, by thermal conductive fluxes directly from the high-temperature current sheet (Oreshina and Somov, 1997).

Fast outflows of 'hot' ($T < 30$ MK) plasma can produce a symmetrical broadening of the optically thin soft X-ray lines observed during solar flares. This broadening is larger than the thermal one. A comparison of the observed profiles of the Fe XXV emission lines with the predictions of the HTTCS model suggests that the presence in the flare region of several small-scale or one (or a few) large-scale curved HTTCS (Antonucci *et al.*, 1996).

It is necessary, however, to differentiate between the super-hot plasma generated in the so-called Hot Thermal (type A) flare and the Impulsive (type B) flare (Tanaka, 1987). The super-hot plasma in type A flares is located at the loop top or above it. Impulsive (type B) flares can also generate a hot plasma which is a direct result of chromospheric heating by powerful beams of accelerated electrons. For example, during the impulsive phase in the flare of 16 December 1991 the Bragg Crystal Spectrometer (BCS) on board *Yohkoh* showed strong blueshifts in the resonance lines of the Ca XIX and Fe XXV ions (Culhane *et al.*, 1994). This clearly indicates the presence of hot plasma upflows, known as the chromospheric evaporation (for a review see Somov, 1992).

4. Primary Origin of Flare Energy

Gold and Hoyle (1960) have shown that the flare energy storage can occur in a particular class of magnetic fields – twisted loops protruding above the photosphere. Two such loops, say two electric currents, attract each other, and the annihilation of the antiparallel components of the magnetic field leads to a sudden release of magnetic energy. *Yohkoh* observations provide substantial evidence of two-loop interaction in active regions and a growing number of wonderful examples of flares due to coalescence of current-carrying loops (Sakai and de Jager, 1996).

Sweet (1969) and Syrovatskii (1981) have developed another concept of the flare energy source. They assumed that an interaction of magnetic fluxes in the solar atmosphere leads to accumulation of an excess energy called the *free* magnetic energy. This is the energy of the magnetic field related to a reconnecting current sheet located along the separator – the field line where the magnetic fluxes interact and reconnect. Recently, Longcope and Cowley (1996) investigated the requirements for current sheet formation at the separator and found how a 3D solution, with a 'current ribbon' (curved current sheet), establishes the total flux available for reconnection, and how much free energy such reconnection will liberate.

Simultaneous observations with *Yohkoh* and the Nobeyama Radioheliograph of several impulsive flares provide information on the relative locations of the microwave, hard X-ray, and soft X-ray sources. Results suggest a loop configuration in which long and short loops interact with each other (Hanaoka, 1997; Nishio *et al.*, 1997). The emerging flux model is most plausible; an emerging small loop collides with a pre-existing large loop and produces accelerated electrons.

These results support the general idea of 3D reconnection in flares. However, which is most important of the two above-mentioned fundamental mechanisms that provide the magnetic energy to a flare remains unclear: the electric currents (loops) generated under the photosphere or the current sheets generated in the corona to screen the interacting magnetic fluxes.

5. Global Models for Solar Flares

The 2D MHD 'standard model' of a two-ribbon flare (Forbes and Priest, 1982; Magara *et al.*, 1996; Ugai, 1996; Yokoyama and Shibata, 1997; see also references in Tsuneta, 1996) explains qualitatively the observed position and upward motion of the above-the-loop-top source as well as the increase of the distance between the footpoint sources.

Following the standard model, the reconnecting current sheet in flares is considered as a 'point' which is responsible for the reconnection process only in a geometrical sense, i.e. the X-type neutral point at which connectivity of field lines in a 2D plane changes. As in a vacuum or in a plasma of very low conductivity, it is supposed that no energy is converted by this point. External hydrodynamics in the vicinity of the X-type point is discussed and is numerically simulated in terms of slow MHD shock waves. They convert the energy of the interacting magnetic fluxes into the thermal and kinetic energy of fast flows of plasma. This means, in fact, that the classical Petschek model of reconnection is used to explain the energetics and dynamics of hydrodynamic plasma flows in solar flares.

More exactly, the standard model assumes and takes as an initial condition that the current sheet *initially* does exist between two homogeneous vertical magnetic fields of opposite directions, but somehow reconnection between them is not allowed to proceed. Thus, in the standard model, the flare energy is initially accumulated in a very large neutral sheet sandwiched by anti-parallel fields. Such an initial condition is very specific and creates strongly collimated plasma flows because the inflow magnetic fields are 'only slightly curved' (Priest and Lee, 1990).

The 2.5D models based on the Woltjer-Taylor minimum-energy principle are more realistic and show that a solar flare can be a transition process between two different states generated as a result of the bifurcation of coronal magnetic fields (Kusano *et al.*, 1995). In general, this implies that reconnection must be accompanied by a flare. Numerical simulations show that the coronal magnetic fields do transit from a higher to a lower energy state via reconnection in a reasonable geometry of plasma flows (Kusano *et al.*, 1997).

6. Local Models for the Reconnection Process

Analytical, numerical, and combined models of the reconnection process exist in very different approximations and with very different levels of self-consistency (for reviews see Biskamp, 1994; Somov, 1994). It becomes more and more obvious that *collisionless* driven reconnection in a high-temperature rarefied plasma such as the solar corona is an important process in considering energetically active phenomena like solar flares. This process was introduced by Syrovatskii (1966) as a *dynamic dissipation* of magnetic field in a reconnecting current sheet and leads to fast energy conversion from field energy to particle energy, as well as a topological change of magnetic field (e.g., Drake and Kleva, 1991; Horiuchi and Sato, 1994; Ono *et al.*, 1996).

Some general properties of such collisionless reconnection can be examined in a frame of a self-consistent model based on the mass, momentum, and energy conservation laws. A particular feature of this model (e.g., Somov, 1992) is that electrons and ions are heated by wave-particle interactions in a different way; contributions to the energy balance are not made either by energy exchange between electrons and ions due to collisions, thermal flux across the magnetic field, nor by energy losses due to radiation. The magnetic-field-aligned thermal flux becomes anomalous and plays the dominant role in the cooling of the electrons in the HTTCS.

These properties are typical for collisionless plasmas in solar flares and can be taken into account by two equations describing the balance of energy for electrons and ions separately:

$$\chi_{\text{ef}}\, \mathcal{E}^{\,in}_{mag} = \mathcal{E}^{\,out}_{t,e} + \mathcal{C}^{\,an}_{\parallel}\,, \tag{1}$$

$$(1 - \chi_{\text{ef}})\, \mathcal{E}^{\,in}_{mag} = \mathcal{E}^{\,out}_{t,i} + \mathcal{K}^{\,out}_{i}\,. \tag{2}$$

Here χ_{ef} is the relative fraction of the heating consumed by electrons from the magnetic energy flux

$$\mathcal{E}^{\,in}_{mag} = \frac{B_0^2}{4\pi}\, v_0\, 4b\,, \tag{3}$$

while the remaining fraction $(1 - \chi_{\text{ef}})$ goes into ions. The sum of the left-hand sides of Equations (1) and (2) represents the direct heating of ions and electrons due to their interactions with waves inside the HTTCS.

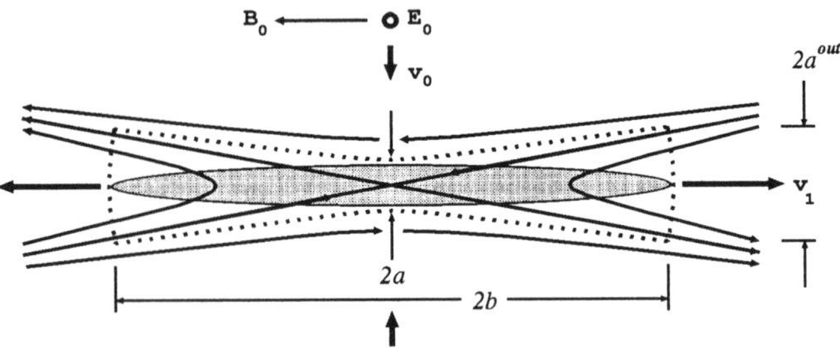

Figure 2. Characteristic scales of the HTTCS: $2a$ is a dissipative thickness, the electric current distribution is shown by shadow; $2b$ is a scale responsible for energy accumulation; the dotted boundary indicates the field lines going through the sheet, so $2a^{out}$ is the scale which determines the outflow of energy and mass.

Because of the difference between the effective temperatures of electrons and ions, the electron and ion enthalpy outflows also differ: $\mathcal{E}^{\,out}_{t,e} \neq \mathcal{E}^{\,out}_{t,i}$. The ion kinetic energy flux $\mathcal{K}^{\,out}_{i}$ is important in (2). The kinetic energy of the electrons is negligible and disregarded in (1). However, electrons play the dominant role in the heat conductive cooling, $\mathcal{C}^{\,an}_{\parallel}$.

In the conditions of flares, the characteristic parameters of such collisionless current sheets are the following. (a) The effective electron temperature $T_e \approx 100 - 200$ MK, the temperature ratio $\theta \equiv T_e/T_i \approx 6.5$. (b) The dissipative thickness $2a \approx 20$ cm is small but the width $2b \approx (1 - 2) \times 10^9$ cm is large, for this reason the linear scale for the outflows of energy and mass $2a^{out} \approx (3 - 6) \times 10^6$ cm is not small (Figure 2). (c) The power of energy release $P_s \approx 3 \times 10^{28} - 2 \times 10^{29}$ erg s^{-1}. The outflow velocity $v_1 \approx 1400 - 1800$ km s^{-1}.

7. Particle Acceleration in Solar Flares

Particle acceleration to high energies in flares is the subject of a great deal of study, but the acceleration mechanism still defies a full theoretical understanding (e.g., Chupp, 1996).

The hard X-ray source located above the top of the flaring loop visible in soft X-rays clearly indicates that some energetic mechanism takes place *outside* the closed loop and works impulsively with efficient acceleration of electrons (Kosugi, 1996). Electron time-of-flight analyses (Aschwanden *et al.*, 1996) confirm that the coronal hard X-ray source discovered by Masuda *et al.* (1994, 1995) is a signature of the electron acceleration site. Plasma inside the above-the-loop-top source seems to be really *collisionless* (Takakura *et al.*, 1993) or marginally collisionless – with very small influence of binary collisions.

As for particle acceleration, two possibilities can be considered. The first is the direct electric field acceleration of electrons and ions to high energies in an non-neutral HTTCS (for a review see Somov, 1996). This seems to be the case for acceleration of relativistic particles in the high corona during the late phase of γ-ray flares. However, in flares with an above-the-loop-top source, some trapping of energetic electrons is unavoidable (Kosugi, 1996).

The second possibility is two-step acceleration (Somov and Kosugi, 1997). Reconnected field lines rapidly move out of the HTTCS, being frozen into super-hot plasma, and form magnetic loops on the upstream side of the fast oblique collisionless shock (FOCS) placed above the soft X-ray emitting loops of a strong magnetic field. The electrons and ions energized and preaccelerated by the HTTCS are trapped in magnetic loops. The top of each loop moves with a high speed towards the FOCS, while its feet penetrate through the shock front. For these reasons, two mechanisms – the adiabatic heating inside the collapsing trap and acceleration by the shock front at two feet of the trap – can efficiently increase the energy of the particles.

The lifetime of an individual collapsing trap can be identified with the observed short time scale (a few seconds) delay in higher energies of hard X-ray and gamma-ray emission. The trapped accelerated electron location can be seen as the coronal hard X-ray above-the-loop-top source. Precipitation of accelerated electrons from the trap through the FOCS into the chromosphere is responsible for the hard X-ray footpoint sources. This model explains the observed relative intensity of the coronal and chromospheric hard X-ray sources and their other physical properties (Somov and Kosugi, 1997).

8. Open Issues of Reconnection in Solar Flares

The current state of theoretical efforts in modeling the reconnection process in flares can be summarized by Figure 3. The *global* 2D MHD models under current use have a very important advantage. They allow us for the first time to make a comparison between *Yohkoh* observations and the models (e.g., Shibata *et al.*, 1995, 1996; Tsuneta, 1995, 1996; Tsuneta *et al.*, 1997). One of the advantages of the *local* models is that they take kinetic plasma effects into account. This allows one to develop 'the basic physics of the reconnection process'. However, the local models are not incorporated in the global consideration of the flare process. For this reason, the local models are less appropriate for comparison with observations.

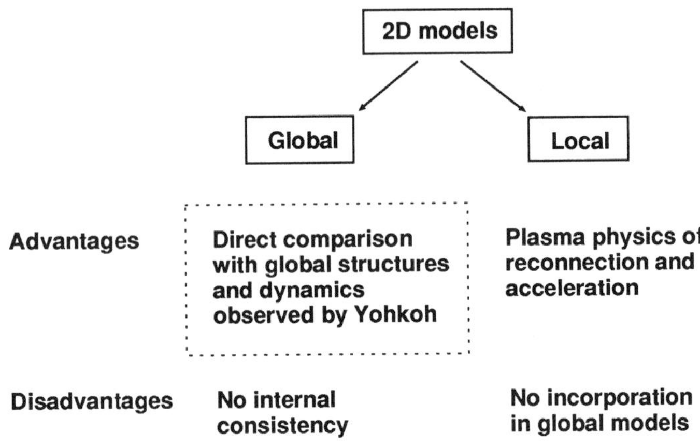

Figure 3. Reconnection models under wide current use.

Future models should join 'global' and 'local' properties of the magnetic reconnection process under solar coronal conditions. For example, chains of plasma instabilities, including kinetic instabilities, can be very important for our understanding of the types and regimes of plasma turbulence inside the collisionless current sheet. In particular it is necessary to evaluate anomalous resistivity and selective heating of particles in the HTTCS. Heat conduction is also anomalous in the high-temperature plasma of solar flares. Self-consistent solutions of the reconnection problem will allow us to explain the energy release in flares, including the open question of the mechanism or combination of mechanisms which explains the observed acceleration of electrons and ions to high energy (Hudson and Ryan, 1995; Somov, 1996).

To understand the 3D structure of actual reconnection in flares is one of the most urgent problems. Actual flares are 3D dynamic phenomenon of electromagnetic origin in a highly-conducting plasma with a strong magnetic field. The following example clearly shows that 3D models of flares (e.g., Gorbachev and Somov, 1990) should be involved in treatment of *Yohkoh* data. Recently, two sub-classes of impulsive flares observed with the HXT have been discovered by Sakao *et al.* (1997). These two sub-classes can be characterized as *more impulsive* (MI) and *less impulsive* (LI) flares. In LI flares, the footpoint separation increases with time. This fact seems to be consistent with the standard model. On the contrary, the flares with decreasing footpoint separation are of the type MI. It does not seem possible to explain these flares in the framework of 2D MHD models.

9. Conclusion

Yohkoh observations with HXT, SXT, and BCS offer us the means to check whether phenomena predicted by solar flare models of a definite type (such as the 2D MHD standard model or the classical 'quadrupole-type model' by Sweet (1969)) actually occur. There are apparent successes of the standard model, for example, in the morphology of flares with cusp geometries. However, some puzzling discrepancies also exist, and further development of more realistic 3D models is urgently required.

References

Acton, L. W. (1996) in Y. Uchida, T. Kosugi, and H. S. Hudson (eds.) *Magnetodynamic Phenomena in the Solar Atmosphere – Prototypes of Stellar Magnetic Activity*, Kluwer Academic Publishers, Dordrecht, 3.

Antonucci, E., Benna, C., and Somov, B. V. (1996) *Astrophys. J.* **456**, 833.

Aschwanden, M. J., Kosugi, T., Hudson, H. S., *et al.* (1996) *Astrophys. J.* **470**, 1198.

Biskamp, D. (1994) in C. Chiudery and G. Einaudi (eds.) *Plasma Astrophysics*, 1.

Chupp, E. L. (1996) in R. Ramaty, N. Mandzhavidze, and X. M. Hua (eds.) *High Energy Solar Physics*, AIP Press, New York, 3.

Culhane, J. L., Phillips, A.T. Inda-Koide, M., *et al.* (1994) *Solar Phys.* **153**, 307.

Drake, J. F. and Kleva, R. G. (1991) *Phys. Rev. Lett.* **66**, 1458.

Forbes, T. G. and Priest, E. R. (1982) *Solar Phys.* **84**, 169.

Gold, T. and Hoyle, F. (1960) *Monthly Not. Royal Astron. Soc.* **120**, 89.

Gorbachev, V. S. and Somov, B. V. (1990) *Adv. Space Res.* **10**, No. 9, 105.

Hanaoka, Y. (1997) *Solar Phys.* in press.

Hirayama, T. (1974) *Solar Phys.* **34**, 323.

Hudson, H. and Ryan, J. (1995) *Ann. Rev. Astron. Astrophys.* **33**, 239.

Horiuchi, R. and Sato, T. (1994) *Phys. Plasmas* **1**, 3587.

Kosugi, T. (1996) in R. Ramaty, N. Mandzhavidze, and X.-M. Hua (eds.) *High Energy Solar Physics*, AIP Press, New York, 267.

Kosugi, T., Makishima, K., Murakami, T., *et al.* (1991) *Solar Phys.* **136**, 17.
Kusano, K., Suzuki, Y., and Nishikawa, K. (1995) *Astrophys. J.* **441**, 942.
Kusano, K., Suzuki, Y., Fujie, K., *et al.* (1997) *Magnetic Reconnection in the Solar Atmosphere*, Astron. Soc. Pacific Conf. Series **111**, 280.
Longcope, D. W. and Cowley, S. C. (1996) *Phys. Plasmas* **3**, 2885.
Magara, T., Mineshige, S., Yokoyama, T., *et al.* (1996) *Astrophys. J.* **466**, 1054.
Masuda, S., Kosugi, T., Hara, H., *et al.* (1994) *Nature* **371**, 495.
Masuda, S., Kosugi, T., Hara, H., *et al.* (1995) *Publ. Astron. Soc. Japan* **47**, 677.
Nishio, M., Yaji, K., Kosugi, T., *et al.* (1997) *Astrophys. J.* in press.
Ono, Y., Yamada,M., Akao, T., *et al.* (1996) *Phys. Rev. Lett.* **76**, 3328.
Oreshina, A.V. and Somov, B.V. (1997) *Astron. Astrophys.* **320**, L53.
Priest, E. R. and Lee, L. C. (1990) *J. Plasma Phys.* **44**, 337.
Sakai, J.-I. and de Jager, C. (1996) *Space Sci. Rev.*, **77**, 1.
Sakao, T. and Kosugi, T. (1996) in Y. Uchida, T. Kosugi, and H. S. Hudson (eds.) *Magnetodynamic Phenomena in the Solar Atmosphere – Prototypes of Stellar Magnetic Activity*, Kluwer Academic Publishers, Dordrecht, 169.
Sakao, T., Kosugi, T., Masuda, S., *et al.* (1992) *Publ. Astron. Soc. Japan* **44**, L83.
Sakao, T., Kosugi, T., and Masuda, S. (1997) in these proceedings.
Schumacher, J. and Kliem, B. (1996) *Phys. Plasmas* **3**, 4703.
Shibata, K. (1996) *Adv. Space Res.* **17**, No. 4/5, 9.
Shibata, K., Masuda, S., Shimojo, M., *et al.* (1995) *Astrophys. J.* **451**, L83.
Shibata, K., Yokoyama, T., and Shimojo, M. (1996) *J. Geomag. Geoelectr.* **48**, 19.
Somov, B.V. (1992) *Physical Processes in Solar Flares*, Kluwer Academic Publ., Dordrecht.
Somov, B. V. (1994) *Fundamentals of Cosmic Electrodynamics*, Kluwer Academic Publ., Dordrecht.
Somov, B. V. (1996) in R. Ramaty, N. Mandzhavidze, and X.-M. Hua (eds.) *High Energy Solar Physics*, AIP Press, New York, 493.
Somov, B. V. and Kosugi, T. (1997) *Astrophys. J.* **485**, in press.
Sweet, P. A. (1969) *Ann. Rev. Astron. Astrophys.*, **7**, 149.
Syrovatskii, S. I. (1966) *Ann. Rev. Astron. Astrophys.*, **19**, 163.
Syrovatskii, S. I. (1981) *Soviet Astronomy – AJ*, **10**, 270.
Takakura, T., Inda, M., Makishima, K., *et al.* (1993) *Publ. Astron. Soc. Japan* **45**, 737.
Tanaka, K. (1987) *Publ. Astron. Soc. Japan* **39**, 1.
Tsuneta, S. (1995) *Publ. Astron. Soc. Japan* **47**, 691.
Tsuneta, S. (1996) *Astrophys. J.* **456**, 840.
Tsuneta, S., Acton, L., Bruner, M., *et al.* (1991) *Solar Phys.* **136**, 37.
Tsuneta, S., Hara, H., Shimizu, T., *et al.* (1992) *Publ. Astron. Soc. Japan* **44**, L63.
Tsuneta, S., Masuda, S., Kosugi, T., Sato, T. (1997) *Astrophys. J.* **478**, 787.
Uchida, Y. (1996) *Adv. Space Res.* **17**, No. 4/5, 19.
Uchida, Y., Fujisaki, K., Morita, S., *et al.* (1997) in these proceedings.
Ugai, M. (1996) *Phys. Plasmas* **3**, 4172.
Yokoyama, T. and Shibata, K. (1996) *Publ. Astron. Soc. Japan* **48**, 353.
Yokoyama, T. and Shibata, K. (1997) *Astrophys. J.* **474**, L61.

<Posters>

ACTIVITY IN NOAA AR7172

B. ANWAR
Watukosek Solar Observatory
Gempol, P.O. Box 4, Pasuruan 67155, East Java, Indonesia

AND

M. AKIOKA
Hiraiso Solar Terrestrial Research Center, CRL
Isozaki 3601, Hitachinaka-shi, Ibaraki 311–12, Japan

Abstract. NOAA Active Region 7172 was observed by the Hα telescope at Hiraiso Solar Terrestrial Research Center during 18-28 May 1992. The region showed high activity during its passage on the solar disk. As many as 33 small flares were detected by the GOES satellite. The *Yohkoh* Soft X–ray Telescope also provided good coverage of data during this period. There were no intense flares observed from the region. This paper presents a comparative study of the activities observed in AR7172 using the Hα, X-ray and magnetic fields data.

1. Introduction

AR7172 was recognized as an active region on 18 May 1992 when the region was located at N23E68 with a sunspot class of *Bxo* (Solar–Geophysical Data 575–Part I, 1992). There was no signature of an active region at the same location during the previous rotation.

The region showed high activity during its appearrance compared to other regions, and therefore was selected as the main target of an international observation campaign. The *Yohkoh* Soft X–ray Telescope (SXT, Tsuneta *et al.*, 1991) produced partial frame images (PFIs) in various X–ray wavelengths as well as white–light images taken by its Aspect Sensor.

AR7172 was also observed with the Hα telescope at Hiraiso Solar Terrestrial Research Center (Akioka, 1996) with high time cadence. This allows us to make a comparative study of the activity in Hα and X–rays. In addi-

T. Watanabe et al. (eds.), Observational Plasma Astrophysics: Five Years of Yohkoh and Beyond, 309–312.

tion, the Heleakala Stokes Polarimeter of Mees Observatory (Mickey, 1985) provided magnetic field maps of the region, which we use to determine polarity of spots and the general magnetic field pattern.

All flares observed by GOES–7 during AR7172's solar disk passage were C–class flares. This allows us to investigave the conditions which lead to the occurrences of small flares in active regions.

2. Evolution of AR7172

2.1. PHOTOSPHERIC EVOLUTION

The Aspect Sensor of SXT began observations of AR7172 with the Narrow-Band filter on 20 May 12:55:06 UT, and produced a sequence of PFIs taken in *Quiet* mode. The white–light observations continued until 26 May 15:05:59 UT, obtaining a total of 85 frames. The time intervals of the sequence are generally short enough compared to the evolution time scale of the active region so that we can follow the photospheric changes.

In the first image of the data sequence, a proceeding spot with negative polarity is the most remarkable feature. It is followed by a small traling spot of positive polarity. Other spots emerged in between these two spots, and the region reached its maximum area on 22 May at about 8 UT, when it had a 'foot–print'–like configuration. The emergence of new spots was accompanied with the disappearance of a spot at the northernmost part of the active region. The region then decayed over the following days and formed a bipolar configuration on 26 May.

2.2. CHROMOSPHERIC EVOLUTION

Hα images taken at about one minute temporal resolution show detailed changes in the chromosphere (Figure 1). In general, the Hα structures of the region did not show any high shear pattern. Many brightenings occurred at the site of the flux emergence. The most prominent bipole emerged very close to the proceeding spot. This bipole had dark loops which eventually became bright. Since the spot located closest to the proceeding spot also had negative polarity, this kind of configuration may not be expected to produce intense flare activity, although an increase in activity may still result from this configuration (Hanaoka, 1994).

2.3. X–RAY EVOLUTION

We have made overlaid X-ray and white-light images to see the correspondence between the coronal loops and features in the photosphere. The *Yohkoh* data analysis software allows the alignment of the two data sets to be achieved accurately. The X-ray loops were almost parallel to the axis of the main spots and perpendicular to the magnetic neutral line. This suggests that the active region had a nearly potential magnetic field.

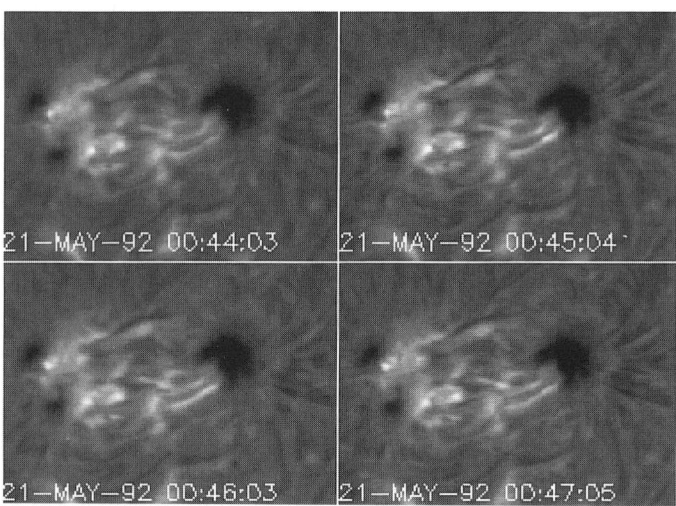

Figure 1. AR7172 seen in Hα on 21 May 1992 with one minute time resolution. The brightening of Hα loops close to the largest spot is evident of Emerging Flux.

3. Discussion

The Hα and SXT images seem to indicate that the main spots as well as the newly emerging spots have untwisted configurations during the observations. This situation is different from cases reported by other authors (*e.g.* Tanaka, 1991; Leka *et al.*, 1994), in which a flux tube was found to be already twisted during its emergence. They also reported that the twisted magnetic configuration produced an intense flare.

It is believed that a potential loop contains less magnetic energy than a twisted magnetic loop. Thus, when a potential loop interacts with a second loop, it will release less energy compared to a non–potential loop. Also, since the newly emerged spots of AR7172 decayed rapidly, sunspot motions during our observation period must not have made a significant contribution to the total magnetic energy.

Measurements show that the inclination angle of the axis connecting the two main spots did not change much during the observation period. This suggests that activity of the region was not due to a magnetic energy build-up by relative motions of the main spots.

Flux emergence was intense during 20 – 23 May, and this was followed by a decay phase, lasting until 26 May when the active region returned to a simple bipolar configuration. We found clear evidence of the merging of two bipoles and their gradual disappearance during the decay phase. From the above facts we believe that the activity in the region was caused mainly by the emergence and cancellation of spots with opposite polarities.

4. Summary

AR7172 showed high activity during its solar disk passage, where only small flares were observed. The activity seems to be caused by an intense flux emergence and the cancellation of spots. Since the region had a simple magnetic configuration during its evolution, as well as a rapid decay, it cannot produce an intense flare.

Acknowledgements

B. Anwar would like to thank the Research Development Corporation of Japan, which supported this work under a Science and Technology Agency (STA) Fellowship. Communication Research Laboratory is his host institute and provides him with the opportunity to work at Hiraiso Solar Terrestrial Research Center. We are grateful to A. Okano for help in preparing this manuscript. Data used here from Mees Solar Observatory, University of Hawaii, are produced with the support of NASA grant NAGW–1542 and NASA contract NAS8–37334. This work is part of a collaboration project between HSTRC and Watukosek Solar Observatory, Indonesia.

References

Akioka, M.: 1996, *J. Geomag. Geoelectr.*, **48**, 5.
Hanaoka, I.: 1994, *ApJ Letters*, **420**, L37.
Leka, K.D., et al.: 1994 in Y. Uchida, T. Watanabe, K. Shibata and H.S. Hudson(eds), *X-Ray Solar Physics From Yohkoh*, Universal Academy Press (Tokyo).
Mickey, D.L: 1985, *Solar Phys.* **97**, 223.
Tanaka, K. : 1991, *Solar Phys.* **136**, 133.
Tsuneta, S. et al.: 1991, *Solar Phys.* **136**, 37.

ACCELERATION OF CHARGED PARTICLES IN COLLISIONLESS MAGNETIC RECONNECTION

P. K. BROWNING AND G. VEKSTEIN
Dept of Physics, UMIST,
P.O. Box 88, Manchester M60 1QD, U.K.

Magnetic reconnection is an important process in both astrophysical and laboratory plasmas, and one of the major contributions of YOHKOH has been to provide observational evidence for this phenomenon in the solar corona. Generation of high-energy particles is one of the major signatures of reconnection; thus, the issue of charged particle acceleration by the inductive electric field associated with reconnection has become an integral part of the study of this process. Many theories of reconnection are concerned with the fluid (MHD) regime, appropriate to collisional plasmas. However, we are concerned here with hot rarefied plasmas, such as the upper corona of the Sun and the Earth's magnetosphere, where the particle mean free path is large compared with the scale length of the magnetic field and a collisionless plasma model must be used.

The basic model used in studies of steady 2D collisionless magnetic reconnection comprises an X-type planar magnetic field combined with a perpendicular uniform electric field. It is thus necessary to investigate the motion of test charged particles in such a configuration. Incoming particles are initially magnetised and the motion is adiabatic; but close to the neutral line the Larmor radius becomes large, the adiabatic approximation breaks down, and particles are directly accelerated by the electric field. Self consistent models such as [1,2] take account of the modification to the magnetic field configuration due to the currents arising from differential acceleration of ions and electrons. Previous publications have concentrated on particle trajectories *inside* the non-adiabatic region since it has been assumed this is the only location where acceleration can occur. Here, we investigate instead particle motion *outside* this "dissipation" region, where the basic equations can be simplified using the adiabatic appoximation.

The trajectory of a particle in the adiabatic region, with conserved magnetic moment μ, is determined by the equation for its motion parallel

T. Watanabe et al. (eds.), Observational Plasma Astrophysics: Five Years of Yohkoh and Beyond, 313–314.

to the magnetic field $\mathbf{B}$

$$\frac{dv_{\parallel}}{dt} = \mathbf{v}_E \cdot (\mathbf{v}_E \cdot \nabla)\mathbf{h} + v_{\parallel}\mathbf{v}_E(\mathbf{h} \cdot \nabla)\mathbf{h} - \frac{\mu}{m}(\mathbf{h} \cdot \nabla)B \qquad (1)$$

where $\mathbf{h} = \mathbf{B}/B$ is a unit vector along the field, together with its electric drift perpendicular to the field $\mathbf{v}_E = \frac{\mathbf{E}\times\mathbf{B}}{B^2}$.

In the regime relevant for fast reconnection, the electric drift motion of particles entering the reconnection site strongly exceeds their thermal speed, and the first two terms on the right hand side of (1) dominate the third. These terms represent coupling between cross field drift and parallel motion in a non uniform magnetic field. We solve this equation numerically for a magnetic field with X-point geometry with uniform perpendicular electric field. We find that as a particle approaches the neutral line, its drift velocity naturally rises; in a non-uniform magnetic field, the perpendicular motion generates parallel motion, and as the particle recedes from the neutral line, the parallel motion persists while the perpendicular motion decays. Particles which enter the site closer to the symmetry axis pass nearer to the neutral line and hence are more strongly accelerated. Thus, particles leave the reconnection site moving mainly parallel to the magnetic field, with an energy spectrum dependent on their initial distance from the symmetry line. This spectrum is found to have a power law form. The maximum kinetic energy achievable by this adiabatic acceleration process is comaprable with that due to the direct acceleration in the non-adiabatic dissipation region. The trajectories differ substantially from pure electric drift, which affects the properties of particles entering the dissipation region.

In order to incorporate the effects of the currents arising due to ions and electrons accelerated in opposite directions near the neutral line, we investigate also particle orbits in a field with uniform perpendicular current. This gives a magnetic field pattern which is compressed in the direction of the particles' inflow and stretched in the outflow direction. For a strong distortion, the particle trajectories consist of a nearly straight drift motion towards the neutral line in the almost uniform magnetic field, followed by a very rapid parallel acceleration near the neutral line which ejects the particle with high velocity along the separatrix. Thus, strong parallel acceleration still arises in a non-potential field, but the final kinetic energy of ejected particles depends more strongly on their initial position.

References

1. Burkhart, G.R., Drake,J.F. and Chen, J.; *J. Geophys. Res.* **A96**, 11539 (1991)
2. Vekstein, G.E. and Priest,E.R.; *Phys. Plas.* **2**, 3169 (1995)

THE FORMATION AND EVOLUTION OF THE CORONAL HOLES ASSOCIATED WITH NOAA REGION 7978

K. L. HARVEY AND H. S. HUDSON

Solar Physics Research Corp., Tucson, AZ, USA

Abstract. We have examined full-disk magnetograms and He I 1083 nm images from the National Solar Observatory/Kitt Peak to follow the development of coronal holes associated with a major active-region complex that emerged in 1996. This complex involved the emergence of six major active regions, the largest of which (NOAA 7978) had sunspot area exceeding 400 millionths. One additional region, important to the development of the coronal holes, had new-cycle magnetic polarity and emerged north of this complex. We follow the development of numerous new connections among these active regions and the polar coronal holes.

1. Introduction

In a four-month period of 1996, within a year of the expected time of minimum between sunspot cycles 22 and 23, a series of solar active regions erupted. Table 1 summarizes these regions, of which NOAA 7978 (emergence July 4, 1996) had the largest sunspot area and most active coronal structure, as observed by the *Yohkoh* soft X-ray telescope (SXT). The emergences occurred within longitude and latitude ranges spanning 44 degrees and 36 degrees, respectively, some 4% of the solar surface area. These active-region emergences within an otherwise extremely quiet Sun provide us with a good opportunity to trace the connectivity of the newly emerging flux. This paper describes one aspect of the connectivity as represented by the development of coronal holes.

T. Watanabe et al. (eds.), Observational Plasma Astrophysics: Five Years of Yohkoh and Beyond, 315–318.

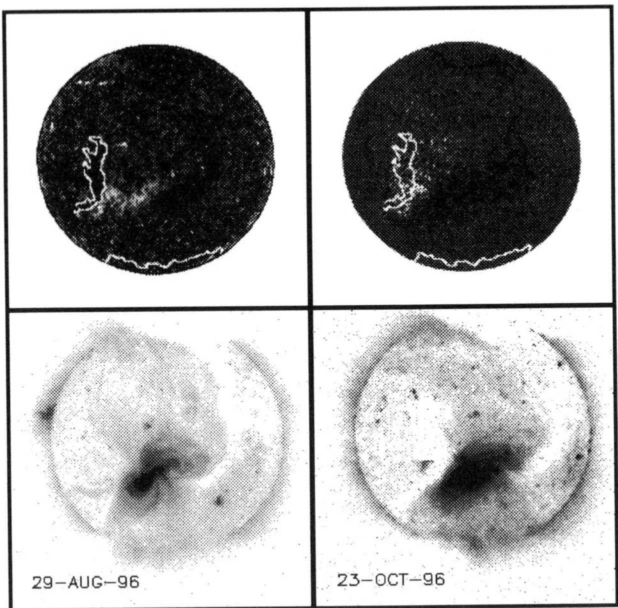

Figure 1. Upper panels, a He I 1083 nm image (left) and a magnetogram (right) from 21 October 1996, with dashed lines indicating the boundary of the magnetic flux domain associated with AR 7978). The black solid lines indicate positive-polarity coronal holes, white lines negative polarity. The lower panels show *Yohkoh* SXT images from about the same date (right) and from two rotations earlier (left), at which time the positive-polarity hole in the west had not yet opened.

2. Development of Connectivity

We followed the spread of magnetic flux associated with each emergence, using Kitt Peak magnetograms and He I 1083 nm images as illustrated in Figure 1. This shows the situation on 21 October 1996, with different magnetic domains and coronal holes outlined. At earlier times during this complex of activity, coronal holes appeared briefly, but the large hole shown crossing the equator persisted for at least three rotations. Figure 1 also shows *Yohkoh* SXT images of this time and two rotations prior to it, when the coronal-hole development was quite different. We have not yet made full comparisons with the SXT images, which should be able to contribute to our understanding of the connectivity. In a complete paper to follow, we track all of the magnetic domains listed in Table 1.

3. Conclusions

Our analysis shows that the 1996 complex of activity around Carrington longitude 250 consisted of the distinct emergence of six major active regions, including one with Cycle 23 polarity, over an 8-rotation time span. Between late April to early August 1996, we estimate that the total magnetic flux that emerged in this complex was at least 10^{23} Mx. Table 1 itemizes the development of coronal-hole areas and fluxes during this interval. The emergence of the new-cycle region resulted in extensive coronal hole development in August, September, and October. We note the following:

1. Isolated coronal holes develop after the emergence of the Cycle 23 region.
2. Some of the follower-polarity fields (black/negative) of the Cycle 23 region connect with nearby polar fields, resulting in a retraction of the north polar hole. This new connection is related to the brightening of X-ray loops (and corresponding He I1083 nm darkening) just outside the edge of the retracted coronal hole. This effect of high-latitude regions on the polar coronal hole boundaries is described by Sheeley *et al.* (1989).
3. The most persistent isolated coronal hole (A) is observed in the leading polarities of the dispersing magnetic fields of ARs 7958 and 7968.
4. By early August 1996, a new coronal hole appeared north of, but extending into the leader of the cycle 23 region. The polar hole also expanded toward the new same-polarity hole.
5. By late August, the leader polarity of the cycle 23 region has become aligned with like-polarity, small coronal holes, which almost connect to form a single unipolar open region extending from the north polar hole to south of the equator.
6. The elongated positive-polarity coronal hole persists for more than one rotation, and a similarly persistent opposite-polarity hole opens in the following region of AR 7978/7982 in September.
7. During the September and October rotations, approximately 17% of the magnetic flux in the activity complex had open configuration.

Acknowledgments. NASA supported this work under contract NAS 8-37334. *Yohkoh* is a project of the Institute of Space and Astronautical Sciences, Japan, with participation from the U.S. and U.K. The NSO/Kitt Peak data used here are produced cooperatively by NSF/NOAO, NASA/GSFC, and NOAA/SEC. This work was done while KLH was a visitor at the National Solar Observatory/National Optical Astronomy Observatories, operated by the Association of Universities for Research in Astronomy, Inc., under contract with the National Science Foundation.

References

Sheeley, N. R., Jr., Wang, Y.-M., and Harvey, J. W.: 1989, *Solar Phys.* **119**, 323.

TABLE 1. Table I. Magnetic Fluxes and Areas of Observed Coronal Holes

Coronal Hole (Location)	Date	Magnetic Flux (10^{20} Mx)	Area (10^{20} cm^2)
South Polar Hole			
	Jul 5	-20.3	11.6
	Jul 8	-18.8	14.8
	Aug 1	-10.5	9.1
	Aug 28	-10.3	7.3
	Sep 26	-16.6	8.2
	Oct 21	-25.3	11.6
	Nov 18	-15.0	6.3
North Polar Hole			
	Jul 5	48.3	26.2
	Jul 8	37.4	22.8
	Aug 1	45.5	27.8
	Aug 28	32.3	21.8
	Sep 26	42.8	20.7
	Oct 21	45.0	18.4
	Nov 18	34.0	18.4
A at S02: Between 7958 leader and 7968 follower			
	Jul 5	8.6	1.3
	Aug 1	11.7	1.0
B at N02: In 7958 follower (observed $<$ 1 rotation)			
	Jul 5	-8.9	1.2
	Jul 8	-8.6	1.3
C at S22: In 7958/7978-B follower (observed $<$ 1 rotation)			
	Aug 1	-8.4	2.0
D at N33: In Cycle 23 AR leader			
	Aug 1	1.5	1.2
E at N32-S20: Cycle 23 AR leader to 7958 leader (A & D)			
	Aug 28	18.9	5.0
	Sep 26	21.5	6.3
	Oct 21	27.9	6.2
	Nov 18	11.6	3.1
F at N07-S33: 7978-B/7978 follower			
	Sep 26	-50.2	4.1
	Oct 21	-29.1	4.4
	Nov 18	-23.7	4.1

HORIZONTAL FLOW FIELD IN THE SOLAR PHOTOSPHERE

R. KITAI, Y. FUNAKOSHI, S. UENO AND S. SANO
Kwasan and Hida Observatories, Kyoto University, Kamitakara, Gifu 506-13, Japan

AND

K. ICHIMOTO
National Astronomical Observatory, Mitaka, Tokyo 181, Japan

Extended Abstract. A real-time frame selector (RTFS), a new observational system, was developed at the Domeless Solar Telescope of Hida Observatory, Kyoto University. RTFS allows automatic selection of a video frame with the highest spatial resolution in a series of live video images during a prescribed time span and stores it in digital format in a hard disk file. We applied the RTFS to observations of solar granulation in the g-band wavelength (λ 4308 Å, passband 20 Å), resulting in a time series of data with a time step of 15 sec. By a modified and simplified version of local-correlation tracking (November and Simon, 1988), we derived the horizontal velocity field using the granulation pattern as tracers. In the quiet photosphere, we clearly found a mesogranulation flow field (November et al., 1981) from a 70 min span of data on May 25, 1996 (Figure 1). Supergranular flow patterns were also detected and identified from comparison with an Hα wing image. Also, using RTFS on a sunspot area of NOAA7981 observed on Aug 2, 1996, we have found new evidence of horizontal inflow from penumbral areas to umbral areas and also the presence of another circumferential inflow from the surrounding photosphere to the penumbral outer edge (Figure 2). The inflow into the umbra is probably due to inflowing motions of penumbral grains found by Muller (1973). The circumferential inflow in the surrounding photosphere may indicate the presence of a circulating doughnut-roll around the spot.

A full version of this paper was submitted to the Publ. Astron. Soc. Japan.

T. Watanabe et al. (eds.), Observational Plasma Astrophysics: Five Years of Yohkoh and Beyond, 319–320.

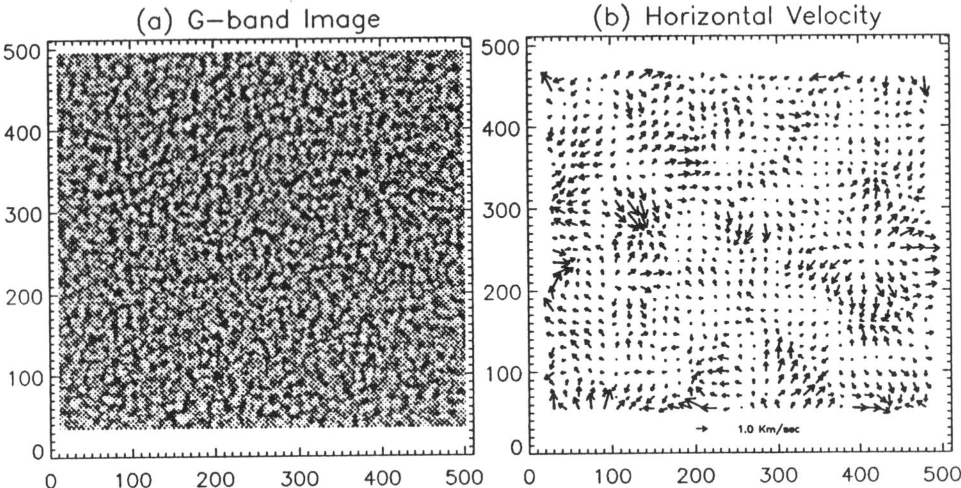

Figure 1. (a) Granulation pattern in a quiet region on May 25, 1996. Coordinates are in pixel unit (= 0.22 arcsec). (b) Horizontal flow field.

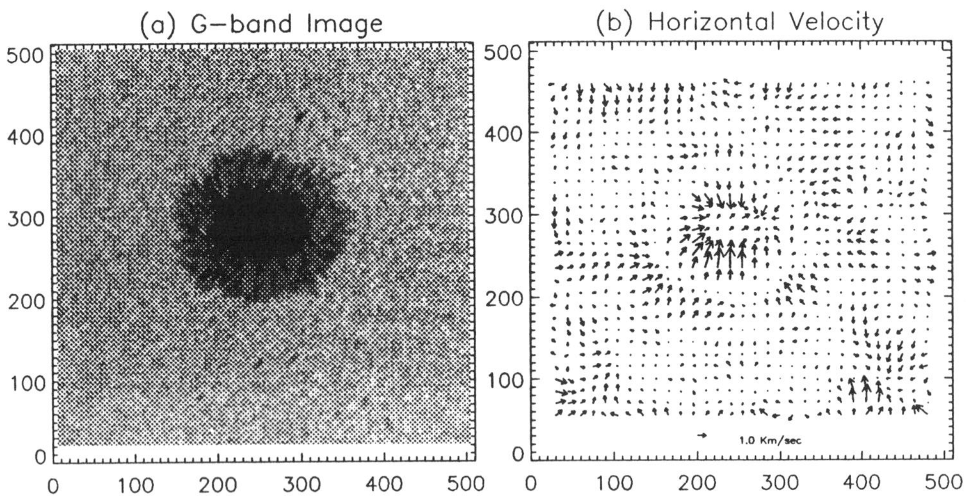

Figure 2. (a) Granules and grains in and around the p-spot of NOAA7981 on Aug 2, 1996. (b) Horizontal flow field.

References

Muller, R. (1973), *Sol. Phys.*, **104**, 287
November, L.J., Toomre, J., Gebbie, K.B., and Simon, G.W. (1981), *Ap. J.*, **245**, L123
November, L.J. and Simon, G.W. (1988), *Ap. J.*, **333**, 427

LOCAL THREE-DIMENSIONAL MHD SIMULATIONS OF THE PARKER INSTABILITY IN DIFFERENTIALLY ROTATING DISKS

T. MATSUZAKI AND R. MATSUMOTO
Department of Physics, Faculty of Science, Chiba University
1-33 Yayoi-Cho, Inage-Ku, Chiba 263, JAPAN

T. TAJIMA
Instutite for Fustion Studies, University of Texas at Austin
Austin, TX 78712, USA

AND

K. SHIBATA
National Astronomical Observatory
Mitaka, Tokyo 181, JAPAN

1. Introduction

Recently, several authors (Hawley *et al.* 1995 ; Matsumoto & Tajima 1995 ; Brandenburg *et al.* 1995) have carried out three-dimensional local magneto-hydrodynamic (MHD) simulations of magnetized accretion disks. In gravitationally stratified disks, buoyant escape of magnetic flux due to the Parker instability (Parker 1966) may limit the strength of magnetic fields inside the disk. Three-dimensional MHD simulations of weakly magnetized Keplerian disks (Stone *et al.* 1996), however, showed little signature of the Parker instability. This is because in high-β disks ($\beta = P_{gas}/P_{mag} \gg 1$), the growth rate of the Parker instability $\gamma_P \sim 0.2-0.5 v_A/H \sim 0.3-0.7\Omega/\beta^{1/2}$ is much smaller than that of the Balbus & Hawley instability $\gamma_{BH} \sim 0.1-0.7\Omega$ (Balbus & Hawley 1991). Here v_A is the Alfvén speed, H is the disk thickness and Ω is the rotational angular speed. Moreover, the Balbus & Hawley instability may cut off the growth of the Parker instability by disrupting the coherence of the azimuthal magnetic field (Stone *et al.* 1996).

T. Watanabe et al. (eds.), Observational Plasma Astrophysics: Five Years of Yohkoh and Beyond, 321–324.

When $\beta \sim 1$, since $\gamma_P \sim \gamma_{BH}$, we expect that the Parker instability couples with the Balbus & Hawley instability. Earlier simulations of the nonlinear evolution of the Parker instability (Matsumoto *et al.* 1988) did not include the effects of differential rotation. In this paper, we report results of 3D MHD simulations of the Parker instability in Keplerian accretion disks and its coupling with the Balbus & Hawley instability.

2. Numerical Method

We consider a local part of the disk as shown in Figure 1. In a frame corotating with the disk at $r = r_0$ from the gravitating center, we take a local Cartesian coordinate with x, y, z in the radial, azimuthal, and vertical directions, respectively. Under the shearing sheet approximation ($|x| \ll r_0$) (Hawley *et al.* 1995), the equilibrium Keplerian shear flow is given by $v_y = -1.5\Omega_k x$ where $\Omega_k = (GM/r_0^3)^{1/2}$ is the Keplerian angular speed. The vertical gravity component is assumed to be $g_z = -GMz/(r_0^2 + z^2)^{3/2}$. The initial state consists of gravitationally stratified, isothermal gas layer threaded by an azimuthal magnetic field. We assume that the initial plasma β $(= \beta_0)$ is uniform. In the following, we use the the normalizations $\rho(z = 0) = C_s = \Omega_k = 1$, where C_s is the sound speed. A characteristic length scale is the scale height defined by $H = C_s/\Omega_k = 1$. We set the thickness of the disk to $H/r_0 = 0.25$. For simplicity, we assume an isothermal equation of state.

A typical size of the simulation box is $(L_x, L_y, L_z) = (0.5H, 18H, 16H)$. We computed both the upper and the lower half space. The azimuthal boundaries are periodic. The radial boundaries are treated by using a sliding periodic condition (Hawley *et al.* 1995). The upper and lower boundaries are free boundaries where waves can transmit. We solved the MHD equations by a modified Lax-Wendroff scheme with artificial viscosity. A typical number of grid points is $(N_x, N_y, N_z) = (42, 62, 165)$. In order to initiate the evolution, a sinusoidal perturbation with amplitude $0.01C_s$ is imposed at $t = 0$ with azimuthal velocity v_y.

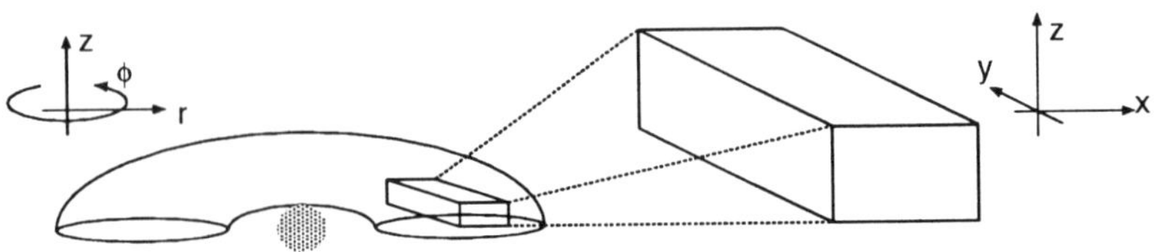

Figure 1. Simulation region and the local Cartesian coordinate.

3. Numerical Results

White curves in Figure 2 show magnetic field lines in a computation with $\beta_0 = 1.0$. We can identify magnetic loops created by the Parker instability. Gray surfaces in Figure 2 show isosurfaces of magnetic field strength. The magnetic fields in the equational region of the disk become turbulent due to the growth of the Balbus & Hawley instability. The turbulent magnetic field structure in Keplerian shear flow is different from more coherent magnetic field structures where only shear motion is included (Shibata *et al.* 1990).

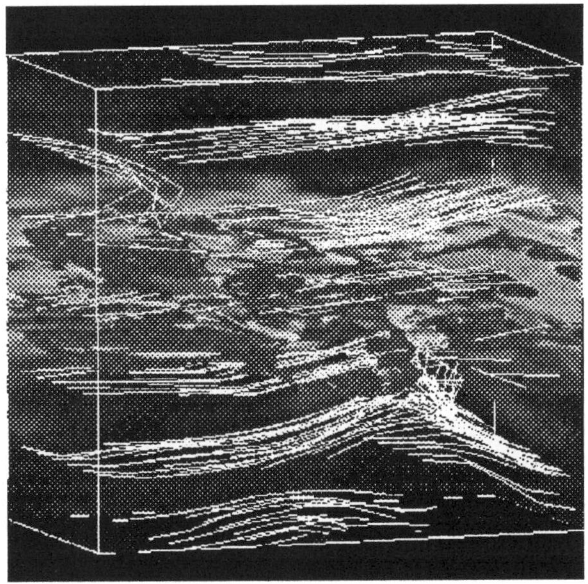

Figure 2. 3D structure of magnetic field lines ($\beta_0 = 1.0$) at $t = 0H/\Omega_k$.

Figure 3 shows the evolution of mean square magnetic field strength $\langle B^2/(8\pi P_0)\rangle$ for $\beta_0 = 1.0$. We found that in mildly magnetized disks ($\beta_0 \sim 1$), the mean magnetic field strength increases for time scale longer than seven rotation periods ($T_{rot} = 2\pi/\Omega_k$). Since the growth time of the Parker instability in a non-rotating disk is $3 - 5H/v_A \sim 0.3 - 0.6T_{rot}$ when $\beta_0 \sim 1$, numerical results indicate that the magnetic field amplification compensates the buoyant loss of magnetic flux.

Figure 4 shows time evolution of magnetic viscosity $\alpha_B(= -\langle B_x B_y\rangle /4\pi P_0)$ for a model with $\beta_0 = 1$. The value of α_B at the saturation stage of the instability is ~ 0.1. This value of α_B is an order of magnitude larger than that in a high-β disk.

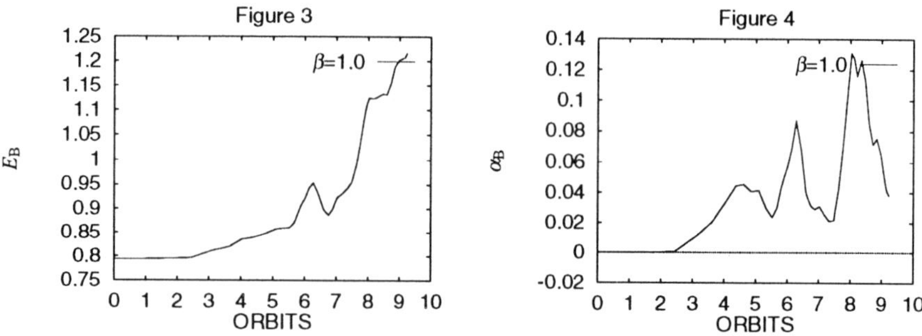

Figure 3. Time evolution of mean square magnetic field strength ($E_B = \langle B^2/8\pi P_0 \rangle$) for a model with $\beta_0 = 1.0$.

Figure 4. Time evolution of effective magnetic viscosity ($\alpha_B = -\langle B_x B_y \rangle / 4\pi P_0$) when $\beta_0 = 1$.

4. Conclusions

We showed that once a mildly magnetised disk ($\beta \sim 1.0$) is formed, the magnetic field is amplified due to the coupling between the Parker instability and the Balbus & Hawley instability and that the disk stays in a low-β state. The buoyant escape of magnetic flux is compensated by the field amplification. Fluctuating magnetic fields created by the Balbus & Hawley instability contribute to the angular momentum transport in the disk. When the magnetic energy stored in such low-β disks is released, the disk may show large amplitude sporadic time variations characteristic of the low-state in black hole candidates. Low-β disks may correspond to low-state disks (Mineshige *et al.* 1995).

Numerical computations were carried out on a VPP300/16R at the Astronomical Data Analysis Center of the National Astronomical Observatory, JAPAN. This work is supported in part by the grant of ministry of education science and culture (07640348).

References

Balbus, S. A., & Hawley, J. F. 1991, ApJ, 376, 214
Brandenburg, A., Nordlund, A., Stein, R., & Torkelsson, U., 1995, ApJ, 446, 741
Hawley, J. F., Gammie, C. F., & Balbus, S. A. 1995, ApJ, 440, 742
Matsumoto, R., Horiuchi, T., Shibata, K., & Hanawa, T. 1988, PASJ, 40, 171
Matsumoto, R., & Tajima, T. 1995, ApJ, 445, 767
Mineshige, S., Kusunose, M., & Matsumoto, R. 1995, ApJL, 445, L43
Parker, E. N., 1966, ApJ, 145, 811
Shibata, K., Tajima, T., & Matsumoto, R. 1990, ApJ, 350, 295
Stone, J. M., Hawley, J. F., Gammie, C. F., & Balbus, S. A. 1996, ApJ, 463, 656

Limits on the Temperature of Impulsive Phase Solar Flare Hard X-ray Emission, and the Amount of Thermal X-rays Present in the Yohkoh HXT-LO (14-24 keV) Channel

J. M. McTiernan

Space Sciences Laboratory, Univ. of California Berkeley, USA

Abstract. We attempt to discover whether the emission seen in the Yohkoh HXT-LO (14-24 keV) channel is thermal, due to the presence of high temperature plasma, or nonthermal, as an extension of the nonthermal emission seen at higher energies (>30 keV). We assume that the HXT-LO emission is thermal, and ask: Under what conditions (i.e., T and EM) will a thermal source be consistent with the HXT-LO emission, and the line emission observed by the Yohkoh BCS? All impulsive events, with HXR rise times ranging from 30 seconds to five minutes, observed by Yohkoh are included.

1. Introduction

To estimate the number and energy of energetic electrons in a solar hard X-ray burst we need the cutoff of the nonthermal electron spectrum. This is taken to be 20 or 30 keV, typically either the low energy limit of the hard X-ray detector used, or the lowest energy where "impulsive" behavior is seen. Stereoscopic observations of partially occulted flares have shown that the nonthermal impulsive component of the flare emission may extend down to 5 keV (Kane *et al.* 1992). Yohkoh SXT observations have shown that there are impulsive footpoints in soft X-ray loops (Hudson *et al.* 1994), leading to speculation that the nonthermal spectrum may extend down to lower energies of 1 to 3 keV. If the nonthermal component extends down into the soft X-ray regime, then the estimates that are made of the energy in accelerated electrons will increase dramatically.

Here we try to determine whether there is nonthermal emission present in the LO channel of the Yohkoh HXT during the impulsive phase of flares. The HXT-LO channel covers an energy range of 14-23 keV, which is a lower energy range than what is typically considered to be nonthermal. We know that thermal emission from flare plasmas can reach temperatures as high as 30 MK, and may dominate the emission spectrum in the range from 10 to 30 keV, but this is typically only true late in the flare (Pike *et al.* 1996). Early in the flare, during the impulsive phase, the temperature is low, and there is little or no thermal emission present in the 10 to 30 keV range.

T. Watanabe et al. (eds.), Observational Plasma Astrophysics: Five Years of Yohkoh and Beyond, 325–326.

2. Data Analysis

The data used includes all flares observed by the Yohkoh BCS and HXT, with impulsive time scales, i.e., the rise time in the HXT-M1 channel was between 30 seconds and 5 minutes. Data from the first 1/3 of the impulsive phase was used, and only flares for which there were enough counts in the HXT-LO and M1 channels to obtain a temperature measurement were kept. Small flares which occurred during the decay phases of large flares were dropped. The resulting sample includes 144 flares.

The procedure for the study follows:

(1) Obtain a temperature from BCS, using CaXIX and FeXXV channel data.

(2) Calculate the response that the thermal component generates in the Yohkoh HXT. From this we see how much HXT-LO emission is due to low-T plasma.

HXT emission, however, may be due to plasma that is much hotter (up to 100 MK) than would be observed in CaXIX and FeXXV. We use the lack of FeXXVI emission to put limits on thermal HXT-LO emission.

(3) We assume that the HXT-LO emission is thermal, and find the expected response of the high-T plasma in the FeXXVI Lyα line. This is a decreasing function of the assumed temperature.

(4) We compare the expected line flux with the fluctuations in the FeXXVI channel. As the assumed T increases, the expected response decreases, and eventually drops below the level of the fluctuations in the FeXXVI background. This gives us a minimum temperature, above which it is possible for the HXT-LO emission to be thermal. We obtain a maximum T by assuming that all of the HXT-LO and M1 emission is thermal. If $T_{max} < T_{min}$ then the emission must be nonthermal.

3. Conclusions

For the sample of flares, we find:

(1) The thermal plasma observed by BCS during the first 1/3 of the impulsive phase is low-temperature plasma, and does not cause much of a response (typically less than 20%) in the HXT-LO energy range.

(2) The HXT-LO emission was nonthermal for 36 out of the 144 flares studied. If the HXT-LO emission was thermal, there would be observable FeXXVI emission. For 54 flares, thermal HXT-LO emission is possible; the HXT-LO emission does not cause an observable effect in FeXXVI at any T. For the remainder of the flares, thermal emission is possible, above a minimum T, which ranges from 30 to 90 MK.

This work was supported by NASA Grant NAGW-5126.

References

Hudson, H.S., *et al.* , 1994 Ap. J. (Letters), 422, L25.

Kane, S.R., *et al.* , 1992 Ap. J., 390, 687.

Pike, C.D., *et al.* , 1996 Ap. J., 464, 487.

HOMOLOGOUS FLARE SERIES OF FEBRUARY 1992

S. MORITA, Y. UCHIDA, K. FUJISAKI AND S. HIROSE
Science University of Tokyo
1-3 Kagurazaka, Shinjyuku-ku, 162 Japan

Abstract.
We describe the series of flares of 21-February-1992, 24-February-1992 and 27-February-1992, which we consider to be a homologous flare series observed with Yohkoh SXT. If the events are homologous, similar features will be seen from different perspectives during the series. Thus, we use the time scale normalized by the total duration for each flare obtained from the GOES data and define flare phases to compare the corresponding features in the series of events. From this analysis, along with the corresponding Kit Peak magnetograms and Mitaka Hα data, we examine the corresponding features seen from different angles, and discuss their three-dimensional structures.

1. Description of the Analysis Procedure

We use the GOES soft X-ray intensity curves to define the total duration (t_0) as the baseline for determining two slopes approximating the rising and declining phases of the flares. The total duration (t_0) of the flares on 21-Feb-1992, 24-Feb-1992 and 27-Feb-1992 are 6.90 hours, 6.62 hours and 6.03 hours, respectively. In this paper, we use timescales normalized by these total durations. We use this timescale to identify relative corresponding times among the three events. If the events are homologous, similar features will be seen at relative corresponding times in each of them.

2. Result and discussion

2.1. MAGNETOGRAMS

Fig.1 is a set of Kitt Peak magnetograms corresponding to each flare. The middle two images are magnetograms at about 7 hours before and 16.5

T. Watanabe et al. (eds.), Observational Plasma Astrophysics: Five Years of Yohkoh and Beyond, 327–330.

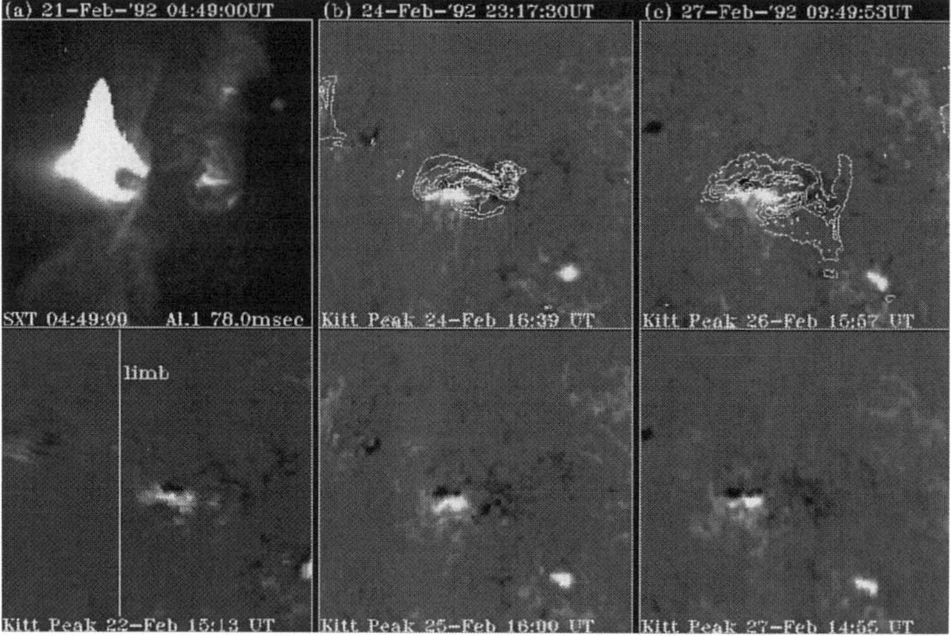

Figure 1. Kitt Peak Magnetograms corresponding to each flare. These images are rotated to the times of the events. The left one, which corresponding to the limb flare is still more rotated to make a top view for a comparison. Contours are SXT images at each times.

hours after the peak time of the 24-Feb-1992 flare. The right two magnetograms are about 18 hours before and 5 hours after the peak of the 27-Feb-1992 flare. These 4 images are rotated to the times of the events, and scaled to the same size as the SXT images. Contours are SXT images at the times of the events. Comparing these 4 images, we cannot see any marked systematic changes, like a footpoint shearing in the restoration phase as expected in some of the models relying upon shearing.

The left magnetogram is about 35 hours later than the peak of the 21-Feb-1992 flare. The magnetogram at the nearest time is rotated to the time of this flare, which was a limb event, and the footpoint–magnetogram relation is examined. From this, we find that this event had a behavior similar to the other disk events of this series. Thus, in addition to the disk events of 24-Feb-1992 and 27-Feb-1992, we can take the 21-Feb-1992 event also as one in this series of homologous events.

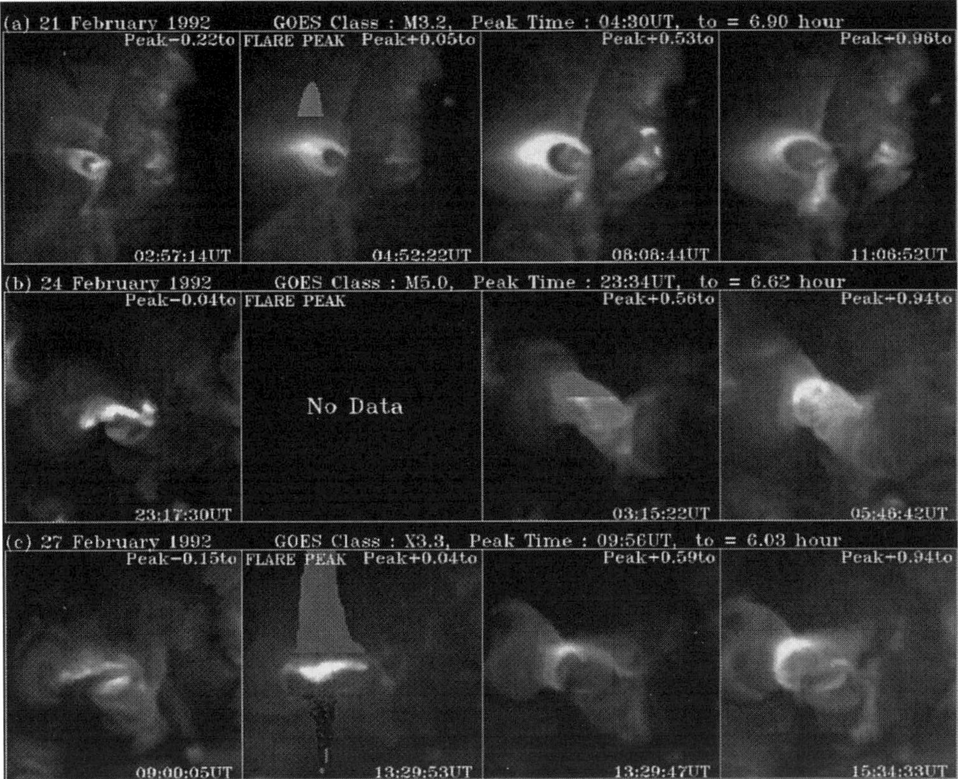

Figure 2. YOHKOH SXT images of the three flares, which we consider to be a homologous flare series. The top, middle and bottom series of images are of the 21-Feb-1992, 24-Feb-1992 and 27-Feb-1992 flares, respectively.

2.2. SXT IMAGES

Fig.2 shows YOHKOH SXT images of the three flares. In this figure, the top, middle and bottom series of images are of the 21-Feb-1992, 24-Feb-1992 and 27-Feb-1992 flares, respectively. Here, the four consecutive panels are at the corresponding scaled-times, at about −15, +5, +55 and +95% of total duration of each event. In comparing the disk events of 24-Feb-1992 and 27-Feb-1992, we find that this flare series's homology is not strict in a mathematical sense, but in a fairly loose sense the homology does hold. The shape and development of the two flares are fairly similar. From this and magnetogram results, we can take the 21-Feb-1992 event also to be one in the same series of homologous events. We take advantage of this to determine the 3D-structure of various features in this homologous arcade flare series, taking each event to be essentially the same structure viewed from largely different angles.

2.3. DETAILED ANALYSIS OF SXT IMAGES AND Hα IMAGES (24-FEB-1992)

We also compared the development of YOHKOH SXT images and Mitaka Hα images of the flare 24-Feb-1992 using the normalized time. The configuration of the SXR arcade changes in time, but its diagonal line keeps its location with respect to the S-shaped X-ray structure, which already existed in the pre-event phase. We can also see the same thing in the Hα images, which show the footpoints of the SXR arcade. It is seen more clearly in the detailed time development of the Mitaka Hα images.

3. Conclusions

The main results are summarized as follows: (i) In comparing the disk events of 24-Feb-1992 and 27-Feb-1992, this flare series is not homologous in a strict sense, but it is so in a fairly loose sense. In SXR-images, the shapes and their development are fairly similar. (ii) We compared the magnetograms for the series of events, but we cannot see any marked systematic changes, like a systematic long-distance footpoint shearing in the restoration phase as expected in some of the models relying upon shearing. (iii) From footpoint–magnetogram analysis, we can take the 21-Feb-1992 event to also be one of this series of homologous events. Thus we determine the 3D-structure of the arcade flare series by comparing the events seen at the limb and on the disk. (iv) In the flare event on 24-Feb-1992, the configuration of an SXR arcade changes in time, but with its diagonal line always overlaying an S-shaped structure which existed in X-rays in the pre-event phase. The corresponding Hα images show the footpoints of the SXR arcade. Thus, it is suggested that the S-shaped structure, which corresponds to the bright structure along the axis of the dark tunnel in the 21-feb-1992 limb event, is quite a fundamental structure for the entire flare process, although it has no corresponding feature in the classical models.

References

Tsuneta, S., Hara, H., Shimizu, T., L. W. Acton, K. T. Strong, H. S. Hudson, and Ogawara, Y. (1992), Observations of a Solar Flare at the Limb with the Yohkoh Soft X-Ray Telescope, *Publ. Astron. Soc. Japan*

SIMULATION STUDY ON MAGNETO-GRAVITY INSTABILITIES IN MAGNETIC SHEAR FIELD

K. MORIYAMA, T. MIYOSHI AND K. KUSANO
Department of Materials Science, Faculty of Science, Hiroshima University, Higashi-Hiroshima 739, Japan

1. Introduction

We performed two-dimensional nonlinear MHD simulations for magnetic buoyancy instabilities in a sheared magnetic field. This work is motivated by research on solar coronal magnetic loop emergence. The magnetic buoyancy instabilities can be classified into the Parker instability and the interchange instability, depending on whether the wavevector is parallel or perpendicular to the magnetic field line. We consider a highly stratified domain which includes the cooler chromosphere and the hotter corona [1]. Initially, the sheared magnetic flux is embedded in the bottom of the chromosphere, and satisfies the magnetostatic equilibrium condition. The TVD scheme is employed to calculate the MHD equations numerically.

2. Simulation Results

First, linear stability analysis is performed. The results show that the direction of the magnetic field lines at the top surface of the magnetic layer strongly dominates the stability of the system. When the magnetic field lines at the top surface are parallel to the wavevector, the Parker-type instability results. When the magnetic field lines are perpendicular to the wavevector at the top of the magnetic layer, the interchange-type instability becomes dominant.

Secondly, we carry out the nonlinear simulation. As a result of the instability the magnetic flux emerges into the corona so that the coronal loops are generated in both the Parker-type instability and the interchange-type instability cases. The nonlinear evolutions, however, are strongly influenced by the distribution of the magnetic shear. The Parker-type instability

T. Watanabe et al. (eds.), Observational Plasma Astrophysics: Five Years of Yohkoh and Beyond, 331–332.

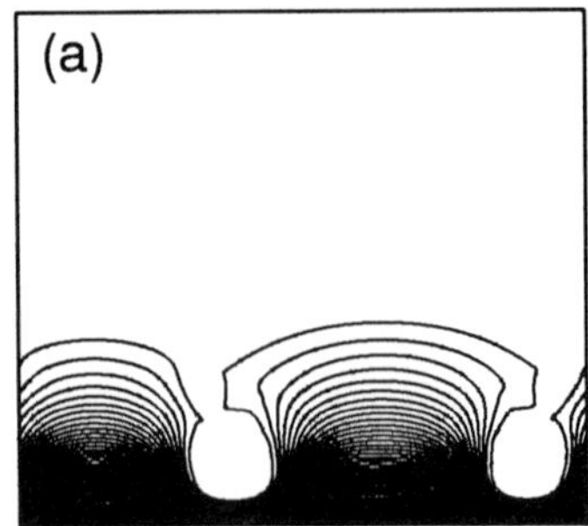

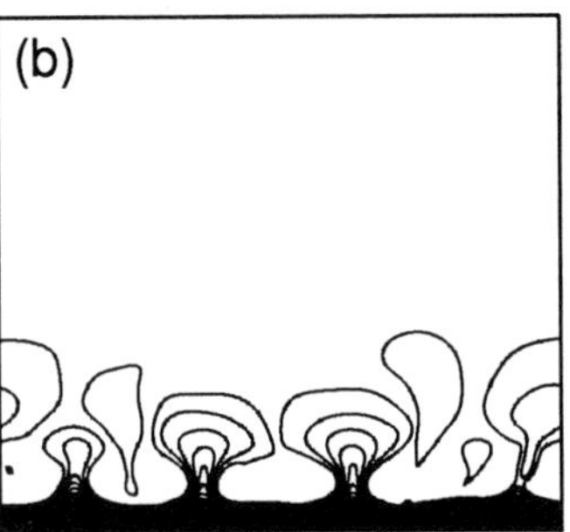

Figure 1. Magnetic field lines generated by (a) the Parker-type instability and (b) the interchange-type instability.

with the magnetic shear forms large-scale coronal loops (Fig. 1a), while the interchange-type instability can drive magnetic reconnection and generates isolated flux tubes (Fig. 1b).

In the nonlinear stage of the interchange-type instability, the lower Fourier modes become dominant through an inverse cascading process, even though in the linear phase the higher modes are the most unstable. Consequently, the large-scale emerging loops can grow. In the pure interchange instability without the magnetic shear, however, the nonlinear growth is not observed. This result suggests that, as a consequence of the magnetic shear, the character of the instability is converted through the nonlinear dynamics. In order to confirm this, we calculated the linear stability for the horizontally averaged component in the simulation result at each time step. We found that the characteristic of the instability evolves from the interchange-type into the Parker-type instability.

These results mean that, when the magnetic field has shear, the Parker instability becomes dominant in the nonlinear stage even if the interchange instability is the more unstable mode in the initial state. This suggests that the size of the coronal loop is always as large as the pressure scale height. We also found that stronger shear stabilizes the short-wavelength mode, so there is a tendency for larger coronal loops to form in stronger shear fields.

References

1. Shibata, K., Tajima, T., Matsumoto, R., Horiuchi, T., Hanawa, T., Rosner, R. and Uchida, Y. (1989) *ApJ*, **338**, 471

X-RAY PLASMA EJECTIONS AND JETS FROM SOLAR COMPACT FLARES OBSERVED WITH THE YOHKOH SOFT X-RAY TELESCOPE

M. OHYAMA, K. SHIBATA, M. SHIMOJO AND T. YOKOYAMA
National Astronomical Observatory
Mitaka 181 Japan

Abstract. Coronal X-ray plasma ejections and jets associated with solar flares have been found with *Yohkoh* soft X-ray observations. We have examined the physical conditions as well as the morphological evolution of X-ray plasma ejections and jets, and report their characteristics. We discuss comparison between X-ray plasma ejections and jets, and also present physical interpretation of these mass ejections.

1. X-ray Plasma Ejections from Compact Impulsive Flares

Before *Yohkoh* observations, it had been thought that impulsive flares might not be explained by the reconnection model for LDE flares. Hard X-ray sources (Masuda *et al.* 1994) and X-ray plasma ejections (*e.g.*, Shibata *et al.* 1995; Ohyama and Shibata 1997a, b, c), however, have been discovered above the soft X-ray flare loops in some impulsive flares. These discoveries suggest that impulsive flares occur through a magnetic reconnection process similar to that for LDE flares.

A C9.7 flare associated with a plasma ejection occurred on 1993 November 11 (Figure 1; Ohyama and Shibata 1997a). An ejected loop started to rise slowly ($\sim$ 10 km s^{-1}) long before the impulsive phase (*the preflare slow rise*), and suddenly accelerated to $\sim$ 130 km s^{-1} just before or at about the onset of the impulsive phase (*the main rise*), with continued acceleration to $\sim$ 200 km s^{-1} during the impulsive phase (Figure 2). We shall discuss the origin of the ejected material. Fortunately, this flare was observed with Yohkoh *before the start of the ejection.* The ejected loop appeared after its footpoint brightnened (Figure 1), and moreover, it was already heated to $\sim$ 11$\pm$4 MK *before the start of the ejection* (Table 1). Furthermore the

T. Watanabe et al. (eds.), Observational Plasma Astrophysics: Five Years of Yohkoh and Beyond, 333–336.

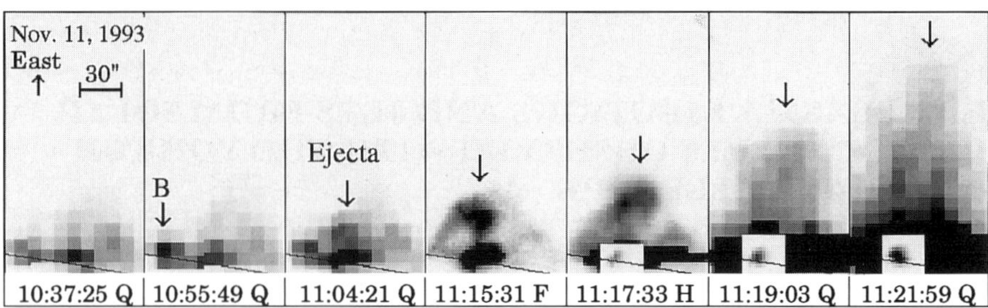

Figure 1. Development of the X-ray plasma ejection associated with the 1993 November 11 flare. The X-ray plasma ejecta is indicated by arrows. 'B' indicates a brightening at the footpoint of the ejected loop. (Ohyama and Shibata 1997a)

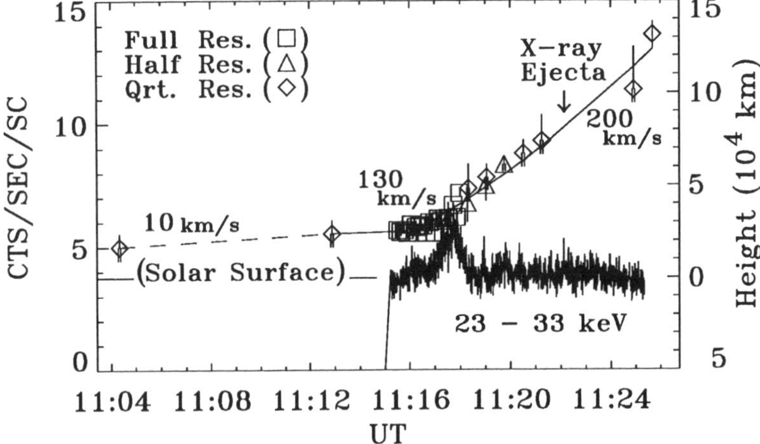

Figure 2. Apparent height of the X-ray ejecta and counting rates from the 23–33 keV channel of HXT for the 1993 November 11 flare. The height of the ejecta is defined as the location of its maximum intensity. The square, the triangle and the diamond indicate data taken with full, half, and quarter resolution, respectively. Error bars indicate the region where the intensity is within 90% of the maximum value of the ejecta.

electron density of the ejected loop was larger than the typical density of the active-region corona, 10^9 cm^{-3}. These results suggest the possibility of chromospheric evaporation due to preflare heating.

Shibata *et al.* (1995) and Ohyama and Shibata (1997a, b, c) have studied some X-ray plasma ejections, and the results are as follows: (1) The shape of the ejecta is a loop. (2) The apparent velocity is 50 – 400 km s^{-1}. (3) The acceleration occurs during the impulsive phase. (4) The ejected material is already heated to 11±4 MK *before the start of the ejection.* (5) The electron density of the ejecta is about an order of magnitude larger than the typical density of the active-region corona. (6) The kinetic energy of the ejecta is smaller than the thermal energy content of the flare loop.

If the ejected plasma stretches the overlying magnetic fields to form a current sheet leading to reconnection (as assumed in some flare models; *e.g.*, Sturrock 1992), its kinetic energy must be larger than the thermal energy content of the flare loop because the magnetic free energy responsible for the flare heating has to be created by the motion of the ejecta. Result (6) does not, however, support this hypothesis (Ohyama and Shibata 1997b, c).

TABLE 1. Physical parameters of X-ray ejecta and flare loop at the preflare phase (11:15:27 – 11:16:21 UT) of the 1993 November 11 flare.

Physical parameter	Flare region	X-ray ejecta
Temperature (10^6 K)	6 – 16.5	4.6 – 15.8 (11.3±4)†
Emission measure (10^{28}cm^{-5})	5.4 – 276	2 – 18
Electron density (10^9cm^{-3})	6.4 – 53	4.5 – 14
Gas pressure (dyn cm^{-2})	14 – 136	10 – 45
Mass (10^{13}g)	~ 10	9.2±0.7
Thermal energy content (10^{29}erg)	2.6±0.2	2.3±0.3
Kinetic energy (10^{27} erg)	—	4.2 – 20

† The value indicates the temperature at the center of the ejecta.

2. X-ray Jets

X-ray jets have been discovered as transitory X-ray enhancements with apparently collimated motion (Shibata *et al.* 1992). Almost all jets are associated with microflares or subflares in active regions or X-ray bright points. Some of these jets are associated with Hα surges.

Shibata *et al.* (1994) and Shimojo *et al.* (1996) have studied the characteristics of X-ray jets. The results are as follows: (1) The length is 10^4 – 5 × 10^5 km. (2) The width is 5 × 10^3 – 10^5 km. (3) The apparent velocity is 10 – 10^3 km s^{-1}. According to preliminary analysis of jets by Shimojo et al., the physical conditions of X-ray jets are as follows: (4) The temperature is 4 – 6 MK. (5) The thermal energy content is 10^{26} – 10^{27} erg. (6) The kinetic energy is 10^{25} – 10^{26} erg. Other characteristics of X-ray jets are reported in detail by Shimojo and Shibata (1997).

3. Comparison Between X-ray Plasma Ejections and Jets

The physical parameters of X-ray plasma ejections and X-ray jets are shown in Table 2. It is noted that there is a tendency that the physical parameters, such as density, mass, thermal energy content, and kinetic energy, of compact X-ray jets are smaller than those of X-ray plasma ejections.

The shape of X-ray jets (*i.e.*, a collimated jet-like structure) is very different from that of X-ray plasma ejections (*i.e.*, plasmoids). Shibata (1996) suggested the following possibility by considering a numerical simulation by Yokoyama and Shibata (1996): A blob-like plasmoid ejected from the current sheet soon collides and reconnects with ambient magnetic fields, and finally disappears. The mass contained in the plasmoid is transferred into the reconnected open flux tube and forms a collimated jet along the tube. Shibata (1996) also suggested that it is difficult to observe the plasmoid ejection in smaller flares because the lifetime of the plasmoid is very short, of the order of $\sim 10 - 100$ sec. We, however, lack the observational knowledge of the relation and the difference between the X-ray plasma ejections and jets. This will be an important subject for future work.

TABLE 2. Physical parameters of X-ray jets and plasma ejections

	X-Ray Jet (preliminary results)	Plasma Ejection
Temperature (10^6 K)	$4 - 6$	$5 - 16$
Density (cm^{-3})	$\sim 10^9$	$10^9 - 10^{10}$
Mass (g)	$10^{11} - 10^{12}$	$10^{13} - 10^{14}$
Thermal Energy Content (erg)	$10^{26} - 10^{27}$	$10^{28} - 10^{29}$
Kinetic Energy (erg)	$10^{25} - 10^{26}$	$10^{27} - 10^{28}$
Thermal Energy of Flare (erg)	$10^{27} - 10^{28}$	$10^{29} - 10^{30}$
Goes Class	B – C class	M class

References

Masuda, S. *et al.* (1994), *Nature*, **371**, 495-497
Ohyama, M. and Shibata, K. (1997a), *PASJ*, **49**, 249-261
Ohyama, M. and Shibata, K. (1997b), *ApJ*, submitted
Ohyama, M. and Shibata, K. (1997c), *ApJ*, submitted
Shibata, K. (1996), *in Magnetodynamic Phenomena in the Solar Atomosphere*, IAU Symp. **153**, ed. Y. Uchida *et al.*, (Kluwer, Dordrecht), p13-20
Shibata, K. *et al.* (1992), *PASJ*, **44**, L173-L179
Shibata, K. *et al.* (1994), *ApJ*, **431**, L51-L53
Shibata, K. *et al.* (1995), *ApJ*, **451**, L83-L85
Shimojo, M *et al.* (1996), *PASJ*, **48**, 123-136
Shimojo, M. and Shibata, K. (1997), in these proceedings
Sturrock, P. A. (1992), *in Eruptive Solar Flares*, ed. Z. Svestka *et al.*, (Springer-Verlag), p397-409
Yokoyama, T. and Shibata, K. (1996), *PASJ*, **48**, 353-376

NONTHERMAL EFFECTS IN FE XXV SPECTRA OBSERVED BY YOHKOH/BCS

A. PHILLIPS
Mullard Space Science Laboratory
University College London, Holmbury St. Mary,
Dorking, Surrey, England

Abstract.

The presence of nonthermal electron energy distributions during impulsive solar flares has been independently observed by the Yohkoh/BCS. Using an FeXXV line ratio diagnostic, departures from a Maxwellian electron energy distribution can be measured. For the flares studied, a significant departure from a purely thermal case is seen co-incident with the hard X-ray burst. For events with a complicated hard X-ray light curve this allows an independent method of separating thermal and nonthermal contributions to the hard X-ray emission seen by the HXT.

1. Introduction

The presence of nonthermal electron energy distributions in solar flares is normally inferred from the shape of the microwave or hard X-ray light curve. Separation of the thermal and nonthermal contributions to X-ray emission in the lower energy bands of the Yohkoh Hard X-ray Telescope (HXT) is however not an easy task for complex hard X-ray bursts. Extrapolation of HXT HI band lightcurves, which are assumed to be purely nonthermal in origin, may not be possible in cases where the electron power law spectrum is particularly soft, or weak enough to fall below the detection threshold of the HXT.

This paper instead applies a FeXXV spectral diagnostic for the presence of nonthermal electron energy distributions originally developed by Gabriel and Phillips (1979) to several flares, in order to determine if there are any measurable departures from a purely thermal electron energy distribution.

T. Watanabe et al. (eds.), Observational Plasma Astrophysics: Five Years of Yohkoh and Beyond, 337–340.

2. Spectral Diagnostic

The diagnostic used here involves the FeXXV dielectronic satellite lines j ($1s^2\ ^1S + e \longleftrightarrow\ 1s2p^2\ ^2D_{5/2}$) and $d13$ ($1s^2\ ^1S + e \longleftrightarrow\ 1s2p3p\ ^2D_{5/2}$) which are excited by electron collisions in a narrow range of energies (less than 10^{-4} keV for each satellite). The ratio of these two satellite lines is temperature sensitive, as is the ratio of each to the resonance line ('w', $1s^2\ ^1S + e \rightarrow\ 1s2p\ ^1P + e$). The resonance line however is collisionally excited by any electron with an energy above 6.701 keV. This leads to a discrepancy between the temperatures returned by different combinations of spectral lines when a population of electrons with a nonthermal energy distribution, which will preferentially excite the w line, is present. There is an implicit assumption that any nonthermal energy distribution has a lower bound above the excitation energies of j and $d13$ (j=4.69 keV, 1.8666Å; $d13$=5.81 keV, 1.8526Å).

Gabriel and Phillips (1979) tabulate the ratios I_j/I_{d13}, I_j/I_w and I_{d13}/I_w as a function of temperature. We can define a measure of the deviation from a purely thermal distribution (D_{nt}) as a simple ratio;

$$D_{nt} = \left(\frac{I_j}{I_w}\right)_{thermal} / \left(\frac{I_j}{I_w}\right)_{observed}$$

Here the ratio $\left(\frac{I_j}{I_w}\right)_{thermal}$ is obtained from the observed j to $d13$ line ratio. Since both the j and $d13$ lines are assumed to be excited by the thermal population only, this observed ratio can be used to calculate the expected j to w line ratio under the condition of no nothermal contribution to the flux in the w line. A value of $D_{nt} = 1$ indicates that the plasma is purely thermal. Values of D_{nt} greater than 1 indicate departures from a Maxwellian distribution.

This technique was applied to Yohkoh/BCS data by means of a simple parametric fit to the three spectral lines and a section of the continuum around 1.83Å. This relatively simple approach was justified, in that the w line intensity (I_w) used to calculate the ratios includes the contribution from all unresolved satellites within 0.001Å of w.

The parametric fit to $d13$ includes a substantial contribution from the $d15$ satellite line at 1.8518Å. The intensity of $d13$ was estimated by using the satellite dependent $F_2(s)$ terms to derive a correction factor. The correction factor included all strong satellites within 0.001Å of 1.8526Å. The contribution of unresolved lines to the intensity derived for the j line over the wavelength range 1.865Å to 1.867Å was estimated to be less than 7%, and thus negligible. A simple parametric fit is however unable to cope with any significant blueshifts caused by directed mass motions of the plasma.

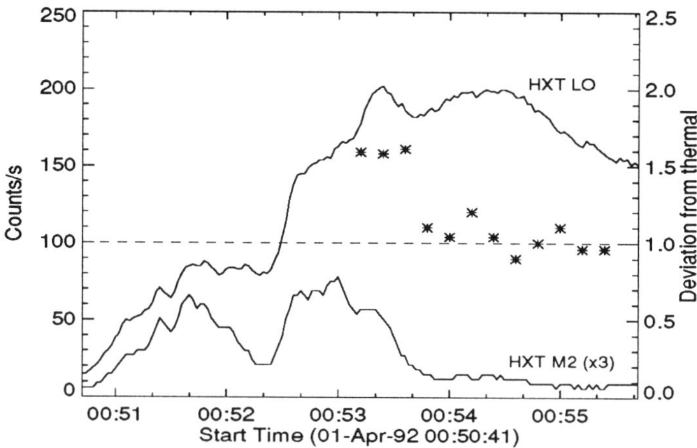

Figure 1. Plot of D_{nt} for the 1 April 1992 flare, with HXT LO and M2 band light curves as a reference. The M2 band lightcurve has been multiplied by the indicated factor for the purposes of display.

This restricts the utility of the technique to those spectra with small or non-existent soft X-ray line blueshifts.

3. Results

Three events were studied, two type B (see Tanaka (1987) for notation) impulsive flares which occurred on 27th September 1993 and 1st April 1992, and one type A flare which occurred on 6th February 1992. Both the impulsive flares show a characteristic spiky time profile. The type A event shows a suppression of this rapidly varying impulsive component (for other studies of this event see Kosugi *et al.* (1994) and Sterling (1994)).

The deviations from a purely isothermal spectrum for the 1st April 1992 flare, and the 6th February 1992 flare, determined by the diagnostic described above, are shown in Figures 1 and 2. In general there is a measurable deviation from the purely thermal case during the impulsive phase (as determined from the M2 energy band light curve), of the type B flares. The type A hot thermal flare, in contrast, displays no evidence of any deviation from a purely thermal electron energy spectrum during the initial stages of the flare. During the gradual phase of both types of flare no significant deviations from thermal are recorded. The result must be regarded as tentative however, as the saturation of the BCS FeXXV channel later in the 6th February flare limited the number of useful spectra.

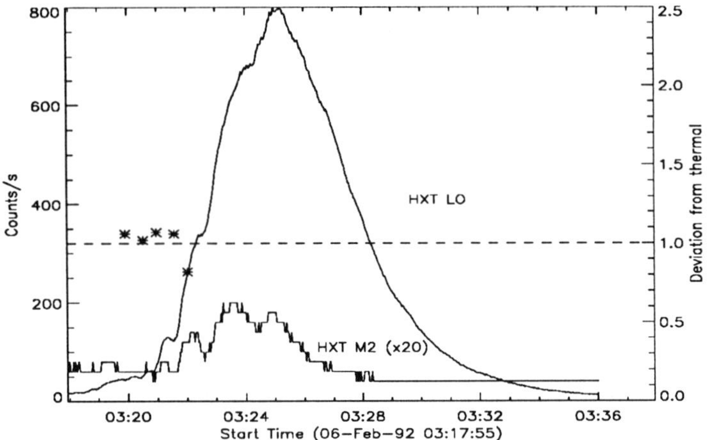

Figure 2. Plot of D_{nt} for the 6th February 1992 flare, with HXT LO and M2 band light curves for reference. There is little evidence for any significant deviation from the purely thermal case in the initial stages of this flare.

4. Discussion

The values of D_{nt} in general confirm the inferences about the existence of a nonthermal population of electrons drawn from the shape and hardness of the light curves alone. The simple parametric fit technique is sufficient to distinguish the existence, or otherwise, of such populations. The usefulness of the technique is limited both by detector effects within the BCS FeXXV channel which affect the local dispersion and line shape (Trow *et al.*, 1994) and by the requirement of a mostly stationary plasma.

Overcoming these problems, and making the procedure more robust, will require a more complex approach involving a spectral synthesis fitting technique. It may then be possible to derive a temperature unaffected by nonthermal contributions. The comparison of this temperature with that derived from a standard satellite to resonance line ratio should act as a more robust indicator of the presence of any nonthermal population of electrons.

References

Gabriel, A. H., Phillips, K. J. H., 1979 *Mon. Not. R. Astr. Soc.*, **189**, 319

Kosugi, T., Sakao, T., Masuda, S., Hara, H., Shimizu, T., Hudson, H. S., 1994 *Proc Kofu Symp. NRO Report*, **360**, 127

Sterling, A. C., 1994 *Proc. Kofu Symp. NRO Report*, **360**, 131

Tanaka, K., 1987 *Publ. Astron. Soc. Japan*, **39**, 1

Trow, M. W., Bento, A. C., Smith, A., 1994 *Nucl. Instr. and Methods*, **348**, 232

USING THE YOHKOH BCS TO CHECK ELEMENT ABUNDANCES AND IONIZATION FRACTIONS

K. J. H. PHILLIPS
Space Science Dept., Rutherford Appleton Laboratory,
Chilton, Didcot, Oxon. OX11 0QX, U.K.

AND

U. FELDMAN
E. O. Hulburt Center for Space Research,
Naval Research Laboratory,
Washington, D. C. 20375, U.S.A.

1. Introduction

During solar flares, the Bragg Crystal Spectrometer (BCS) on *Yohkoh* observes the resonance ($1s^2\,{}^1S_0 - 1s2p\,{}^1P_1$, symbolised w) lines of He-like S, Ca, and Fe ions and neighbouring spectral regions, and exceptionally, during large flares, the Ly-α lines of H-like Fe. The contribution functions $G(T_e)$, describing the emissivity as a function of electron temperature T_e, peak at about 14, 35, and 70 MK (1 MK = 10^6 K). The $G(T_e)$ curves cover a very broad range of temperature and overlap each other considerably. Spectra from the *Yohkoh* BCS and other spacecraft instruments that have been flown in recent years have been analyzed to give temperatures T_e from the ratio of nearby dielectronic satellite lines to the resonance line. The theory is described by Gabriel (1972). Standard software packages (Morrison 1994) are available for doing this analysis.

Some attention has been directed to flares which have detectable X-ray emission lasting several hours, the 'Long Duration Events' (LDE's). They are of interest for this work, as such events in their late stages have emission which is not far from isothermal, i.e. the temperatures as measured by the *Yohkoh* BCS from S, Ca, and Fe spectra (T_e(S XV), T_e(Ca XIX), and T_e(Fe XXV)) are nearly equal. This allows the accuracy of some input parameters into the analysis to be examined.

T. Watanabe et al. (eds.), Observational Plasma Astrophysics: Five Years of Yohkoh and Beyond, 341–344.

2. Analysis

We selected *Yohkoh* BCS data from seven LDE's, some well known from previous analyses, but most identified from BCS light curves. We found that emission from even the Fe XXV lines could be detected at very late stages in some of these flares, when values of T_e(Fe XXV) was as low as 10 MK and T_e(Ca XIX) as low as 7 MK. Thus, these are among the lowest-temperature flare plasmas still emitting Fe XXV and Ca XIX line emission yet detected.

The emission flux F_w (photons cm^{-2} s^{-1}) of each resonance line (w) observed by the BCS is given by

$$F_w = \text{const.} \times G(T_e)\ N_e^2 V$$

where $N_e^2 V$ is the emission measure (V is the emitting volume) and the constant includes factors expressing the earth–sun distance. The contribution function $G(T_e)$ is given by

$$G(T_e) = f(T_e) \times [N(\text{E})/N(\text{H})] \times C_{\text{tot}}$$

where $f(T_e)$ is the fraction of He-like ions of the element E in question at temperature T_e, $N(\text{E})/N(\text{H})$ is the relative abundance of E, and C_{tot} is the total excitation rate coefficient, which includes collisional excitation from the ground state (dominant), cascades from higher levels, and recombination, as well as contributions due to unresolved satellite lines. Fig. 1 shows the fractions $f(T_e)$ for the ions of interest here.

In this work, we used the standard *Yohkoh* BCS software to derive temperatures and emission measures for spectra integrated over time intervals long enough (from 5 to 40 minutes) to obtain sufficient statistical quality. (At late stages in the LDE's development, i.e. several hours after their onset, the temperatures were changing only very slowly.) We departed in one respect from the procedure used in the software in that we used the element abundances relative to hydrogen of Feldman (1993), i.e. [S/H] $= 2.1 \times 10^{-5}$ (instead of 8.7×10^{-6}), [Ca/H] $= 6.9 \times 10^{-6}$ (4.5×10^{-6}), and [Fe/H] $= 6.5 \times 10^{-5}$ (4.5×10^{-5}), i.e. S:Ca:Fe = 0.32:0.11:1 (instead of 0.19:0.10:1).

As indicated, we found that for most of the spectra analyzed, the temperatures T_e(S XV), T_e(Ca XIX), and T_e(Fe XXV) were equal to within observational uncertainties, which are computed by the software based on photon count statistics. This is indicated by Fig. 2, a plot of T_e(Ca XIX) against T_e(S XV) for all the spectra analyzed for the seven LDE's. Nearly all the points fall within two standard deviations of the equal-temperature line.

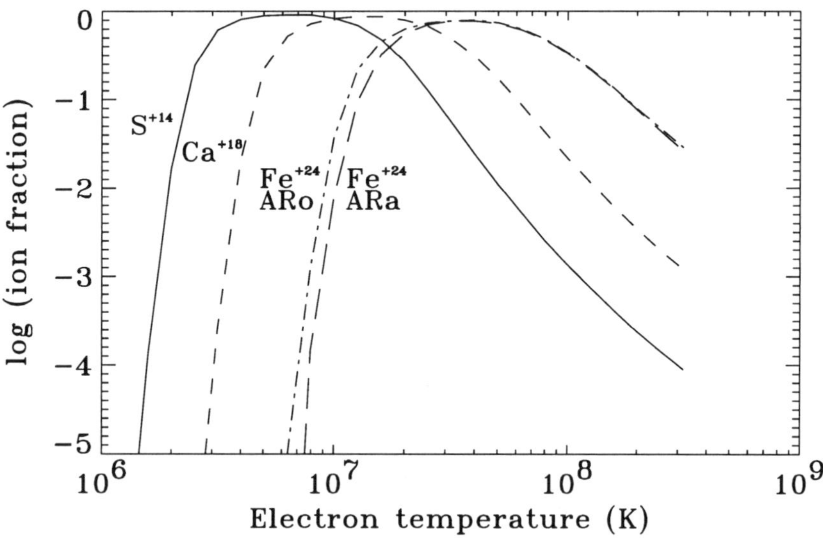

Figure 1. Ionization fractions of He-like S, Ca, and Fe. The S, Ca, and Fe curves (marked ARo) are based on Arnaud & Rothenflug (1985), the Fe curve marked ARa is based on Arnaud & Raymond (1992).

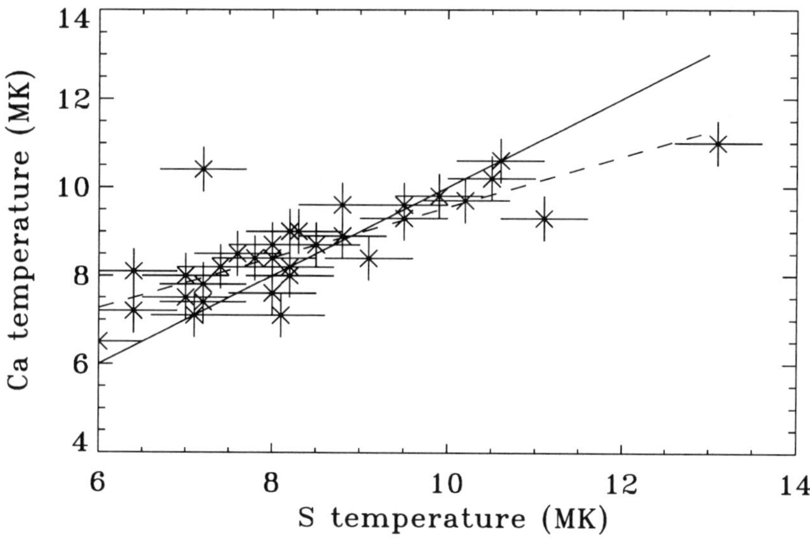

Figure 2. T_e(Ca XIX) plotted against T_e(S XV) for 32 spectra during the decay stages of seven LDE's. The solid line indicates equal temperatures, the dashed line least-squares linear fit to the observed temperatures.

3. Discussion

We formed emission measure ratios $R_1 = EM(\mathrm{Fe})\ /EM(\mathrm{Ca})$, $R_2 = EM(\mathrm{Fe})\ /EM(\mathrm{S})$, and $R_3 = EM(\mathrm{Ca})\ /EM(\mathrm{S})$ and searched for trends with temperature, either T_e(S XV) or T_e(Ca XIX) or (for R_3) the mean of the two.

We found the following:

(*a*) There were no significant trends with temperature for either the R_1 or the R_2 ratios.

(*b*) The mean value of $R_1 = 1.3 \pm 0.3$ s.d. and the mean value of $R_2 = 1.8 \pm 0.5$ s.d., i.e. within two standard deviations of unity.

(*c*) For R_3, we found that there was a slight trend, with R_3 taking values less than unity (~ 0.5) for temperature < 8 MK and more than unity for temperature > 9 MK, but is about one for a temperature of 8.5 MK.

Point (*c*) shows that R_3 is around unity for a temperature nearly equal to the maximum of $f(T_e)$ of both He-like S and Ca, when f is almost 1. This must mean, then, that the abundance ratio we took is the correct value to within about 50%. The trend with temperature may imply that the fractional abundance of He-like Ca ions is about 50% too large for temperatures 6–8 MK. This is in accord with the fact that He-like Fe fractions were revised downwards for temperatures less than the peak in the Arnaud & Raymond (1992) calculations compared with the earlier Arnaud & Rothenflug (1985) work.

Points (*b*) and (*c*) give us confidence that the both the ionization fractions are within a factor of 2 of the correct value over the temperature range considered and that the relative element abundances we used in this analysis are equally correct.

Our final values for the relative abundances, with estimated uncertainties, are S:Ca:Fe = 0.32±50% : 0.11±30% : 1. These are consistent with the results of Fludra et al. (1993), who find these same ratios to be 0.22±23% : 0.09±11% : 1.

This work was partly supported financially by British Council/KBN (Poland) collaborative grant no. WAR/992/057.

References

Arnaud, M., & Raymond, J. 1992, ApJ, 398, 394
Arnaud, M., & Rothenflug, R. 1985, A&AS, 60, 425
Feldman, U. 1992, Physica Scripta, 46, 202
Fludra, A, et al. 1993, Adv. Space Res., 13(9), 395
Gabriel, A. H. 1972, MNRAS, 160, 99
Morrison, M. 1994, Yohkoh Analysis Guide (LMSC-P098510)

MAXIMUM ENERGY OF FORCE-FREE MAGNETIC FIELDS

T. SAKURAI
National Astronomical Observatory
2–21–1 Osawa, Mitaka, Tokyo 181, Japan

Abstract. Force-free magnetic fields attain more and more energy as they are stressed, but the energy is never able to exceed an upper limit set by the boundary condition (Aly, 1984). This property of the force-free fields has been discussed based on analytic studies using integral relationships (virial relations) about the magnetic field, and also based on numerical simulations. In this study, we use a simple solution for the force-free fields derived by Low and Lou (1990) and show the above-mentioned property most directly.

1. Introduction

Aly (1984) showed that the energy of any force-free field is bounded from above by the energy of the open magnetic field having the same distribution of magnetic flux on the boundary. His argument is based on integral relationships for the force-free magnetic fields. The aim of this paper is to demonstrate this property directly by using a simple force-free field model formulated by Low and Lou (1990). A similar study has been made recently by Wolfson (1995).

2. Formulation

The model by Low and Lou (1990) is an axisymmetric solution for the force-free field equation

$$(\nabla \times \boldsymbol{B}) \times \boldsymbol{B} = 0,$$

where the field $\boldsymbol{B}$ is assumed to have a form

$$\boldsymbol{B} = \frac{1}{r \sin\theta} \left[\frac{1}{r} \frac{\partial A}{\partial \theta} \hat{\boldsymbol{r}} - \frac{\partial A}{\partial r} \hat{\boldsymbol{\theta}} + Q(A) \hat{\boldsymbol{\phi}} \right],$$

T. Watanabe et al. (eds.), Observational Plasma Astrophysics: Five Years of Yohkoh and Beyond, 345–348.

Figure 1. Dipolar field with moderate shear ($n = 0.8$); front view (left) and side view (right).

in spherical polar coordinates. Here $(\hat{\boldsymbol{r}}, \hat{\boldsymbol{\theta}}, \hat{\boldsymbol{\phi}})$ represent the unit vectors in (r, θ, ϕ)-directions, respectively. The vector potential A then satisfies

$$\frac{\partial^2 A}{\partial r^2} + \frac{1-\mu^2}{r^2}\frac{\partial^2 A}{\partial \mu^2} + Q\frac{dQ}{dA} = 0 \qquad (\mu = \cos\theta).$$

We assume a particular form for A and Q as

$$A = A_0 \frac{P(\mu)}{(r/a)^n}, \quad Q = k\frac{A_0}{a}\left|\frac{A}{A_0}\right|^{1+\frac{1}{n}},$$

where a, k, n and A_0 are constants. Then the radial dependence of the solutions is automatically satisfied, and the angular dependence of the vector potential is described by the following ordinary differential equation

$$(1-\mu^2)\frac{d^2 P}{d\mu^2} + n(n+1)P + \frac{n+1}{n}k^2 P|P|^{\frac{2}{n}} = 0.$$

For the solutions to be regular at $\mu = \pm 1$, P should satisfy the boundary condition $P(\pm 1) = 0$. We adopted the normalization for P as $P'(-1) = 10$ following Low and Lou (1990).

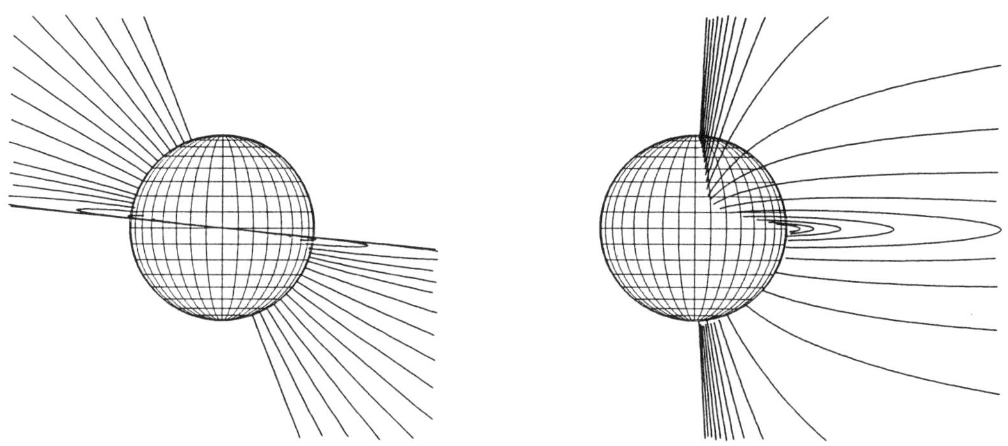

Figure 2. Dipolar field with large shear ($n = 0.1$); front view (left) and side view (right).

3. Results

For a given value of n, the eigenvalue k is determined (Sakurai and Chae 1996). The solutions characterized by $k = 0$ and $n = 1, 2, 3, \ldots$ are current-free fields in the form of multipoles (dipole: $n = 1$, quadrupole: $n = 2$, octopole: $n = 3$). The field deviates from the current-free state as k deviates from zero. However, the basic geometry of the field is conserved along each eigenvalue; therefore, the solutions starting from $k = 0$ and $n = 1, 2, 3$ can be called dipolar, quadrupolar, and octopolar fields even if $k \neq 0$.

As the field deviates from a current-free multipole solution, first k grows to a certain value and then decreases asymptotically to zero. The value of n monotonically decreases to zero along the equilibrium sequence. The magnetic field decays as $B \sim r^{-n-2}$. Therefore, small values of n indicate that the field decays very slowly in radial distance, namely a spatially extended field geometry.

Figures 1 and 2 show the field lines for dipolar force-free fields. For current-free multipole fields, the field lines are contained in the meridional plane. Deviation of the field lines from the meridional plane (namely, shear in field lines) is due to force-free currents. The case of Figure 1 ($n = 0.8$) has only moderate shear because it is close to the current-free case ($n = 1$). On the other hand Figure 2 shows results for a very small value of n ($n = 0.1$). In that case the field lines are elevated, and a structure like a neutral sheet

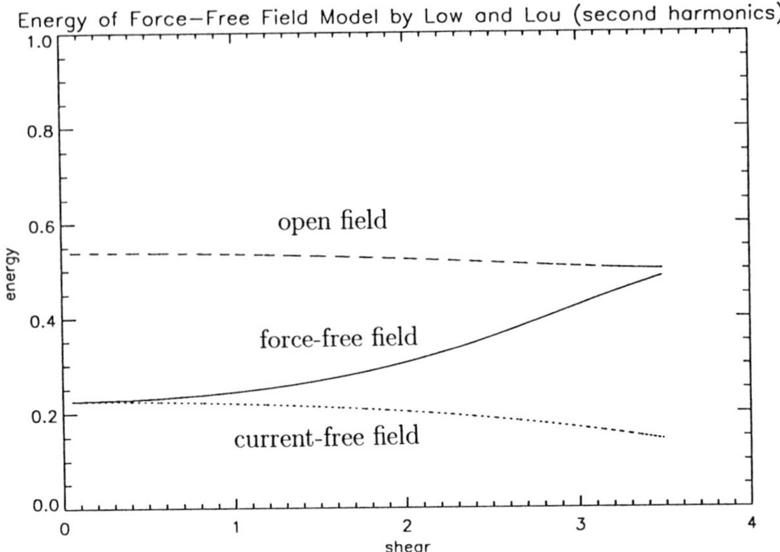

Figure 3. Energies of the open field, the force-free field, and the potential field (from top to bottom) as a function of magnetic shear. The total magnetic flux is held fixed.

is formed.

Figure 3 shows the evolution of magnetic energies as a function of shear for the quadrupolar model. Other multipoles show a similar behavior. Here, the shear is defined as

$$\text{shear} = \int_0^{\pi/2} \frac{B_\phi}{B_\theta \sin\theta} d\theta.$$

This quantity gives the maximum excursion in longitude of the trace of the horizontal magnetic vectors on the solar surface. The energies are evaluated by using the virial theorem. The open fields are generated by replacing the radial components of the field with their absolute values. The energies are re-normalized in such a way that the total magnetic flux is kept constant. As the shear increases, the energy of the force-free field approaches (but never exceeds) the energy of the open magnetic field.

References

Aly, J.J. (1984) *Astrophys. J.*, **283**, 349.
Low, B.C. and Lou, Y.Q. (1990) *Astrphys. J.*, **352**, 343.
Sakurai, T. and Chae, J.C. (1996) in *Magnetodynamic Phenomena in the Solar Atmosphere*, ed. Y.Uchida, T.Kosugi, and H.S.Hudson, Kluwer Academic Publishers, p.579.
Wolfson, R. (1995) *Astrophys. J.*, **443**, 810.

A STUDY OF PREFLARE X-RAY BRIGHTENINGS

FRANTIŠEK FÁRNÍK
Astronomical Institute of the Academy of Sciences, Ondřejov, Czech Republic

AND

SERGE K. SAVY
Institute of Space and Astronautical Science, Sagamihara, Japan

1. Extended Abstract

Preliminary results of an extended study of the characteristics of preflare brightenings are presented. It was shown recently (Fárník et al., 1996) that SXT observations place the preflare emission source roughly at the same altitude level as the susequent flare, close to the main flare source but not necessarily at the same position, and that the incidence of significant X-ray brightening in coronal magnetic structures before they flare is rather low ($< 25\%$). In an attempt to extend our knowledge of this phenomenon we studied in detail 29 flares using *Yohkoh* data.

The role of the preflare process in relation to the driving flare mechanism is poorly understood. Only a few studies of the topic have been undertaken so far: Bumba and Křivský (1959) studied preflares in optical data, Kai,Nakajima and Kosugi (1983) made a detailed study of 97 bursts in radio region, some other papers were based on data from the SMM satellite. The spatial resolution of the SXT instrument onboard *Yohkoh* is much better compared with the SMM instruments and enables a more detailed study.

Using *Yohkoh* satellite data from the third year of its operation (October 93 - September 94), we examined all the solar flares *Yohkoh* observed with peak emission greater than a total of 1000 counts/sec in the BCS CaXIX channel, (this corresponds approximately to $>$ GOES class C6). Only 29 of these events had SXT images of the flaring region during the flare and at least several minutes before the onset. Table 1 shows the list of selected

T. Watanabe et al. (eds.), Observational Plasma Astrophysics: Five Years of Yohkoh and Beyond, 349–352.

events with their basic characteristics (available to us). A classification scheme we used is described below.

To see if any of the flares in the set exhibited preflare activity, the location of each flare in the SXT images was searched for identically shaped structures before the onset of the flare. One problem with this technique is that magnetic structures in the line of sight, but at a different altitude to the flare, can brighten, giving the appearance of preflare activity at the same location as the subsequent flare. However, we note that in most SXT preflare and flare images, distinct X-ray sources have a unique combination of size, shape and orientation. On this basis we believe that it is reasonable to assume that sources at the same position in the image plane, and with the same size, shape and orientation, but viewed at different times, are associated with the same magnetic structure. As we pointed out in the Introduction, we primarily focused on studying the spatial relation between preflaring and flaring areas. We grouped all the studied events (29 in total) into 3 categories according their shape, orientation, location and size of the bright pattern:

I. The main flare bright pattern is identical with the shape and location of the core part of the whole preflare (intensity contour at 50 % of maximum) - in Table 1 shown as 'Coincident'.

II. The flare pattern lay over a part of the core preflare pattern - in Table 1 shown as 'Adjacent/Overlapping'.

III. The flare is separated from the pre-flaring area (inside the same active region) and no connection is observed in soft X-rays, in Table 1 shown as 'Distant'.

One example of the 'Adjacent/Overlapping' type is shown in Figure 1. We found evidence that, of the 29 events we examined, 7 coronal structures that flared were bright in X-rays at least several minutes before the start of the flaring and the remaining 22 flares coincided fully or partially with X-ray bright areas before flare onset, but the bright regions had a significantly different size and shape from the flare loops that subsequently appeared near the same location. In several cases it appeared that only a small area of the flare loop structure was bright significantly before flare onset.

2. Acknowledgments

F.F.'s work on the topic was supported by the grant (No.205/96/0167) of the Grant Agency of the Czech Republic (GACR). Both authors would like to thank H.Hudson and J.Khan for very stimulating discussions.

TABLE 1. List of selected events (flares with preflares)

Date:	Peak Time in XR	Flare Class	Flare Location	CaXIX Preflare	Preflare Location
9 Oct 93	08:10	M1.1	N11W78	No	Coincident
27 Oct 93	06:46	C9.4	N09E39	No	Adjacent/Overlapping
30 Nov 93	06:07	C9.2		Yes	Adjacent/Overlapping
6 Dec 93	20:42	C7.5		Yes	Coincident
7 Dec 93	00:45			Yes	Coincident
7 Dec 93	02:16	C4.2	S21E12	No	Adjacent/Overlapping
25 Dec 93	04:29	C6.2	N07E12	Yes	Adjacent/Overlapping
28 Dec 93	16:58	M1.1	N07W37	No	Coincident
29 Dec 93	15:42	C9.9	N12W43	Yes	Coincident
30 Dec 93	16:12			Yes	Adjacent/Overlapping
30 Dec 93	17:12	C7.7	N10E58	No	Adjacent/Overlapping
5 Jan 94	07:02	M1.0	S13W23	No	Adjacent/Overlapping
6 Jan 94	04:07	C4.9		No	Adjacent/Overlapping
16 Jan 94	23:16	M6.1		Yes	Adjacent/Overlapping
17 Jan 94	09:20	C9.3	N06E65	No	Adjacent/Overlapping
19 Jan 94	03:03	C4.0	N05E40	No	Adjacent/Overlapping
29 Jan 94	11:25	M2.4		No	Adjacent/Overlapping
27 Feb 94	09:12	M2.8		No	Adjacent/Overlapping
3 Mar 94	02:10	C6.2		No	Coincident
3 Mar 94	12:35			Yes	Coincident
4 Mar 94	09:10			Yes	Adjacent/Overlapping
14 Aug 94	17:35	M3.9	S12W08	No	Distant
17 Aug 94	01:05	M1.5	S15W35	No	Adjacent/Overlapping
18 Aug 94	03:04	M1.1	S08W53	No	Adjacent/Overlapping
30 Aug 94	19:50	C6.2	S09E83	No	Distant
31 Aug 94	07:10	C2.8	S07E69	No	Adjacent/Overlapping
3 Sep 94	15:55	C6.8	S09E76	No	Adjacent/Overlapping
5 Sep 94	05:36	C6.0	S09E04	Yes	Adjacent/Overlapping
6 Sep 94	00:58			Yes	Adjacent/Overlapping

3. References

Bumba, V. and Křivský, L.: 1959, *Bull.Astron.Inst.Czech.* **10**, 221.

Fárník, F., Hudson, H., and Watanabe, T.: 1996, *Solar Phys.* **165**, 169.

Kai, K., Nakajima, H., and Kosugi, T.: 1983, *Publ.Astron.Soc.Japan* **35**, 285.

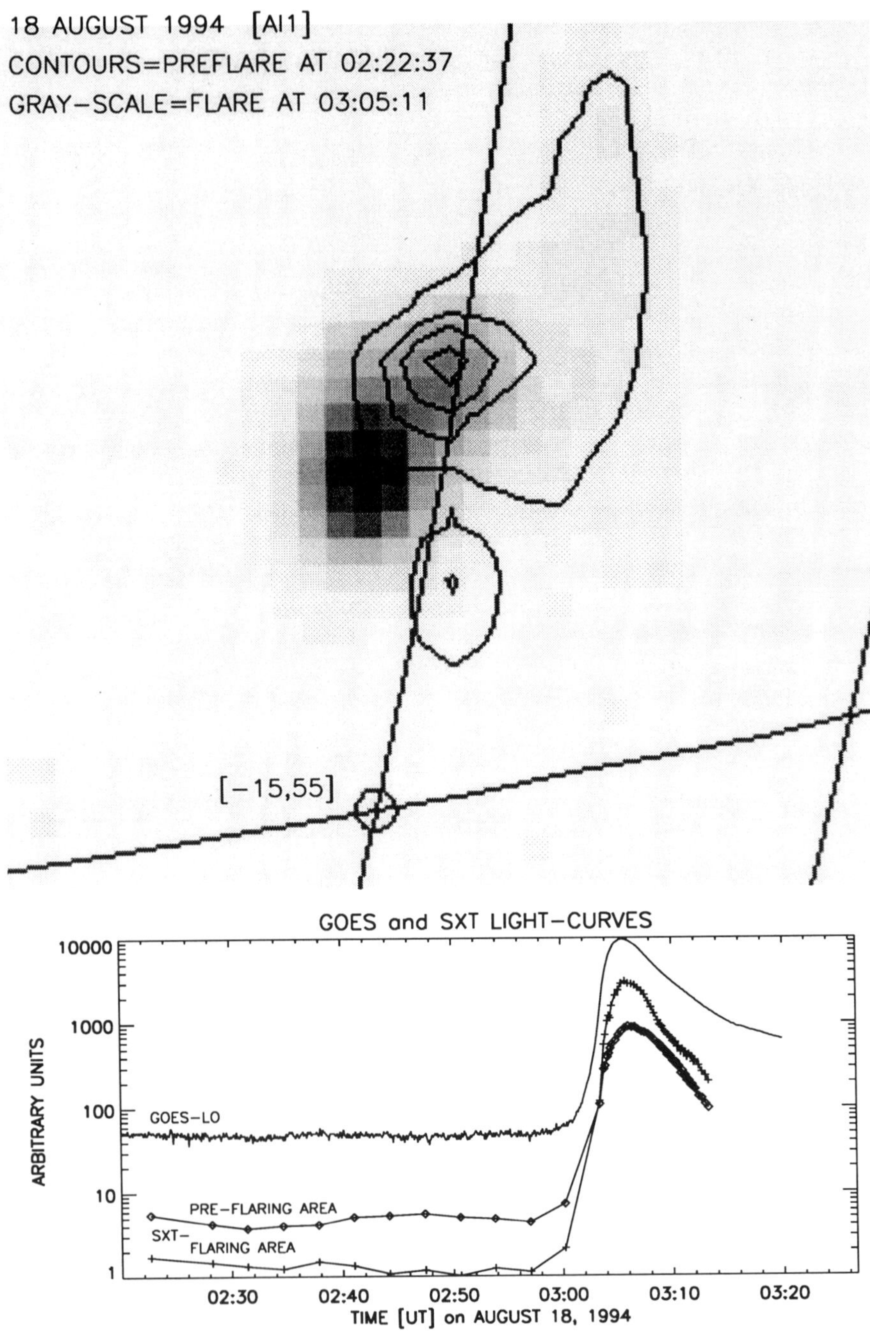

Figure 1. An example of the preflare-flare images of the event of 18 Aug 94.In Table 1 this event is classified as 'Adjacent/Overlapping'. The location of the event is given by solar grid and coordinates. The lower part, the GOES and SXT light curves, shows how relatively weak the preflare emission may be.

NON-LINEAR EVOLUTION OF ERUPTING CORONAL MAGNETIC FIELDS

BAMBANG SETIAHADI AND BACHTIAR ANWAR
Watukosek Solar Observatory
Gempol, P.O. Box 04, Pasuruan 67155, East Java, Indonesia
MAKI AKIOKA
Hiraiso Solar Terrestrial Research Center, CRL
3601 Isozaki, Hitachinaka, Ibaraki 311-12, Japan
AND
TAKASHI SAKURAI
Solar Physics Division, National Astronomical Observatory
Mitaka, Tokyo 181, Japan

Abstract. We performed two-dimensional numerical simulations of erupting coronal magnetic fields. Initial condition is unsheared cylindrically symmetric magnetic arcade in magnetohydrostatic balance. The initial pressure profile and magnetic topology are in quasi static configuration. Our results show that such an arcade will turn into an erupting coronal magnetic field closely resembling a loop-type coronal mass ejection (see e.g. Filippov *et al.*, 1994). We also studied the non-linear evolution of various physical parameters, such as pressure, density, and kinetic energy, during the simulated eruption.

1. Introduction

Early images of the solar coronal magnetic fields taken during the rare occurrences of total solar eclipses give an impression of static structures. As more data was compiled from ground-based and space-based platforms covering various wavelength regimes (e.g. *HAO, YOHKOH, P78-1, OSO 7* and *SKYLAB*), it becomes obvious that the solar coronal magnetic field is more dynamic than previously thought. It can change frequently and abruptly. During the eruptions a considerable amount of mass (about 10^{15}

T. Watanabe et al. (eds.), Observational Plasma Astrophysics: Five Years of Yohkoh and Beyond, 353-356.

grams) of plasma with speeds ranging from 100 km s^{-1} up to 1500 km s^{-1} is expelled into interplanetary space (Jackson and Howard, 1993). These ejections are called coronal mass ejections (CMEs). They are expected to produce significant changes throughout the solar atmosphere and shocks in the corona (Setiahadi *et al.*, 1995).

In this paper we try to make a plausible model of an assumed coronal loop which might be pre-existing before the launch of a CME. Using two-dimensional time-dependent MHD numerical simulations we demonstrate that the loop shape can be reproduced from our model, and we derive two-dimensional non-linear behaviour of some basic physical parameters during eruption.

2. Initial Configuration

The initial configuration of our model is an unsheared arcade with cylindrically symmetric, purely azimuthal magnetic fields with the axis lying on the dense photosphere (see e.g. Steele and Priest, 1990). The footpoints of the arcade are line-tied in the photosphere. The arcade is in the xy-plane and the base of the arcade lies along the horizontal x-axis which runs parallel to the solar surface. The vertical y-axis is directed upward perpendicular to the solar surface with $y = 0$ at the solar surface. We apply fixed boundary conditions along the arcade base to mimic the dense photosphere, but let the plasma penetrate from below the bottom boundary with a relatively long time scale. Other (top, left, and right) boundaries, are free boundaries where the plasma and magnetic fields can pass freely. The plasma pressure and the magnetic pressure balance the magnetic tension and satisfy the equation of magnetohydrostatic equilibrium.

We assume that the base of the arcade contains high beta plasma and that the evolution of the arcade is controlled by plasma pressure. By further assuming that the pressure profile drops exponentially in the radial direction, we have a solution for the azimuthal magnetic structure.

3. Results

In this paper our initial configuration is in a critical state. Below this critical state any small perturbation on the base of the arcade will not produce an eruption, but rather the pressure and the magnetic field perturbations will slowly and constantly propagate and decay toward interplanetary space. Evolution from the quiet state to the critical state is assumed to be instantaneous.

We have run our simulation up to 1000 cycles, which is equivalent to 512 seconds of real time. The MHD fast shocks were excited 50 seconds into the simulation at about 5.0×10^4 to 5.5×10^4 km above the solar surface. We can study the system's evolution using the profile of the vertical velocity component (V_y) along the axis of symmetry. At t = 50 s V_y exhibits speeds larger than the ambient solar wind speed at a height of about 3.0×10^4 to 6.0×10^4 km above the base. This is the signature of the beginning of a shock in the corona. The place of shock excitation is not very localized, but is extended over a region of about 10,000×10,000 km^2. The shock front, on the other hand, may be as sharp as 1,000 km and moves out with a speed much larger than the ambient solar wind speed. In our simulation the fast-mode shock front moves at a speed of about 390 to 400 km s^{-1}. This is in accordance with the speeds reported by several authors (see e.g. Wang *et al.*, 1995).

Within 100 to 150 seconds the shocks begin to show a loop-like shape. The loop shape became more obvious after t = 200-450 s and closely resembles the white light data from *HAO* for events of 14 April 1980, 8 Augusts 1980, and 24 October 1989.

The dynamics at heights below 5.0×10^4 km cannot be fully inferred from the above events because of obscuration by the occulting disk of the coronagraph. However, from the above events we can clearly see a darker part trailing behind the moving loop structure. This part may be the upper part of the low density and low pressure region (or hollow) developed after the excitation of the MHD fast shocks as a consequence of material depletion by the moving loop. According to our simulation the region may have lower temperature, since the adiabatic pressure drops faster than the density after t = 50-100 s. Our simulation shows that the temperature may drop from 1.0×10^6 K to 0.8×10^6 K and the plasma density on the base may drop to 68 % of its initial state.

We analyse the evolution of kinetic energy density along the axis of symmetry. At t = 50 s the kinetic energy is concentrated in the lower part of the arcade. We can watch the beginning of kinetic energy transfer following the excitation of the fast shocks. The kinetic energy is continuously transfered to higher and wider regions, leaving a hollow of very low kinetic energy density of about 0.08×16.3 erg cm^{-3}.

4. Discussion

Pressure evolution may still be decisive in controlling the dynamics of the lower part of the corona during CME eruptions. As the pressure evolves, the magnetic fields may follow the evolution by slowly adjusting their topology. If there is no magnetic field enhancement during this passive evolution, the magnetic field will attain its critical state, at which time the magnetic topology cannot withstand further pressure build-up. Eruption is then the only way to adjust the pressure and magnetic topology back into a quiet

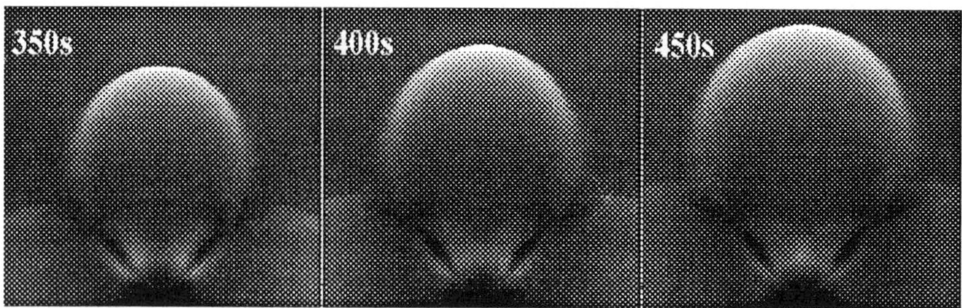

Figure 1. Result of MHD simulation of the solar CME at 350s, 400s, and 450s.

condition.

The above scheme might be valid for a clean CME, i.e., an ejection without any subsequent low-lying events occuring after the initial eruption. If the ejection is followed by other low-lying phenomena then the interpretation of the physics becomes a difficult task. A low-lying mature active region may transfer additional energy to the arcade and may accelerate the process leading to the critical state. As the critical state is attained the arcade will erupt and form the hollow. The low kinetic energy and pressure content of the hollow would easily be compensated by an energy flow from the mature active region.

All of computations presented in this paper were performed on an *IBM-PC AT* with Intel *486DX4-100MHz* and the *UNIX* operating system. A typical time needed for a simulation running up to 1000 cycles was 250 to 300 minutes, depending on the behaviour of the fast-mode shocks developed during each cycle.

References

Filippov, M., Golbraikh, E., Steinitz, R. (1994), *IAU Coll.* **no 144** , p. 101.
Jackson, B.V., Howard, R.A. (1993), *Solar Physics*, **148**, p. 359.
Setiahadi, B., Anwar, B., Sakurai, T. (1995), The Ninth International Symposium on Equatorial Aeronomy (ISEA), Bali, Indonesia, March 20-24, 1995
Steele, C.D.C., Priest, E.R. (1990), *Solar Physics*, **127**, p. 65.
Wang, A.H., Wu, S.T., Suess, S.T., Poletto, G. (1995), *Solar Physics*, **161**, p. 365.

STUDY OF SOLAR X-RAY JETS
OBSERVED BY THE YOHKOH SOFT X-RAY TELESCOPE

M. SHIMOJO AND K. SHIBATA
National Astronomical Observatory,
2-21-1, Osawa, Mitaka, Tokyo, 181, Japan

Abstract. We have found 100 X-ray jets in the database of full Sun images taken with the Soft X-ray Telescope (SXT) aboard Yohkoh during the period from 1991 November through 1992 April. We report on the results of a statistical study of X-ray jets and a study of the magnetic field properties at the sites of X-ray jets. These results suggest that magnetic reconnection produces the jets.

1. Introduction

The soft X-ray telescope (SXT) aboard Yohkoh has better temporal coverage and resolution than previous solar X-ray imagers, and has revealed many previously unknown dynamic coronal phenomena. Among the various newly discovered dynamic phenomena, one of the most interesting findings is the common occurrence of X-ray jets (Shibata *et al.* 1992, Strong *et al.* 1992). According to Shibata *et al.* (1992), these X-ray jets are transitory X-ray enhancements with an apparent collimated motion.

The purpose of this paper is to report on a study of the statistical properties of X-ray jets which were observed during the six months from 1991 November through 1992 April and a study of the magnetic field properties at the sites of the X-ray jets.

2. Statistical Study of Solar X-Ray Jets

For this study, we used only full-frame image (FFI) data taken from 1991 November through 1992 April. We studied events satisfying the following conditions:

T. Watanabe et al. (eds.), Observational Plasma Astrophysics: Five Years of Yohkoh and Beyond, 357-360.

1. The aspect ratio (length/width) of the jet is larger than 3.
2. The time interval between the pre-jet image and the first jet image is less than 1 hour, or morphological changes occur within 1 hour.

First, we made a descriptive list of the jets and stored their representative X-ray images in a database. A statistical study for these jets results in the following chracteristics (Shimojo *et al.* 1996): 1) Most of the jets are associated with small flares (microflare – subflare) at their footpoints. 2) The lengths of jets range from a few $\times$ 10^4 km to a few $\times$ 10^5 km. 3) The apparent velocities range from $\sim$ 10 km/s to 1000 km/s and a typical apparent velocity is about 200 km/s. 4) The lifetimes range from a few min to $\sim$ 10 hours and the distribution of lifetimes is a power law. 5) We found that 43% of jets have constant widths, and 33% of the jet appear to be converging. 6) About two thirds (64%) of the jets appear in or near active regions (AR). When a jet is ejected from a bright point-like feature in an AR, it occurs to the west 86% of the time. 7) A clear gap ($> 10^4$ km) is seen between the exact footpoint of the jet and the brightest part of the associated flare in 27% of the jets. 8) The X-ray intensity along an X-ray jet often shows an exponential decrease with distance from the footpoint. 9) This exponential intensity distribution holds from the early phase to the decay phase.

3. Magnetic Field Properties of Solar X-ray Jets

From a list of X-ray jets, we selected events, for which there were magnetic field data from NSO/Kitt Peak. Using co-aligned SXT images and magnetograms, we examined the magnetic field properties of X-ray jets (Shimojo *et al.* 1997). We classified the magnetic field properties of the jet footpoints into the following four types:

Single Pole (8%) : Only one observed magnetic polarity.
Bipole (12%) : A bipole structure.
Mixed Polarity (24%) : There are some or many poles with mixed polarities.
Satellite Polarity (48%) : A mixed polarity type area near sun spots.

Examples of these magnetic field configurations are shown in Figure 1. We found that the most common magnetic field property of jet-producing regions is the satellite polarity type. Since we consider the satellite polarity as physically the same as the mixed polarity type, we can say that 72% of jets occurred in (general) mixed polarity regions.

We also investigated the magnetic evolution of jet-producing areas in active regions NOAA 7067, NOAA 7270 (Figure 2) and NOAA 7858. It is found that X-ray jets favored regions of evolving magnetic flux (increasing or decreasing).

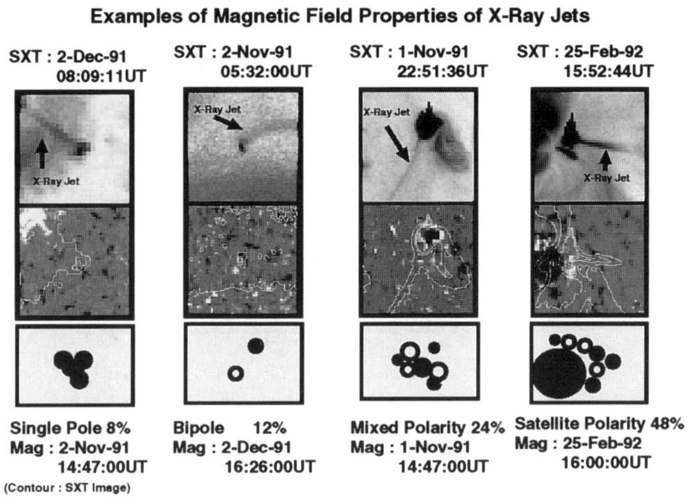

Figure 1. Examples of magnetic properties of X-ray jets

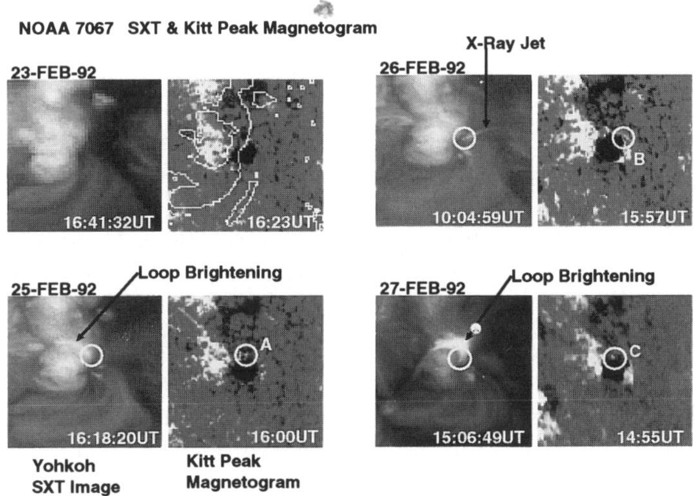

Figure 2. Time variation of magnetic flux with polarity opposite to that of the nearby sunspot. The solid curve and + indicate the magnetic flux of satellite polarity. Four dots-dashed vertical lines show the times of X-ray jets.

4. Discussion : Solar X-Ray Jet Build Up

A model of X-ray jets was been suggested by Shibata *et al.* (1992), and numerically simulated by Yokoyama and Shibata (1995). Their model suggests that magnetic reconnection occurs between emerging flux and overlaying

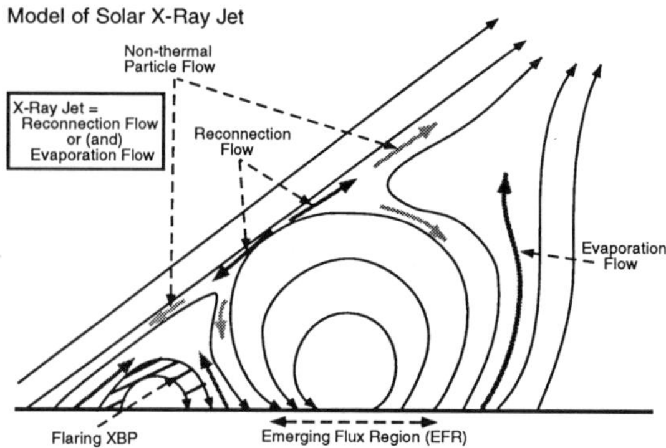

Figure 3. Model of Solar of X-ray jets

existing magnetic fields, resulting in hot plasma flows (X-ray jets). Our results (such as the gap between the jet and the brightenings, and that the jet-producing region is usually a mixed polarity region with the increasing magnetic flux) supports this model. However, our results also show that X-ray jets occurred in canceling magnetic structures.

We can explain our results of X-ray jets occurring at the sites of both emerging flux and the canceling magnetic bipole using the following emerging flux model: 1) A bipole emerges from the sub-photosphere. 2) The emerging flux makes contact with an overlaying or adjacent oppositely-directed pre-exsting field. 3) Magnetic reconnection between the emerging flux and the overlaying or adjacent field produces an X-ray jet and reconnected loops. 4a) If the rate of emerging flux in the emerging bipole is larger than the rate of reconnection, the total observed magnetic flux increases. 4b) If the rate of emerging flux is smaller than the rate of reconnection, magnetic flux decreases in the jet-producing region at the photosphere, because there is a chance that the reconnected flux submerges under the photosphere ("converging flux model" by Priest *et al.* 1994). The difference between an observed increase and decrease in the flux in the jet-producing region comes simply from the difference in the rate of emergence of magnetic flux.

References

Priest, E.R., Parnell, C.E., Martin, S.F. : 1994, *Astrophys. J.*,**427**, 459
Shibata, K., *et al.* : 1992, *Publ. Astron. Soc. Japan*, **44**, L147
Shimojo, M., *et al.* : 1996, *Publ. Astron. Soc. Japan*, **48**, 123.
Shimojo, M., Shibata, K. and Harvey, K.-L. : 1997, *Solar Phys.*, to be published.
Strong, K. T., *et al.* : 1992, *Publ. Astron. Soc. Japan*, **44**, L161
Yokoyama, T. and Shibata, K. : 1995, *Nature*, **375**, 42

DECONVOLUTION OF YOHKOH SOFT X-RAY IMAGES

J. SHIN AND T. SAKURAI
National Astronomical Observatory
Mitaka, Tokyo 181, JAPAN

AND

N. MIURA
Kitami Institute of Technology
165 Koen-cho, Kitami 090, JAPAN

Abstract.

We examine the effects of deconvolution of *Yohkoh* SXT images using a Blind Iterative Deconvolution algorithm and a modified PSF/noise suppression method. Though the results show clear enhancement of peak intensities in X-ray structures, some critical problems are found during the calculation, which makes it hard to deconvolve the SXT images with general algorithms. The difficulty comes from the undersampling of SXT images, and therefore more detailed consideration is required for the deconvolution.

1. Introduction

The *Yohkoh* soft X-ray telescope (SXT) has taken numerous images of active, quiet, and coronal hole structures on the Sun for more than five years, which makes it possible to estimate various physical quantities in the solar corona. However, the observed images are smeared to some degree by the finite resolving power of the telescope. Further, a large pixel size of the CCD makes the image substantially undersampled. Without proper calibration of these effects, it is not guaranteed that we can accurately describe features of coronal activity from the observed images. In this report we introduce various methods of the deconvolution of SXT images and discuss problems found during the calculation.

T. Watanabe et al. (eds.), Observational Plasma Astrophysics: Five Years of Yohkoh and Beyond, 361–364.

2. Deconvolution of Yohkoh SXT Images

2.1. PROBLEMS IN THE DECONVOLUTION OF SXT IMAGES

With the point spread function (PSF) given from the laboratory experiments (Martens et al. 1995), we examined the effect of image deconvolution using a sample SXT image (Figure 1(a)): First we simply deconvolved the image without considering the noise included in the image (Figure 1(b)). The effect of the deconvolution is shown clearly at the bright areas, where the intensities in the peaks become much enhanced. But after the deconvolution we found a critical problem in the result: the reconstructed image shows a negative structure, composed of a group of negative values, on the image, especially where the intensity gradient is very steep. This is quite different from negative values due to noise which could be seen at random places in the image. This negative structure forms as a group of negative values and is always located just beside the brightest area where there is an abrupt change in the intensity distribution. Further, this problem is not a special case for this particular image, but generally appears for other SXT images. Here we consider two possible candidates that can produce this negative structure: a broad PSF and the noise produced by the CCD. Now we will focus our interest on the effect of the shape of PSF and the noises of CCD.

2.2. BLIND ITERATIVE DECONVOLUTION

The main advantage of the Blind Iterative Deconvolution (BID) algorithm (Miura et al. 1992 and references therein) is that it considers both PSF and the reconstructed image simultaneously. Using the BID method we estimated the shape of the PSF and the "true" SXT image directly from the observed images. However, the PSF deduced from the BID method was far from that determined from pre-launch experiments although it has the same basic core/wing structure. The negative structure does not appear in the deconvolved image, because the BID algorithm always assumes a positive solution. However, when we re-convolved the result with the PSF, it did not adequately match with the observed image within noise level, which indicates that the deconvolution was not satisfactory. We may conclude that the BID method is not suitable for the deconvolution of SXT images: a steep variation of intensity distribution can lead to an incorrect prediction of the shape of the PSF.

2.3. FILTERING OF THE NOISE

If the negative structure problem is due to a broad PSF, then we may remove it by modifying the shape of the PSF. We slightly reduced the

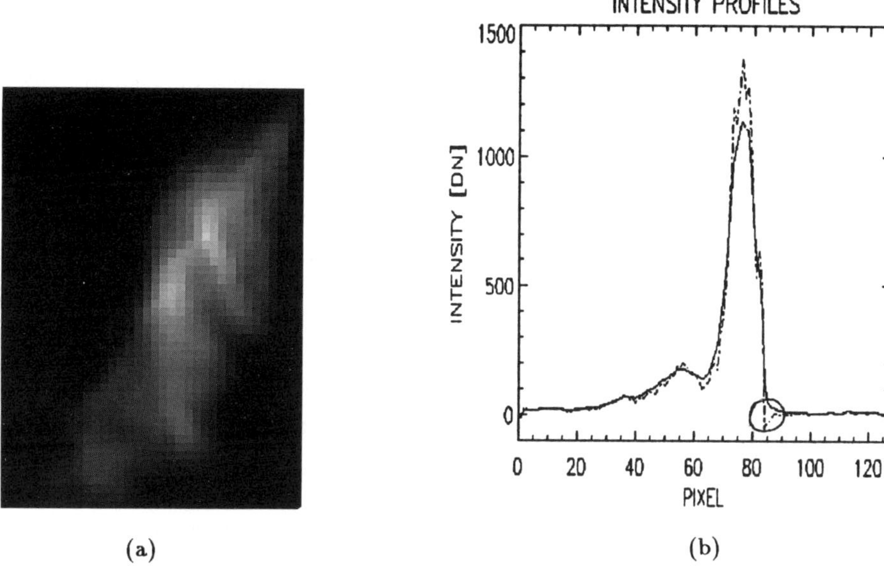

(a) (b)

Figure 1. (a) A sample SXT image on the 14th of August, 1992. (b) The intensity profiles of the observed (solid line) and the deconvolved (dotted line) images when they are cut horizontally at the peak area. The negative structure is clearly seen on the right side of the bright region (circled).

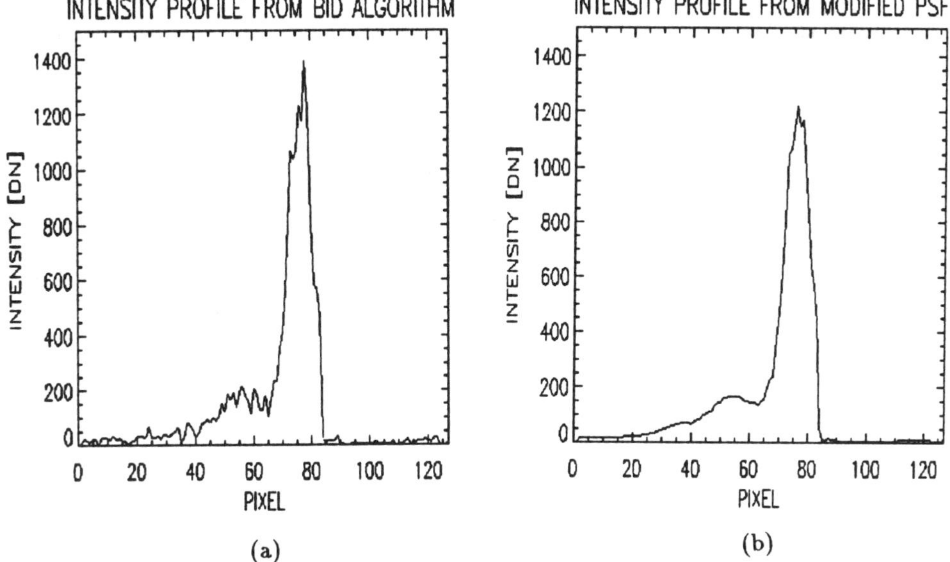

(a) (b)

Figure 2. (a) The profile obtained from the BID algorithm. (b) The profile obtained from the modified PSF/noise suppression method.

width of the PSF by changinig the coefficients of the Moffat function of Martens et al. (1995), which mathematically represents the PSF, and used it in our calculation. Also we simply suppressed the spiky component of the low intensity locations, which might be due to noise in a single pixel, by using a median filtering method within the noise level. Of course, this kind of simple smearing of the structure might cause a loss of real information, but we allow this loss here in order to investigate the effect of noise in the image. We iterate suppressing the spikes and modifying the PSF until the negative structure disappears on the image. The result of this deconvolution is in Figure 2(b). It shows a clear enhancement of brightness at the peak area. We could remove the negative structures from the reconstructed image with only a slight modification of the PSF. This means that the result of deconvolution of the SXT image is very sensitive to the shape of the PSF. Therefore, careful consideration is needed for the deconvolution of SXT images.

3. Summary

With a slightly modified PSF and a noise supression method, we examined the effect of deconvolution on the *Yohkoh* SXT images. The deconvolved image shows a clear difference in its intensity distribution, especially in the intensities of the peaks. However, we do not insist that a merely sharpened PSF can cure the problems in deconvolution, even though the results of Martens et al. (1995) show large scattering in the shape of PSF. We suggest that all the problems found during the deconvolution of *Yohkoh* SXT images are strongly related to the undersampling effect of the SXT CCD. The core diameter of the PSF is almost the same as the pixel size, which must be another reason why the BID method fails in estimating the shape of the PSF. Thus the rapid change in brightness between adjacent CCD pixels due to undersampling might be a significant obstacle in deconvolving SXT images with a general deconvolution procedures. Consequently, careful consideration of the undersampling effect on the SXT images is necessary. We are currently investigating several methods to overcome this difficulty.

References

Martens, P.C., Acton, L.W. and Lemen, J.R. (1995) *Solar Physics*, **157**, 141

Miura, N., Baba, N., Isobe, S., Noguchi, M. and Norimoto, Y. (1992), *J. Modern Opt.*, **39**, 1137

POST-FLARE STRUCTURES RISING WITH DECREASING OR CONSTANT SPEED

ZDENĚK ŠVESTKA
CASS, UCSD, La Jolla, CA 92093-0111, U.S.A. and SRON, Utrecht, The Netherlands

AND

FRANTIŠEK FÁRNÍK
Astronomical Institute of the Academy of Sciences, Ondřejov, Czech Republic

Abstract. As a rule, post-flare loop systems rise with a decreasing speed, while those structures which rise with constant speed are post-flare giant arches. However, there are exceptions when post-flare loop systems rise with constant speed for many hours (including three cases observed by SXT on *Yohkoh*). These exceptions imply that the Kopp and Pneuman interpretation of post-flare loops as sequentially reconnecting open field lines cannot be generally valid. The most likely interpretation is that all post-flare loop systems start with the Kopp and Pneuman process, but in some of them later-formed loops begin to be continuously heated; thus they cease to cool and begin to expand into the corona.

1. Two Kinds of Post-Flare Coronal Structures

The Hard X-ray Imaging Spectrometer (HXIS) on board the Solar Maximum Mission (SMM) spacecraft discovered that there are two different coronal structures associated with major flares: the commonly known post-flare loop system, and a much higher large-scale structure which was named 'post-flare giant arch'. While the systems of post-flare loops rise with gradually decreasing speeds, in all moving giant arches the rise speed was found constant over the whole observing period of several hours (e.g., Švestka, 1984). This constancy of speed of rising giant arches has been recently con-

T. Watanabe et al. (eds.), Observational Plasma Astrophysics: Five Years of Yohkoh and Beyond, 365–369.

firmed by SXT observations on *Yohkoh* (Švestka *et al.*, 1995). Figure 1(a,b) shows two examples of the different behavior of the two structures.

Bruzek (1964) discovered that the growth of post-flare loop systems is not caused by expansion of the loops, but rather by successive formation of progressively higher loops while the lower loops decay. In the initial phase the loop system grows rapidly, but the rise gradually slows down - as can also be seen in the **L** curves in Figures 1(a,b). Bruzek's observations led to the interpretation by Kopp and Pneuman which has been generally accepted as the most likely interpretation of the observed post-flare loop systems.

In contrast, in all cases when post-flare giant arches expanded and the expansion speed could be measured, the rise speed was surprisingly constant. Eventually, after many hours of observation, very high in the corona the speed begins to decrease (example in Figure 1(c))), but in the extreme case on 15–17 March 1993 the constant rise continued for at least 24 hours (Figure 1(d)). It is not possible that such a rise with constant speed could be caused by successive reconnection of progressively higher loops, because the force driving the reconnection should decrease with the altitude. Therefore, Švestka *et al.* (1995) suggested that in such cases there is actual expansion of loops which survived a coronal mass ejection process and rise into the plasma and magnetic-field vacuum left behind by the CME.

2. Exceptions

Švestka (1996) collected 31 published events of post-flare loops and post-flare giant arches in which both the altitude and the rise speed were measured. This set of data has confirmed that post-flare giant arches reach much higher altitudes (up to 250 000 km) than post-flare loops (maximum altitude 140 000 km) and rise with constant speed, while the rise speed of post-flare loops begins to decrease with altitude shortly after the onset of the event. However, there are exceptions to these general rules:

The post-flare loops of 5 April 1960 rose with a constant speed of about 2 km s^{-1} for more than 8 hours up to an altitude of 88 000 km. That structure could not have been a giant arch, because the event was observed in the Hα line and its description convincingly represents a post-flare loop system in papers by Slonim (1963) and Bruzek (1964). The loops of 21 August 1973 rose with a constant speed of about 1.8 km s^{-1} for 7 hours to an altitude of about 70 000 km. X-ray pictures of this event (Smith *et al.*, 1977; MacCombie and Rust, 1979) show an arcade of post-flare loops on the limb. The post-flare loops of 26 June 1992, observed both in X-rays and Hα (Figure 1(f)) showed a decrease in rise speed for the first 3–4 hours, but thereafter the speed stayed constant at 1.5 km s^{-1} for about 14 hours up to an altitude of 110 000 km (Van Driel - Gesztelyi *et al.*, 1997).

Yohkoh SXT data show several other examples of anomalous behavior of post-flare loop systems. On August 28–29, 1992 the rise speed of post-flare loops, observed on the east limb of the Sun, remained constant for 10 hours (Figure 1(e)). And on October 28–30, 1992, close to the east limb, the rise speed of post-flare loops slightly slowed down in the onset phase, but thereafter stayed almost constant for 38 hours. During that time the loops reached an altitude of at least 230 000 km (Švestka *et al.*, 1997). In both cases the limb structures looked in soft X-ray images like 'ordinary' post-flare loop systems (cf., Figure 4 in Švestka, 1996).

3. Discussion

Kopp and Pneuman's (1976) model fits very well the behavior observed in ordinary post-flare loop systems which grow in the corona with decreasing rise speed. However, it cannot explain the growth of giant arches and anomalous post-flare loops with constant speed for many hours.

If the Kopp and Pneuman model is wrong, one has to suppose that all post-flare loop systems are formed by expanding loops, of which a few expand with constant speed for many hours, while the majority begin to decelerate within a short time after loop formation. However, this interpretation completely contradicts the well-known Hα observations made by Bruzek (1964), confirmed, among others, by Big Bear Hα movies with high spatial resolution (Martin, 1979; 1989). Thus instead it seems that there are two different processes which can form post-flare loop systems in the solar corona.

Some of the altitude plots (see, e.g., Figure 1(f)) seem to show that the rise speed actually declined in the beginning phase, but thereafter continued to stay constant. Thus it is possible that even in those events with apparently constant speed the Kopp and Pneuman model can describe the initial evolution (i.e., sequentially forming new loops), but it was eventually replaced by another process involving loop expansion. One can suppose that while in 'regular' post-flare loop systems new, progressively higher loops sequentially form and then cool and eventually disappear, in some events, after a short period of the 'regular' Kopp and Pneuman process, newly formed loops begin to be continuously heated from below. One should not forget that other structures, which exist in the quiet corona, live for a long time and we do not know yet what heats them. Thus we suggest that some active, newly formed loops are heated by some uknown process, thereby preventing their cooling for a long period and driving their expansion.

Acknowledgements

ZŠ's contribution to this paper was partly supported by Contract No. ATM–9312023 with the U.S. National Science Foundation. The authors thank Paul Hick, Petr Heinzel, and Sara Martin for helpful and stimulating discussions. FF was partly supported by the grant No.A3003707 of the Grant Agency of the Academy of Sciences of the Czech Republic.

References

Bruzek, A.: 1964, *Astrophys. J.* **140**, 746.
Hick, P. and Švestka, Z.: 1987, *Solar Phys.* **108**, 315.
Kopp, R.A. and Pneuman, G.: 1976, *Solar Phys.* **50**, 85.
MacCombie, W.J. and Rust, D.M.: 1979, *Solar Phys.* **61**, 69.
Martin, S.F.: 1979, *Solar Phys.* **64**, 165.
Martin, S.F.: 1989, *Solar Phys.* **121**, 215.
Slonim, J.M.:1963, *Soln. Dann.* No. 4, 67.
Smith, J.B., Speich, D.M., Wilson, R.M., Tandberg-Hanssen, E., and Wu, S.T.: 1977, *Solar Phys.* **52**, 379.
Švestka, Z.: 1984, *Solar Phys.* **94**, 171.
Švestka, Z.: 1996, *Solar Phys.* **169**, 403.
Svestka, Z., Fárník, F., Hudson, H.S., Uchida, Y., Hick, P., and Lemen, J.R.: 1995, *Solar Phys.* 161, 331.
Svestka, Z., Fárník, F., Hick, P., Hudson, H.S., and Uchida, Y.: 1997, *Solar Phys.* to be submitted.
Van Driel-Gesztelyi, L., Wiik, J.E., Schmieder, B., Tarbell, T., Kitai, R., Funakoshi, Y., and Anwar, B.: 1997, *Solar Phys.* **174**, in press.

Figure 1. (next page).
(a) Projected altitudes of post-flare loops (**L**), observed both in Hα and by SMM instruments, and a rising post-flare giant arch ($\mathbf{A}_T$, $\mathbf{A}_B$) observed by HXIS close to the solar limb on 6 November 1980. $\mathbf{A}_T$ shows the altitude of the maximum temperature in the arch, while $\mathbf{A}_B$ is the altitude of the brightest location on the arch.
(b) Altitude of the maximum brightness in the loop (**L**) and arch (**A**) structures observed by *Yohkoh* on 21 February 1992. While the rise speed of the loops decreases with time and the loops end their rise at an altitude of 120 000 km, the arch continues to rise with a constant speed, eventually reaching twice that altitude (straight line).
(c) Growing altitude of the giant arch of 27–28 April 1992 observed by SXT on board *Yohkoh* showing a constant rise speed for 18 hours, and then a slow down at an altitude of 170 000 km.
(d) Growing altitude of the giant arch of 15–17 March 1993 observed by SXT, with the rise speed staying constant for more than 24 hours, up to an altitude of at least 200 000 km. Solar rotation has been taken into account for this event (and the event in (c)), assuming vertical extension of the loops.
(e) Altitudes of the post-flare loops of 28–29 August 1992, with solar rotation taken into account under the assumption of vertical extension of the loops (SXT observations). Circles and crosses correspond to two different parts of the loop system.
(f) Projected altitudes of the post-flare loops of 25–26 June 1992 observed in X-rays (squares, SXT data) and in Hα (diamonds, Hida Observatory). ((a) after Švestka (1984) and Hick and Švestka (1987), (b–d) after Švestka *et al.* (1995), (f) after Van Driel-Gezstelyi *et al.* (1997).)

height (10^4 km)
6 NOVEMBER 1980
(a)
A_T
A_B
L
time (UT)
(b)
A
L
ALTITUDE [km]
20-22 FEBRUARY 1992
(c)
27-28 APRIL 1992
(d)
15-17 MARCH 1993
Proj.alt. of H-alpha X-ray loops above the limb in 10^4 km.
(e)
ALTITUDE [10^4 km]
(f)
LOOP 3
X-ray (Yohkoh SXT)
Hida
Start Time : 25-Jun-92 20:11:00 UT

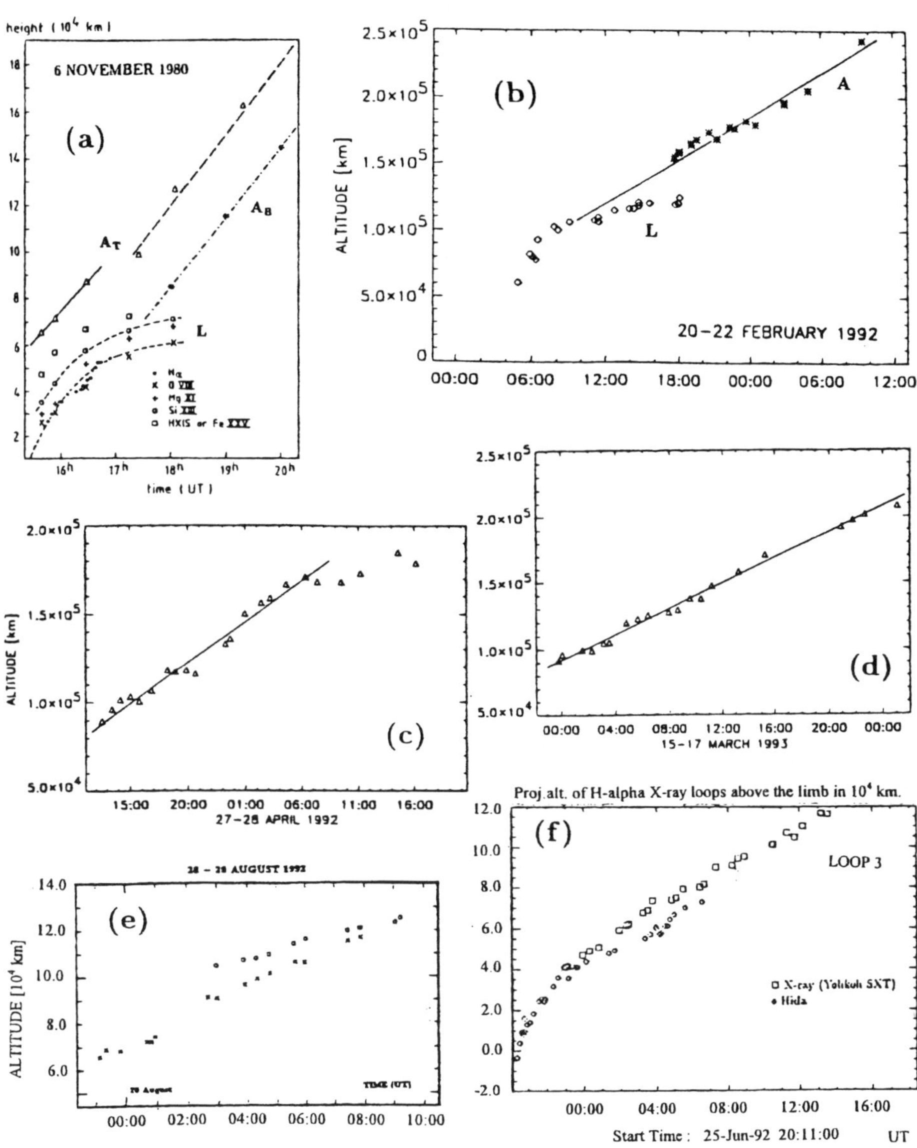

Figure 1.

SOFT X-RAY FEATURES OF PROMINENCE ERUPTION AND DISAPPEARANCE

H. TONOOKA, R. MATSUMOTO AND S. MIYAJI
Department of Physics, Chiba Univ., Chiba 263, Japan
S. F. MARTIN
Helio Research, La Crescenta, CA 91214, USA
R. C. CANFIELD
Physics Department, Montana State Univ., Bozeman, MT 59717-0350, USA
K. REARDON
Univ. of Hawaii, Honolulu, HI 96822, USA
A. MCALLISTER
NOAA Space Environment Lab., Boulder, CO 80303, USA
AND
K. SHIBATA
National Astronomical Observatory, Mitaka, Tokyo 181, Japan

1. Introduction

We investigated the mechanism of the disparition brusque, i.e., the sudden disappearance of a prominence (e.g., Tandberg-Hanssen 1995), observed with an Hα limb filtergraph by comparing *Yohkoh* SXT observations and Hα observations.

Using prominence data from the Hα limb filtergraph of the Mees Solar Observatory, it has been verified that there exist at least three categories of prominence eruptions or disappearances (e.g., Tonooka *et al.* 1995). The most common category is the *eruptive prominence* in which some of the filament mass is injected into the high corona and is considered to escape the Sun while the remainder of the mass falls back to the chromosphere. The second category is the *quasi-eruptive prominence* in which the prominence begins to rise bodily, similar to a complete eruption, but all of its observable

T. Watanabe et al. (eds.), Observational Plasma Astrophysics: Five Years of Yohkoh and Beyond, 371–374.

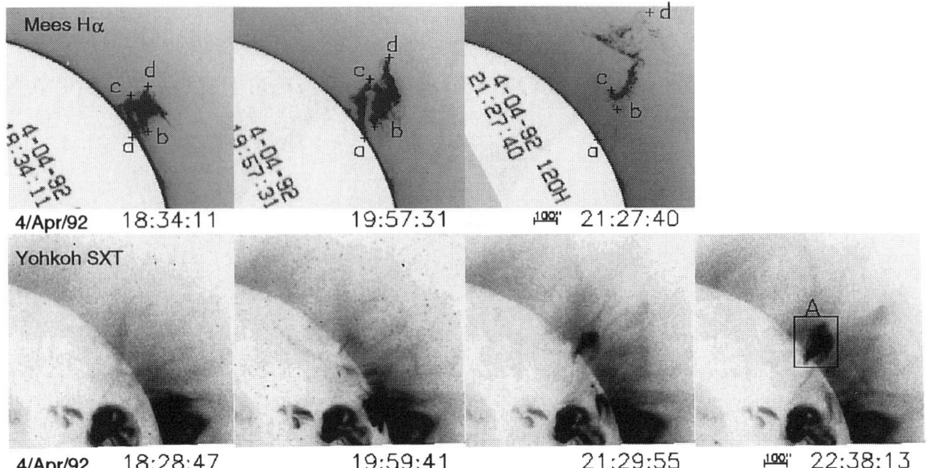

Figure 1. *Eruptive prominence* of 4/Apr/1992. Top three images are taken by Mees Hα coronagraph and bottom four images are by *Yohkoh* SXT. All images are negative (i.e. black is higher intensity). Hα images are co-aligned with the SXT images of which north is up. The symbol "+" and the letters (a–d) in the top images are measuring points of distance variation (Figure 3:bottom). The box with "A" in the bottom right image of SXT (22:38:13 UT) is the region that intensity variation is measured (Figure 3:top).

mass in Hα flows back to the chromosphere near or beyond the apparent ends of the prominence. The third is the *disappearing prominence* in which the mass drains out of the prominence without any evidence of either slow ascent or eruption. The latter two categories are thought to be relatively rare.

2. Analysis

We examined 7 events which have good temporal and spatial resolution out of 12 disparition brusque events observed in Hα at Mees Solar Observatory between December 1991 and May 1994. We categorize 3 of them as *eruptive prominence*, 3 as *quasi-eruptive prominence*, none as *disappearing prominence* and 1 as unknown case.

We found the following features by overlaying the Hα images onto Soft X-ray images observed by *Yohkoh* SXT for both *eruptive prominence* and *quasi-eruptive prominence.*

- Eruptive prominence (Figure 1)
    - In all *eruptive prominence* events, filament structure and cusp shaped loop in soft X-ray were seen just below the erupting Hα prominence.

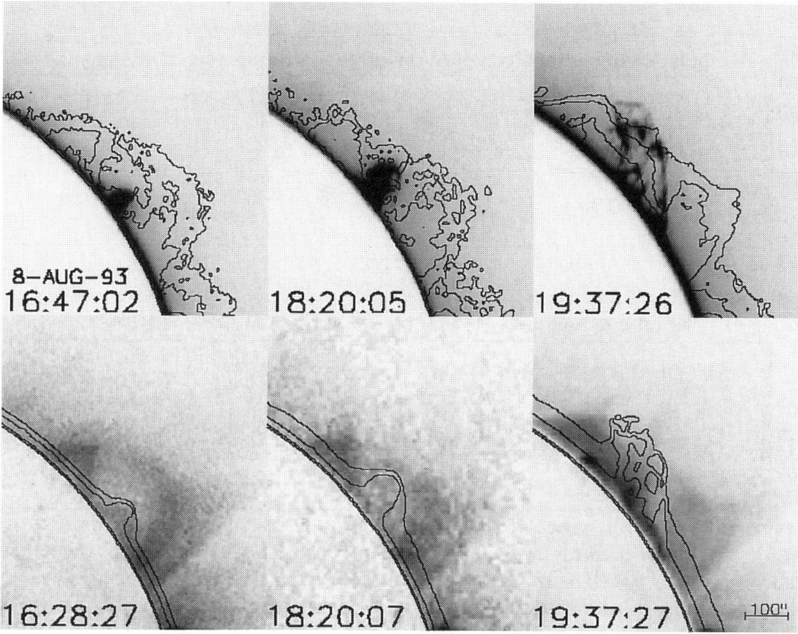

Figure 2. Quasi-erputive prominence of 8/Aug/1993. Coaligned SXT AlMg and Mees Hα coronagraph images are shown. Top images are the Hα intensity with soft X-ray contours, bottom images are the opposite. The sequence shows that a system of overlying soft X-ray loops (16:28:27 UT) disappeared as the Hα prominence erupts into them. In the final stage (19:37:27 UT) two new system of soft X-ray loops were seen on both sides of the prominence.

    - Each prominence eruption occurred not in a soft X-ray bright region, but in the region of faint X-ray loops.

- Quasi-eruptive prominence (Figure 2)
    - Soft X-ray loop structures were seen above the rising prominence. The rising motion of the prominence was stopped as if it was obstructed by the previously seen X-ray loop. Finally, the prominence matter fell down. These feature were seen in all 3 *quasi-eruptive prominence* events.
    - Intensity increase caused by rising prominence or descending matter was not seen in 2 events out of 3. In one event, however, a cusp shaped loop located on both sides of the prominence brightened in soft X-rays before and after the prominence rise.

Temporal variation of soft X-ray intensity of two *eruptive* and a *quasi-eruptive prominence* events showed that the soft X-ray intensity peaked 1 to 3 hours after Hα morphology changed (see Figure 3).

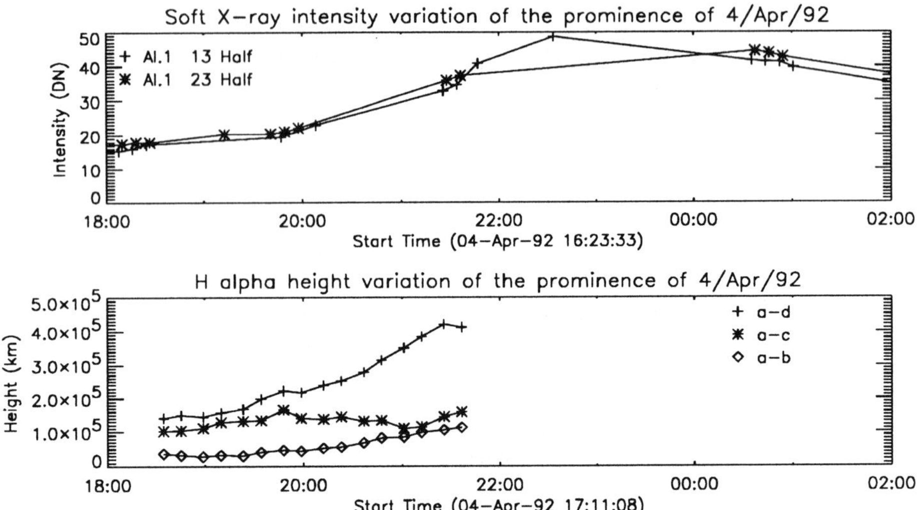

Figure 3. The temporal variation of soft X-ray intensity and Hα height in the prominence of 4/Apr/92, the same event as shown in Figure 1. Top panel is the Soft X-ray intensity variation inside box A in Figure 1. The bottom panel is the temporal variation of the distance between a (root of the prominence) to b (a-b), a to c (a-c), and a to d (a-d) in Hα (Figure 1:top). The distance a-d began to increase just before 20:00 UT. The soft X-ray intensity started to rise almost simultaneously. The Soft X-ray intensity maximum is about 3 hours later (after 22:30 UT). These variation of X-ray intensity and Hα height are common in *eruptive prominences* and some of the *quasi-eruptive prominences.*

3. Conclusion

We found that *eruptive prominences* often lead to the formation and brightening of a soft X-ray cusp structure below the Hα prominence. In *quasi-eruptive prominences*, a soft X-ray loop which often preexists above the Hα prominence seems to block the rise of the prominence. In both cases, the morphological change of the Hα prominence precedes the soft X-ray intensity peak by 1 to 3 hours.

References

Tonooka, H., *et al.*, 1995, in *Magnetodynamic Phenomena in the Solar Atmosphere - Prototypes of Stellar Magnetic Activity*, ed. Uchida, Y., *et al.*, Kluwer Acad. Publ. pp.493

Tandberg-Hanssen, E., 1995, *The Nature of Solar Prominences*, Kluwer Acad. Publ.

YOHKOH OBSERVATIONS AND INTERPLANETARY DISTURBANCES

SHINICHI WATARI
Communications Research Laboratory
4-2-1 Nukuikita, Koganei, Tokyo 184, JAPAN
YUKIO KOZUKA
Solar-Terrestrial Environment Laboratory
3-13 Honohara, Toyokawa, Aichi 442, JAPAN
AND
TAKASHI WATANABE
Ibaraki University
2-1-1 Bunkyo, Mito, Ibaraki 310, JAPAN

Abstract: Interplanetary disturbances during the five-year *Yohkoh* era are reviewed. Coronal holes have been observed by *Yohkoh* as source of recurrent disturbances. They have changed their size and latitudinal distribution according to decrease of solar activity. Large-scale coronal disturbances (LCDs) have been observed in association with coronal mass ejections, or source of non-recurrent disturbances.

There are two types of interplanetary disturbances: recurrent and non-recurrent. Recurrent disturbances are associated with corotating interaction regions (CIRs). CIRs are formed by the interaction between high-speed solar wind from coronal holes and slow-speed solar wind. Non-recurrent disturbances are associated with coronal mass ejections (CMEs). Interplanetary disturbances passing the Earth interact with the geomagnetic field and increase geomagnetic activity, so geomagnetic activity is an indicator of interplanetary disturbances. In Figure 1 we show contour maps of Ap-index (a linear scale index to describe the planetary geomagnetic activity) in order to summarize the geomagnetic disturbances, or the interplanetary disturbances, during the five-year *Yohkoh* era. The vertical axis of the figure is labeled both with Bartels rotation number (the serial number assigned to 27-day periods since January 1833), and the starting date for each rota-

T. Watanabe et al. (eds.), Observational Plasma Astrophysics: Five Years of Yohkoh and Beyond, 375–378.

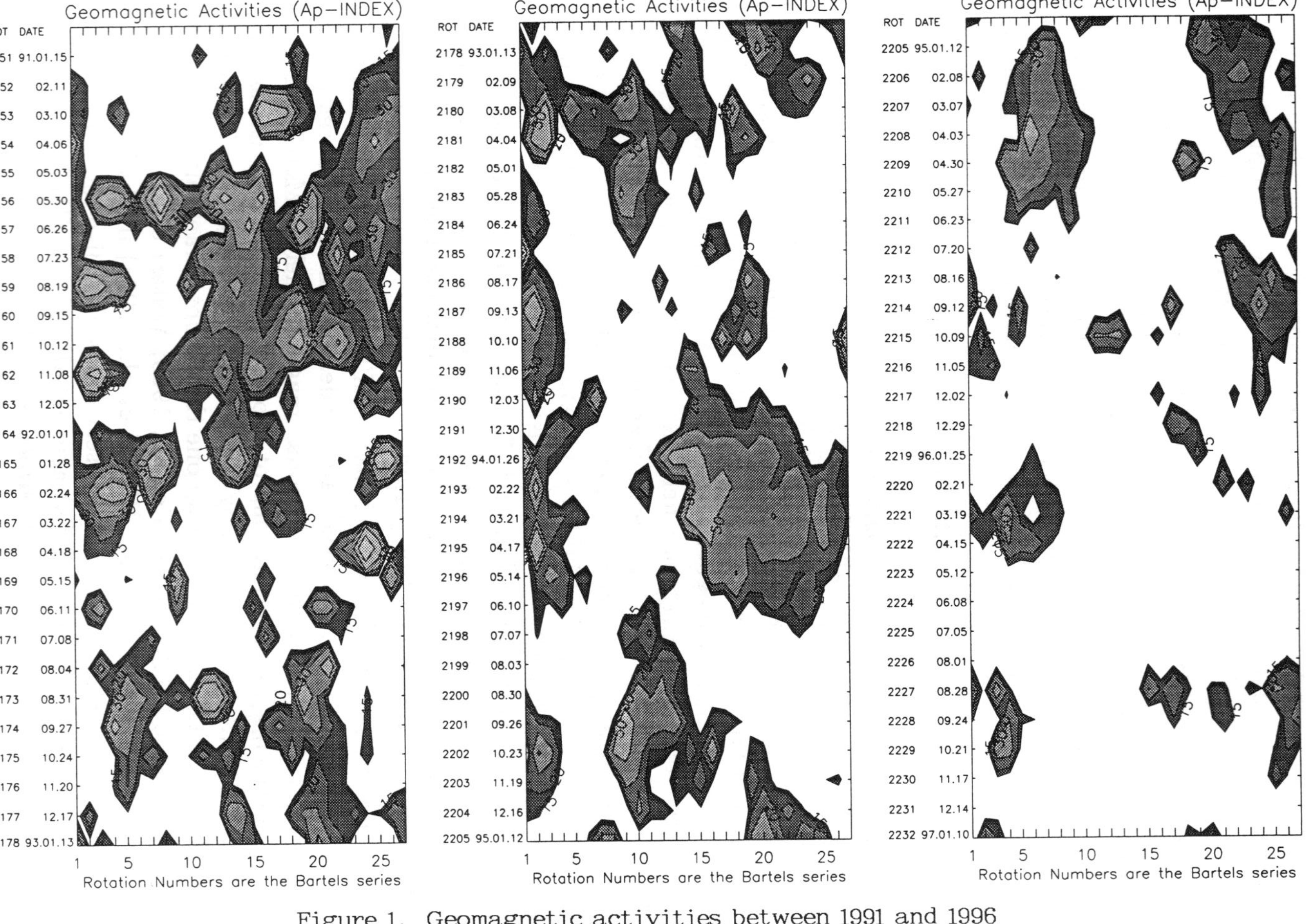

Figure 1. Geomagnetic activities between 1991 and 1996

tion. The horizontal axis of the figure starts on the first day of each rotation and ends on the 27th day. Larger values of Ap-index correspond to larger geomagnetic disturbances, or larger interplanetary disturbances.

Coronal holes are observed as dark areas in soft X-ray images because of their low density. Watari et al. (1995) pointed out that the relation between the sizes of soft X-ray coronal holes and the solar wind speeds from the coronal holes observed near the Earth is more complicated than the relation reported by Nolte et al. (1976). Watari et al. (1995) also stressed the contribution of the high-latitude coronal holes to high speed-solar wind observed near the Earth.

Between 1991 and the middle of 1992, there were many active regions around the solar equator. Figure 1 shows that during this period there were many large non-recurrent geomagnetic disturbances. Between the middle of 1992 and 1995, large coronal holes extended from the poles to the equator and recurred several times on the Sun. Geomagnetic disturbances recurred in association with these coronal holes. The recurrent disturbances were dominant during this period. The patterns of the disturbances (see Figure 1) reflected the global magnetic field of the Sun, or the sector structures of the interplanetary magnetic field. Kozuka et al. (1994 and 1996) and Hara (1996) studied the long-term changes of the Sun's global magnetic field as revealed by the *Yohkoh* / soft X-ray telescope (SXT) observations. Around solar minimum (~1996), the coronal holes mainly appeared around the poles. The number of recurrent disturbances associated with coronal holes decreased as shown in Figure 1. However, the recurrent pattern could still be seen in the solar wind observations near the Earth. The number of disturbances associated with CMEs also decreased because of the low activity of the Sun during this period.

It seems to be difficult for the *Yohkoh*/SXT to detect CMEs directly. The *Yohkoh*/SXT has often observed large-scale coronal disturbances (LCDs), or large-scale restructuring of coronal magnetic fields. McAllister et al. (1996) fortunately found that one of those was associated with an interplanetary disturbance observed by Ulysses. It is, however, hard to say how a geoeffective LCD is different from one that is not. Transient coronal holes (Watanabe et al., 1992; Kozuka et al., 1995; Watari et al., 1995) or soft X-ray dimming (Hudson, 1996; Watari et al., 1996a) might be associated with geoeffective CMEs. The SXT observations identified the long-duration soft X-ray emission events (LDEs) associated with the CMEs as the emission from the soft X-ray arcades formed after the eruptions. These arcades are also associated with the solar radio continuum emission (Watari and Isobe, 1994). Soft X-ray arcades are often formed in association with filament eruptions (Watanabe et al., 1992; Hanaoka et al., 1994; Kozuka et al., 1995, Watari et al., 1996b). Watari et al. (1996b) recently noted that filament eruptions are part of large scale eruptions along the inversion lines of the solar magnetic field.

Acknowledgments

The authors thank the *Yohkoh* team, NASA/NSSDC, NOAA/NGDC, and HAO/MLSO for offering their data.

References

Hanaoka, Y., H. Kurokawa, S. Enome, et al. (1994) Simultaneous observations of a prominence eruption followed by a coronal arcade formation in radio, soft X-rays, and H-alpha, *PASJ*, **46**, pp. 205-216.

Hara, H. (1996) Magnetic activity cycle in the X-ray coronal structures, in *Magnetodynamic Phenomena in the Solar Atmosphere - Prototypes of Stellar Magnetic Activity*, edited by Y. Uchida, T. Kosugi, and H. S. Hudson, Kluwer Academic Publishers, Dordrecht, pp. 321-328.

Hudson, H. S. (1996) Yohkoh observations of coronal mass ejections, in *Magnetodynamic Phenomena in the Solar Atmosphere - Prototype of Stellar Magnetic Activity*, edited by Y. Uchida, T. Kosugi, and H. S. Hudson, Kluwer Academic Publishers, Dordrcht, pp. 89-96.

Kozuka, Y., Ta. Watanabe, M. Kojima, et al. (1994) Large-scale coronal and solar-wind structures, in *X-ray Solar Physics from Yohkoh*, edited by Y. Uchida, Te. Watanabe, K. Shibata, and H. S. Hudson, University Academy Press, Tokyo, pp. 301-304.

Kozuka, Y., Ta. Watanabe, M. Kojima, et al. (1995) The dynamical characteristics of a disappearing-filament associated interplanetary disturbance observed in 1992 early May, *PASJ*, **47**, pp. 377-382.

Kozuka, Y., M. Kojima, and T. Saito (1996) Solar cycle variation of the rotation of the large-scale magnetic field of the sun, in *Magnetic Phenomena in the Solar Atmosphere - Prototypes of Stellar Magnetic Activity* edited by Y. Uchida, T. Kosugi, and H. S. Hudson, Kluwer Academic Publishers, Dordrecht, pp. 627-628.

McAllister, A. H., M. Dryer, P. McIntosh, and H. Singer (1996) A large polar crown coronal mass ejection and a "problem" geomagnetic storm: April 14-23, 1994, *J. Geophys. Res.*, **101**, pp. 13497-13515.

Nolte, J. T., A. S. Krieger, A. F. Timothy, et al. (1976) Coronal holes as sources of solar wind, *Solar Phys.*, **46**, pp. 303-332.

Watanabe, Ta., Y. Kozuka, M. Ohyama, et al. (1992) The structure of the coronal soft X-ray source associated with the dark filament disappearance of 1991 September 28 using the Yohkoh soft X-ray telescope, *PASJ*, **44**, pp. L199-204.

Watari, S., T. Isobe, and Yohkoh SXT Team (1994) Soft X-ray feature in active regions associated with meter wavelength solar radio emission, in *New Look at the Sun with Emphasis on Advanced Observations of Coronal Dynamics and Flares*, edited by S. Enome and T. Hirayama, NRO Report No. 360, pp. 417-420.

Watari, S., Y. Kozuka, M. Ohyama, and Ta. Watanabe (1995) Soft X-ray coronal holes observed by the Yohkoh SXT, *J.Geomag.Geoelectr.*, **47**, pp. 1063-1071.

Watari, S., Z. Smith, H. A. Garcia, T. Detman, and M.Dryer (1996a) Coronal change at the south-west limb observed by Yohkoh on 9 November 1991, and the subsequent interplanetary shock at pioneer Venus orbiter, *Solar Phys.*, **167**, pp. 357-369.

Watari, S., T. Tedman, and J. A. Joselyn (1996b) A large arcade along the inversion line, observed on May 19, 1992 by Yohkoh, and enhancement of interplanetary energetic particles, *Solar Phys.*, **169**, pp. 167-179.

EVOLUTION OF CORONAL ACTIVE REGIONS OBSERVED WITH THE YOHKOH SOFT X-RAY TELESCOPE

S. YASHIRO
Department of Astronomy, University of Tokyo, Hongo, Bunkyo, Tokyo 133, Japan

AND

K. SHIBATA AND M. SHIMOJO
National Astronomical Observatory of Japan, 2-21-1, Osawa, Mitaka, Tokyo 181, Japan

Abstract. We study the early evolution of active regions in the corona by analyzing 56 emerging flux regions (EFRs). We find that the initial apparent velocity of the expansion of the EFRs is 0.4 – 4.6 km/s, which is much lower than that inferred from Hα observations of arch filament systems or from theory.

1. Introduction

Newly emerged solar active regions (ARs) are referred to as emerging flux regions (EFRs) (Zirin 1972). EFRs are one of the key phenomena for understanding solar activity and the origin of solar magnetic fields. In the chromosphere, EFRs are seen as arch filament systems (AFSs) in Hα, and it is known that the rise velocity of arch filaments is ~10 km/s (Bruzek 1967, 1969). However, the properties of EFRs in the corona have not been studied well, since the spatial and temporal resolution of previous soft X-ray observations were not adequate for a detailed study of EFRs.

The Soft X-ray telescope (SXT) aboard *Yohkoh* has provided us with the first opportunity to observe emerging flux regions in the corona with high spatial and temporal resolution. We present initial results of a systematic study of EFR evolution in the solar corona. (See Yashiro et al. 1997 for more details.)

T. Watanabe et al. (eds.), Observational Plasma Astrophysics: Five Years of Yohkoh and Beyond, 379–382.

2. Analysis

All data for this study were obtained using the AlMg filter in the Full Frame Image (FFI) mode of the *Yohkoh*/SXT. We define two subregions for each active region, which we call the *extended region* and the *core region.* The *extended region* is defined as the portion of the active region with intensity greater than some critical boundary value. (see Yashiro et al. 1997 for more exact definition.)

We define the *core region* as the portion of the active region radiating the brightest 80% of the total integrated soft X-ray intensity of the EFR. (In some cases the core region consists of two or more subregions inside the EFR.) We define the size of an EFR as the square root of the area of either the extended region or the core region, and derive the apparent velocity of expansion of an EFR from the time variation of its size.

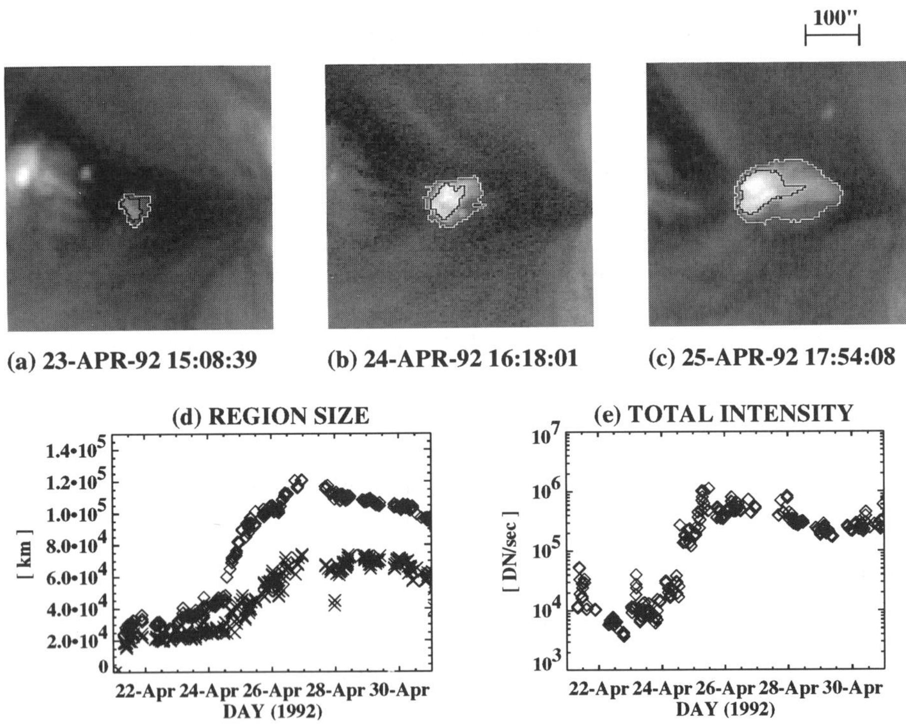

Figure 1. An example of a coronal emerging flux region (NOAA 7145). Images of (a)-(c) show the boundaries of the *core region* (black curve) and *the extended region* (white curve). (d) Time evolution of the region size. Diamonds show the extended region and crosses show the core region. (e) Time evolution of the total soft X-ray intensity.

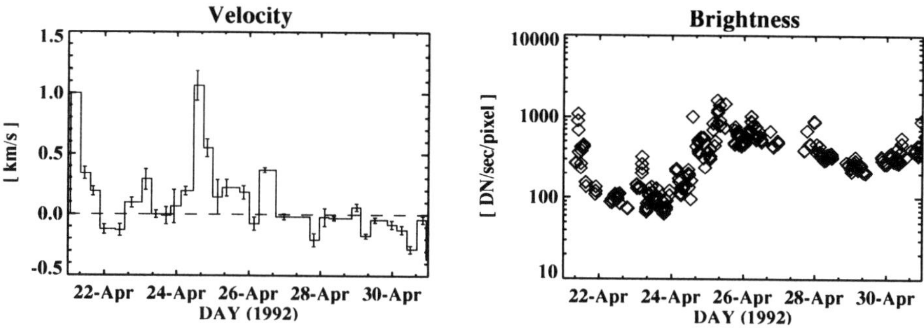

Figure 2. The left figure shows the time variation of the apparent velocity of the expansion of EFR 7145. The right figure shows the surface brightness (I_{total}/A) for EFR 7145, where I_{total} is the total soft X-ray intensity of the extended region.

3. Results

We analyze 56 emerging flux regions which appeared in coronal holes or quiet regions and examine the time variation of the region size and the total X-ray intensity. Figures 1(d) and (e) show the time development of region size and total intensity of NOAA 7145. We can see that the variation of size of the extended region is the same as that of the core region.

Figure 2 shows the time variation of the expansion velocity of extended region of EFR 7145. Here, the velocity is defined as the slope of the least square fitting straight line on the time variation curve of the region size over 6 hours. The velocity is relatively high (1 km/s) just after the birth of the EFR, and then decreases to 0.1 – 0.3 km/s, but increases again to 0.5 – 1.1 km/s on 24 – 25 Apr.

Figure 3 shows the distribution of the initial expansion velocity of 56 EFRs. We find that most of the initial velocities are less than 4 km/s. It is interesting to note that the the initial expansion velocity is comparable to the separation velocity of the two footpoints of the emerging flux (Harvey & Martin 1973).

4. Discussion

The apparent EFR rise velocity is about half of its expansion velocity because the apparent shape of a EFR is roughly hemispherical. This implies that the rise velocity is less than 2 km/s, which is much slower than the rise velocity of Hα arch filaments ($\sim$ 10 km/s). Two dimentional magnetohydrodynamic (MHD) numerical simulations and theory of Shibata et al. (1989, 1990) also predict a faster rise velocity for EFR loops ($\sim$ 10 km/s).

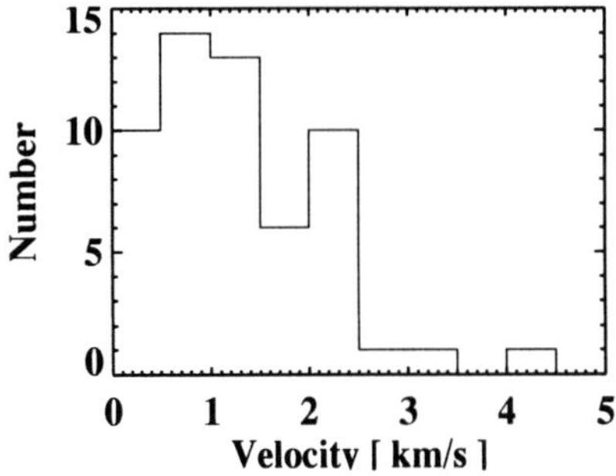

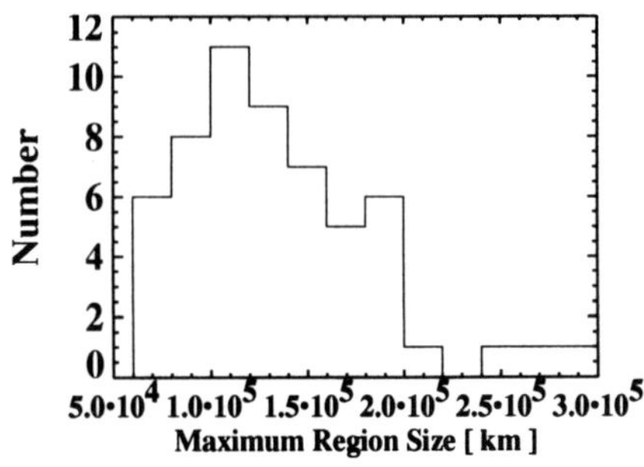

Figure 3. The histogram of the initial expansion velocity and the maximum region size.

We consider two possible explanations for the apparent discrepancy between the expansion velocities we find for coronal EFRs from soft X-ray observations and those inferred from Hα observations and MHD simulation/theory. One possibility is that the emerging magnetic flux may have slowed down during a very early phase of its evolution (less than 10 – 30 min, which is the time resolution in FFI mode). The other possibility is that the apparent expansion of EFRs in soft X-rays does not reflect the actual rise motion of magnetic loops. That is, magnetic loops might be actually moving upward at $\sim$ 10 km/s, but this motion may not be visible in SXT images. In order to detect such motions, we require Doppler shift measurements of coronal plasmas, which may be possible with the SOHO spacecraft and/or the next Japanese solar mission (Solar B).

The authors would like to thank all of the *Yohkoh* team members, and also thank A. Sterling and S. Savy for useful comments. The *Yohkoh* satellite is a Japanese national project, launched and operated by ISAS, and involving many domestic institutions, with multilateral international collaboration with the US and UK.

References

Bruzek, A. 1967, Solar Phys., 2, 451.
Bruzek, A. 1969, Solar Phys., 8, 29.
Harvey, K. L. & Martin, S. F. 1973, Solar Phys., 32, 389.
Shibata, K., Tajima, T., Steinolfson, R. & Matsumoto, R. 1989, Astrophys. J., 345, 584.
Shibata, K., Nozawa, S., Matsumoto, R., Sterling, A. & Tajima, T. 1990, Astrophys. J., 351, L25
Yashiro, S., Shibata, K., Shimojo, M. 1997, submitted to ApJ.
Zirin, H. 1972, Solar Phys. 22, 34.

XUV DOPPLER TELESCOPE ABOARD SOUNDING ROCKET

T. YOSHIDA, R. KANO AND S. NAGATA
Department of Astronomy, School of Science,
The University of Tokyo,
Bunkyou-ku, Tokyo, 113 JAPAN

AND

H. HARA, T. SAKAO, T. SHIMIZU AND S. TSUNETA
National Astronomical Observatory,
2-21-1, Osawa, Mitaka, Tokyo, 181 JAPAN

Abstract. We present an overview of an ongoing Japanese sounding rocket project with the Solar XUV Doppler Telescope (XDT) scheduled to be launched in January, 1998. The telescope employs a pair of high-wavelength-resolution multilayer mirrors and a back-thinned CCD. It is designed to observe the coronal velocity field of the whole Sun by measuring line-of-sight Doppler shifts of the Fe XIV 211 Å line (1.8 MK). The detection limit is estimated to be about 60 km/s.

1. Introduction

1.1. TEMPERATURE STRUCTURE OF SOLAR CORONA

There are two types of temperature structures in the solar corona: transient structures with short lifetimes ($<$ a few hours), and quasi-steady structure with long lifetime ($>$ one day) (Yoshida and Tsuneta, 1996). The loop-top temperatures of the transient loops (such as flares and microflares) exceed 5 MK, and those of the quasi-steady loops range from 2 to 5 MK (Figure 1.1).

How these temperatures are formed is an important problem (Figure 1.2). Yohkoh-SXT (Tsuneta et al. 1991) reveals that the heating mechanism of the transient loops is magnetic reconnection (Tsuneta 1996, Shimizu 1995). There is, however, no general understanding of the heating mechanism of the quasi-steady loops, and detailed observations of all temperature

T. Watanabe et al. (eds.), Observational Plasma Astrophysics: Five Years of Yohkoh and Beyond, 383–390.

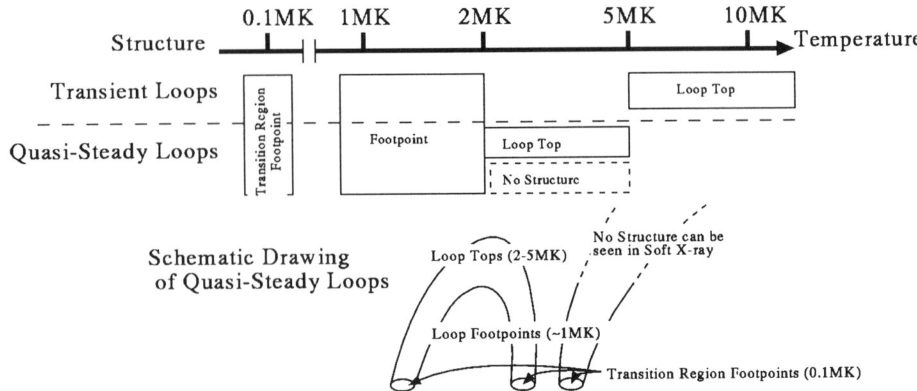

Figure 1.1. Temperature structure of coronal loops.

components are important to reveal the heating mechanism of the quasi-steady loops.

There are two types of solar coronal imaging telescopes: normal incidence optics and grazing incidence optics. Yohkoh-SXT (Tsuneta et al. 1991) is a grazing incidence telescope and observes continuum and emission lines in soft X-rays. It is sensitive to temperatures above 2 MK (Figure 1.3), whereas normal incidence telescopes observe emission lines (from $<$ 0.1 to 2 MK) in XUV and UV.

The XUV Doppler Telescope (XDT) (Sakao et al. 1996, Hara et al. 1997) reported on here is a normal incidence telescope. It will observe pure Fe XIV images with high-wavelength-resolution multilayers.

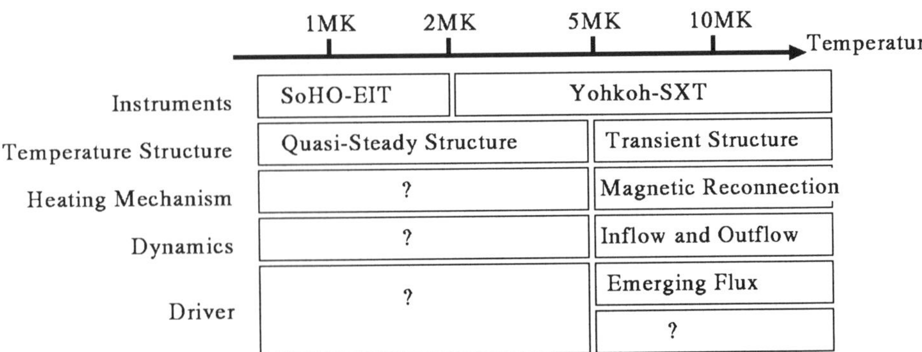

Figure 1.2. Coronal heating problems.

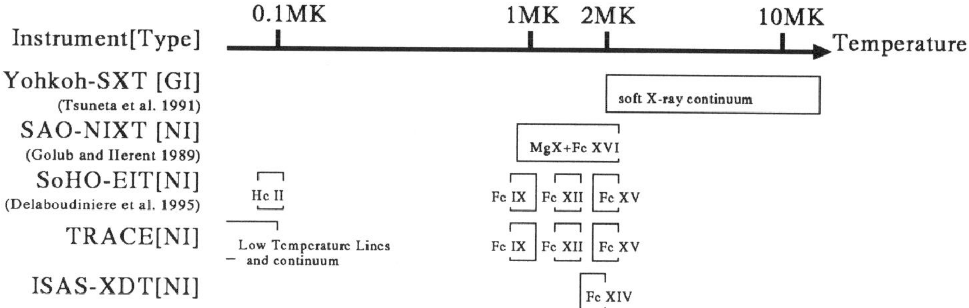

Figure 1.3. Temperature responses of instruments; 'NI' = normal incidence optics, 'GI' = grazing incidence optics.

1.2. VELOCITY MEASUREMENT OF SOLAR CORONA

So far, coronal imaging observations have been done with several instruments. However, there is no simultaneous velocity observation of the whole Sun. Velocity observations in a small part of the Sun with low temporal resolution have been done with scanning spectrometers. We have few observations of the dynamics of the coronal heating.

The XUV Doppler Telescope has two multilayer optical systems. One is tuned to the blue wing of the Fe XIV emission line, and the other to the red wing as shown in Figure 1.4. We can obtain a line-of-sight velocity map of the whole Sun by taking the ratio (Figure 1.5) of the red and the blue wing images. This will be the first simultaneous observations of line-of-sight velocity maps of the whole Sun.

Figure 1.6 shows the velocity distribution of coronal structures. The detection limit of the velocity with the XUV Doppler Telescope is about 60 km/s. We will be able to observe cool X-ray jets (Shimojo et al. 1996), fast flows along loops (Hara 1995), continuous expansion of active regions

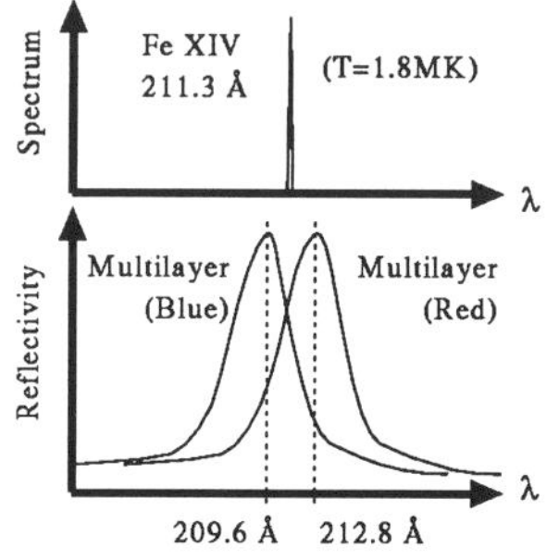

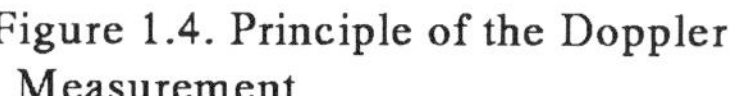

Figure 1.4. Principle of the Doppler Measurement

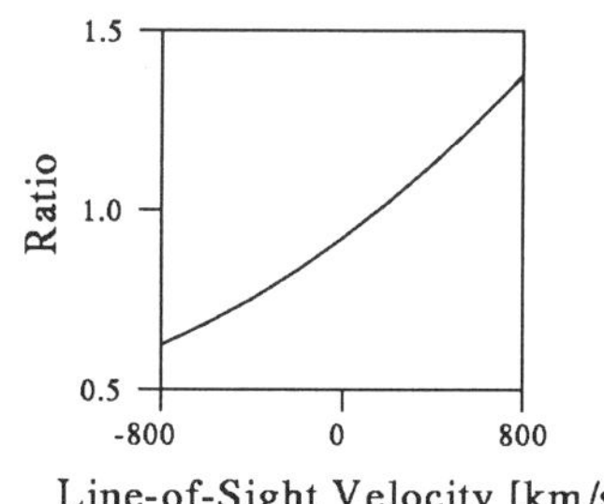

Figure 1.5. Ratio - Velocity Diagram.

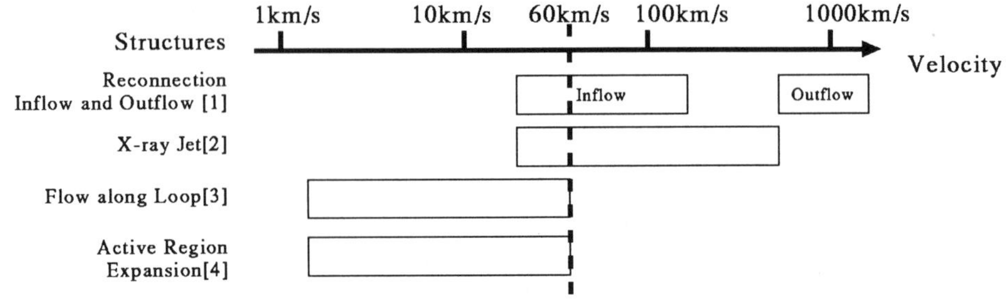

Figure 1.6. Velocity of coronal structures. References are as follows; [1] Tsuneta (1996), [2] Shimojo et al. (1996), [3] Hara (1995), and [4] Uchida et al. (1992)

(Uchida et al. 1992), inflows and outflows caused by magnetic reconnection (Tsuneta 1996), and other flows associated with coronal dynamical phenomena.

2. Velocity Measurement with Multilayers

2.1. HIGH-WAVELENGTH-RESOLUTION MULTILAYER

We need high-wavelength-resolution multilayers to detect velocities as low as possible. Wavelength resolution is characterized by the quantity $\lambda/\Delta\lambda_{\rm FWHM}$, where $\Delta\lambda_{\rm FWHM}$ is the FWHM value of the reflectivity profile of the multilayer and λ is the peak wavelength of the reflectivity.

Generally, Mo–Si multilayers are used in the wavelength range around 200 Å. However, the wavelength resolutions of the Mo–Si multilayers are not high, because the absorption coefficient of the material Mo is high and the number of the layers which contribute to reflection is not large.

We have developed high-wavelength-resolution multilayers with MoSi as reflector material and Si as spacer material (Nagata et al. 1997a, 1997b). The material MoSi has a lower absorption coefficient than Mo. We expect that we can achieve high wavelength resolution by utilizing thin MoSi lay-

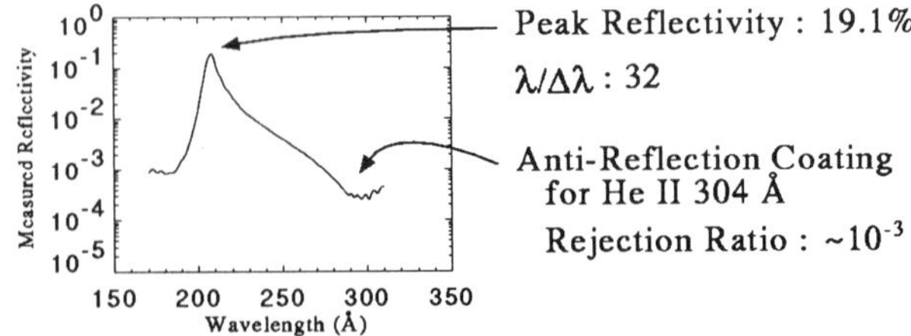

Figure 2.1. Measured Reflectivity of Multilayer MoSi-Si

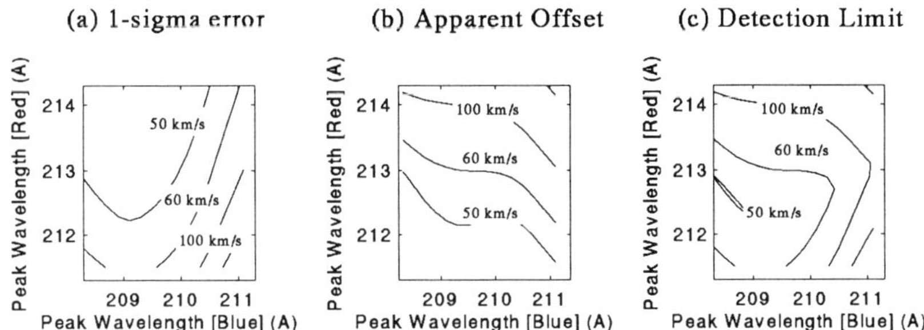

Figure 2.2. (a) Contour map of 1-sigma error of velocity for 10 k photons as a function of the peak wavelength of the blue and the red wing multilayers. (b) Contour map of the apparent velocity offset caused by contamination from nearby emission lines. (c) Maximum of (a) and (b).

ers as reflector layers in place of Mo layers. Figure 2.1 shows the measured reflectivity of a test piece of the MoSi–Si multilayers. The wavelength resolution is about 30, which is much higher than those of Mo–Si multilayers.

2.2. DETECTION LIMIT OF THE VELOCITY

If we can observe pure Fe XIV images, and if there is no contamination of nearby lines, we can uniquely determine the line-of-sight velocity from the ratio of the blue and red wing images, as shown in Figure 1.5. The detection limit of the velocity depends only on photon noise of the images. Figure 2.2-(a) shows a contour map of one-sigma error in velocity for 10k photons (half of CCD full well) as a function of the peak wavelength of the blue and red wing multilayers.

However, the observed Fe XIV images are slightly contaminated from other lines, and the contaminated signals depend on the differential emission measure of the observed plasma. The blue and red images have slightly different contamination from nearby lines. Thus, the signals of the two images can be different even without any velocity, depending on the differential emission measures. Hence, we have an uncertainty in determining the velocity. To examine this uncertainty, we simulate two solar spectra (Mewe et al. 1985,1986) from the observed differential emission measures of an active region and a quiet region obtained with SERTS (Brosius et al. 1996), and derive the difference of the blue-red ratio in the two cases. Figure 2.2-(b) shows the apparent velocity offset derived from the ratio difference.

We evaluate the detection capability of the coronal velocity by considering (1) one-sigma detection limit for 10k photons [figure 2.2-(a)], and (2) apparent velocity offset caused by the contamination of nearby emission lines [figure 2.2-(b)]. Figure 2.2-(c) shows the larger of the two error sources. This finally determines the detection limit. We can detect velocity as low as 60 km/s, when the peak wavelength of the blue wing multilayer is 212.8 Å and that of the red wing multilayer is 209.6 Å, as shown in Figure 2.2-(c).

3. The Telescope

3.1. OVERVIEW OF THE TELESCOPE

The XUV Doppler Telescope will be launched by ISAS in January 1998, with the 22nd S520 rocket (S520CN-22). The spatial resolution will be about 5 arcsec, which is limited by spherical aberration of the telescope. We use a back-thinned CCD (512$\times$512) as the detector. The field of view covers the whole sun. The observations will continue for $\sim$ 5 minutes. The typical exposure time is 5 seconds, and the number of the exposures will be about 40. Table 3.1 summarizes the key design parameters of the telescope. The details of the telescope design are presented in Sakao et al. (1996) and Hara et al. (1997).

Figure 3.1 shows a schematic drawing of the telescope. The telescope has two multilayer optical systems tuned on both sides of the Fe XIV line at 211.3 Å. The red and the blue wing observations are alternately done with the selector shutter at the aperture of the telescope. The secondary mirror is actively controlled in a closed-loop system to stabilize the focal plane images. A Sun sensor [position sensitive diode (PSD)] is used for this purpose.

3.2. CCD

We use a back-thinned CCD fabricated by SITe. The measured quantum efficiency of the CCD in the wavelength range from 100 to 300 Å is about 0.5, which is much higher than that of the front-illuminated CCD. The full well of the CCD is about 350 k electrons (20 k photons at 211 Å). To

Table 3.1. Telescope design

Telescope Type	Cassegrain
Aperture	150 mmø
Focal Length	944.9 mm
CCD Pixel Size	5.2"/pixel
CCD Field-of-View	44'x44'
CCD	Back Illuminated CCD 512x512 (24 μm pixel)

Launch Date	Jan. 31, 1998
Observation Period	300 sec.
Typical Exposure Time	5 sec.
Number of Exposures	~40
Velocity Detection Limit	~60 km/s

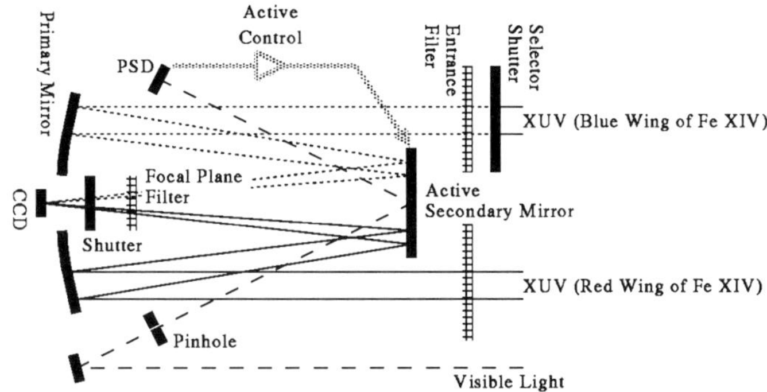

Figure 3.1. Schematic drawing of the telescope

suppress the dark current of the CCD, the chip is thermoelectrically cooled to 0 – 10 $°C$ and the multi-pinned phase mode (MPP) operation is adopted.

3.3. REJECTION OF UV AND VISIBLE LIGHT

We use thin metal foils made by LUXEL to reject visible light: aluminum with thickness 1800 Å at the aperture of the telescope and aluminum 1800 Å and carbon 500 Å at the focal plane. The transmittance in the wavelength range longer than 1000 Å is less than 10^{-15} of that at 211 Å. UV light is rejected with these filters and with the primary and the secondary multilayers. The total transmittance (or reflectivity) from 400 Å to 700 Å is about 10^{-4} of that at 211 Å. The intense He II line at 304 Å is rejected with an anti-reflection coating on the multilayers as shown in Figure 2.1 (Nagata et al. 1997b). The total reflectivity at 304 Å is less than 10^{-5} of that at 211 Å. To avoid acoustic shocks on the foil filters at the time of launch, the telescope tube is designed to act as a vacuum vessel. A vacuum door at the top of the tube is to be opened after launch.

Acknowledgments

We acknowledge Profs. Y. Ogawara, I. Nakatani, and S. Nakajima of ISAS for their support of this sounding-rocket program. We thank Prof. K. Yamashita for his useful comments in developing the multilayer mirrors. This program was financially supported by a joint development program of the National Astronomical Observatory, Japan, and the Grant-In-Aid for Scientific Research of the Ministry of Education, Science, Sports and Culture (Contract No: 07454037 to S. Tsuneta). T. Yoshida, T. Shimizu, R. Kano, and S. Nagata were supported by Research Fellowships of the Japan Society for the Promotion of Science for Young Scientists.

References

Brosius, J., Davila, J., Thomas, R., and Monsignori-Fossi, B. (1996) *ApJ Suppl.*, **106**, 143
Delaboudiniere, J. -P., Artzner, G. E., Brunaud, J., Gabriel, A. H., Hochedez, J. F., Millier, F., Song, X. Y., Au, B., Dere, K. P., Howard, R. A., Kreplin, R., Michels, D. J., Moses, J. D., Defise, J. M., Jamar, C., Rochus, P., Chauvineau, J. P., Marioge, J. P., Catura, R. C., Lemen, J. R., Shing, L., Stern, R. A., Gurman, J. B., Neupert, W. M., Maucherat, A., Clette, F., Cugnon, P., and Van Dessel, E. L. (1995) *Sol. Phys.*, **162**, 291
Golub, L. and Herant, M. (1989) *SPIE*, **1160**, 629
Hara, H. (1995) *Ph. D. Thesis*, The University of Tokyo
Hara, H., Kano, R., Nagata, S., Sakao, T., Shimizu, T., Tsuneta, S., Yoshida, T.,and Kosugi, T. (1997) *SPIE* **3113**, in press
Mewe, R., Gronenschild, E. H. B. M.,and van den Oord, G. H. J. (1985) *A& AS* **62**, 197
Mewe, R., Lemen, J., and van den Oord, G. H. J. (1986) *A& AS* **65**, 511
Nagata, S., Tsuneta, S., Sakao, T., Yoshida, T., Hara, H., Kano, R., Ishiyama, W., Murakami, K, and Oshino, M. (1997a) *Applied Optics*, Vol. 36 No. 13, 2830
Nagata, S., Hara, H., Sakao, T., Shimizu, T., Tsuneta, S., Yoshida, T., Ishiyama, W., Murakami, K, and Oshino, M. (1997b) *SPIE* **3113**, in press
Sakao, T., Tsuneta, S., Hara, H., Kano, R., Yoshida, T., Nagata, S., Shimizu, T., Kosugi, T., Murakami, K., Wasa, W., Inoue, M., Miura, K., Taguchi, K., and Tanimoto, K. (1996) *SPIE* **2804**, 153
Shimizu, T. (1995) *Ph. D. Thesis*, The University of Tokyo
Shimizu, T. (1996) *PASJ*, **47**, 251
Shimojo, M., Hashimoto, S., Shibata, K., Hirayama, T., Hudson, H., and Acton, L. (1996) *PASJ*, **48**, 123
Tsuneta, S., Acton, L., Bruner, M., Lemen, J., Brown, W., Caravalho, R., Catura, R., Ogawara, Y., Hirayama, T., Owens, J. (1991) *Solar Physics*, **136**, 37
Tsuneta, S. (1996) *ApJ*, **456**, 840
Uchida, Y., McAllister, A., Strong, K., Ogawara, Y., Shimizu, T., Matsumoto, R., and Hudson, H. (1992) *PASJ*, **44**, L63
Yoshida, T., Tsuneta, S., Golub, L., Strong, K., and Ogawara, Y. (1995) *PASJ*, **47**, L15
Yoshida, T., and Tsuneta, T. (1996) *ApJ*, **459**, 342

MAGNETIC RECONNECTION IN THE ACTIVE REGION INFERRED BY HOMOLOGOUS SOFT X-RAY FLARES IN FEBRUARY 1992

H.Q. ZHANG
Beijing Astronomical Observatory, Chinese Academy of Sciences, Beijing 100080, China

T. SAKURAI, K. SHIBATA AND H. SHIMOJO
National Astronomical Observatory, Mitaka, Tokyo 181, Japan

H. KUROKAWA
Kwasan and Hida Observatories, Kyoto University, Kyoto 606, Japan

AND

S. MORITA AND Y. UCHIDA
Faculty of Science, The Science University of Tokyo, 1-3, Kagurazaka, Shinjuku 162, Japan

Abstract
In this paper, we provide the possible three-dimensional configuration of the magnetic field in a solar active region during the homologous soft X-ray (long duration event) flares in February 1992.

One of the biggest discoveries by the soft X-ray telescope (SXT) aboard Yohkoh satellite is that of cusp-shaped loop structures in long duration events (LDE) flares and large-scale arcade loops associated with filament eruption or coronal mass ejections (CME) (Tsuneta and Lemen, 1993; Shibata, et al., 1995). Due to the difficult for the measurement of magnetic field in the corona, the most of observations of the magnetic field are concentrated in the solar photosphere and some of them in the chromosphere. The questions are how the basic spatial configuration of the magnetic field and the trigger mechanism in the solar LDE (or classical two-ribbon) flares in solar atmosphere (Carmichael, 1964; Strurrock, 1966; Hirayama, 1974; Kopp and Pneuman, 1976). In normal, we belive that the homologous flares

T. Watanabe et al. (eds.), Observational Plasma Astrophysics: Five Years of Yohkoh and Beyond, 391–396.

in a solar active region are triggered by the same mechanism (Tandberg-Hanssen and Emslie, 1988). As the homologous flares in the active region occur in the different positions of the solar disk due to the solar rotation, we can infer the real spatial configuration of the soft X-ray flares by its perspective effect in the image plane.

A series of homologous flares occurred in the active region (NOAA 7070) in February 1992. One of flares occurred near the solar east limb on February 20-21, 1992 and the others occurred near the center of solar disk on February 24-25 and 27, 1992. These give us a chance to analyze the flares at different angles of view. As we analyse the extrapolation of the photospheric magnetic field in the corona and compare with the project effect of the homologous soft X-ray flares in the image plane, we can construct the possible spatial configuration of the magnetic field and its reconnection process during the flares in the solar active region. Figure 1 shows a time sequence of soft X-ray images on Feb. 20-21, 1992. We can seen that a small "helmet streamer"-like structure formed near the solar surface and a loop-like feature overlaid on it at 23:52:42 UT on Feb, 20, 1992 in Figure 1. Figure 2 shows a vector magnetogram in this active region at 02:21 UT on Feb. 25, 1992. We can see that the large-scale magnetic field in the active region is the normal distribution of the magnetic polarity. However, the magnetic main pole S_1 of negative polarity was located in the north relative to the magnetic main pole N_1 of positive polarity. The transverse magnetic field is almost parallel to the magnetic inversion line L_0 between the magnetic main poles N_1 and S_1. The magnetic main pole S_2 of negative polarity separated away from the magnetic main pole N_2, and the transverse components of the magnetic field almost connected both the main poles. We can find that there is a newly growing bipole of magnetic flux in comparison with a series of photospheric vector magnetograms (Kurokawa, et al., 1994). The homologous flares were triggered by the interaction between the emerging magnetic flux and the pre-existed magnetic field in the solar atmosphere (Zhang, 1995 and 1996). As one overlaps the soft X-ray image on the longitudinal magnetogram in the Figure 3, we can find that the initial bright loop-like structure in the pre-flare in the soft X-ray image connected the magnetic main pole N_1 of positive polarity and enhanced network magnetic field of negative polarity. In the both ends of soft X-ray feature L_1, the transverse component of the photospheric magnetic field is almost parallel to the soft X-ray loop features L_1. This means that the magnetic lines of force extend along this soft X-ray loop L_1 to connect the magnetic structures of opposite polarity. We also find that the soft X-ray loop L_1 partly separates the magnetic main poles (N_1 and S_1) of opposite polarity and the newly growing bipole (N_2 and S_2). The small "helmet streamer"-like structure can not be found well in the image obtained at

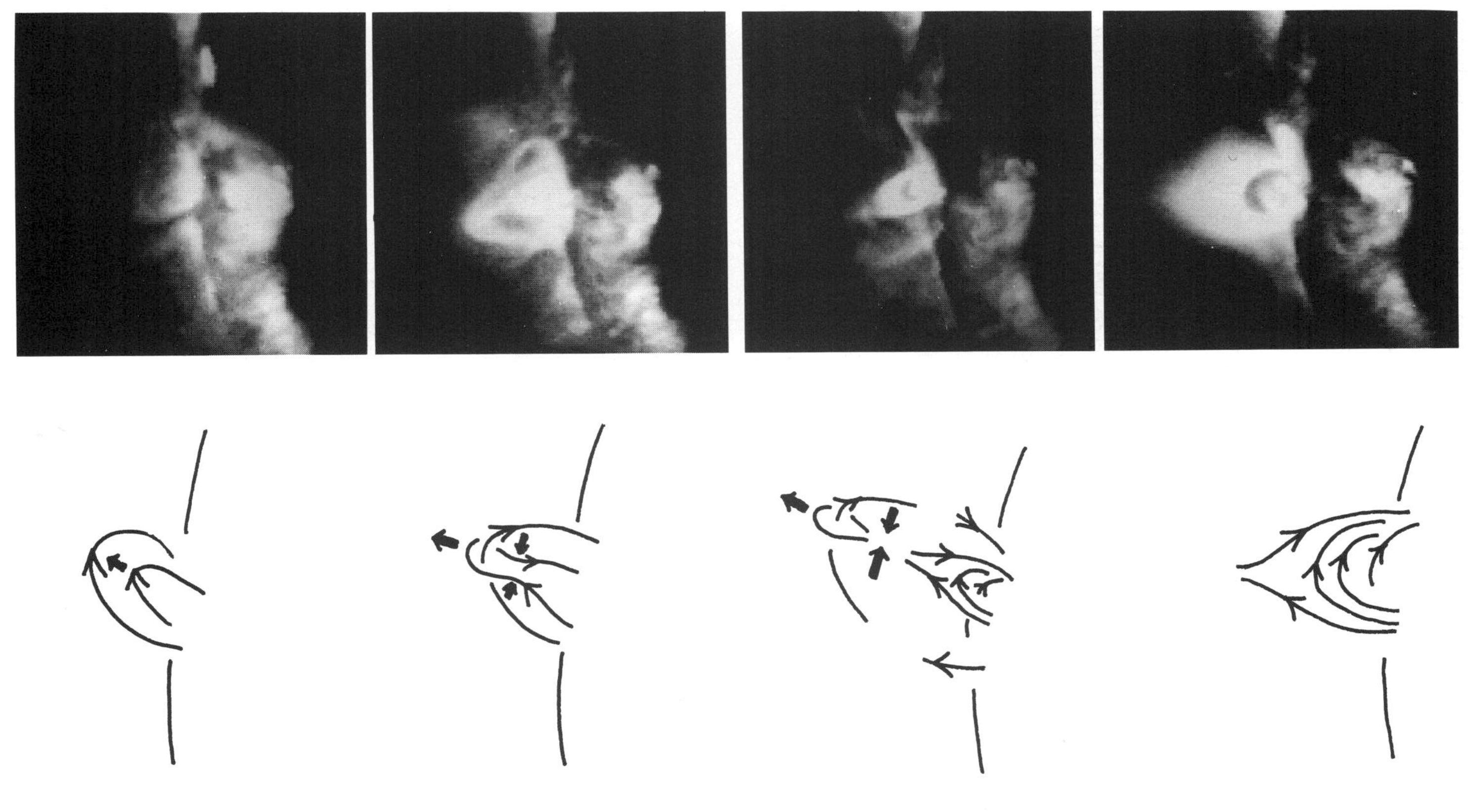

Figure 1. A time sequence of soft X-ray flare on 20-21 Feb. 1992 (top). The reconnection process of the magnetic lines of force in the period of the soft X-ray flare is inferred by homologous flares in the active region in February 1992 (bottom). The arrow lines show the magnetic lines of force above solar photosphere in the solar active region. The short arrows indicate the moving direction of magnetic lines of force.

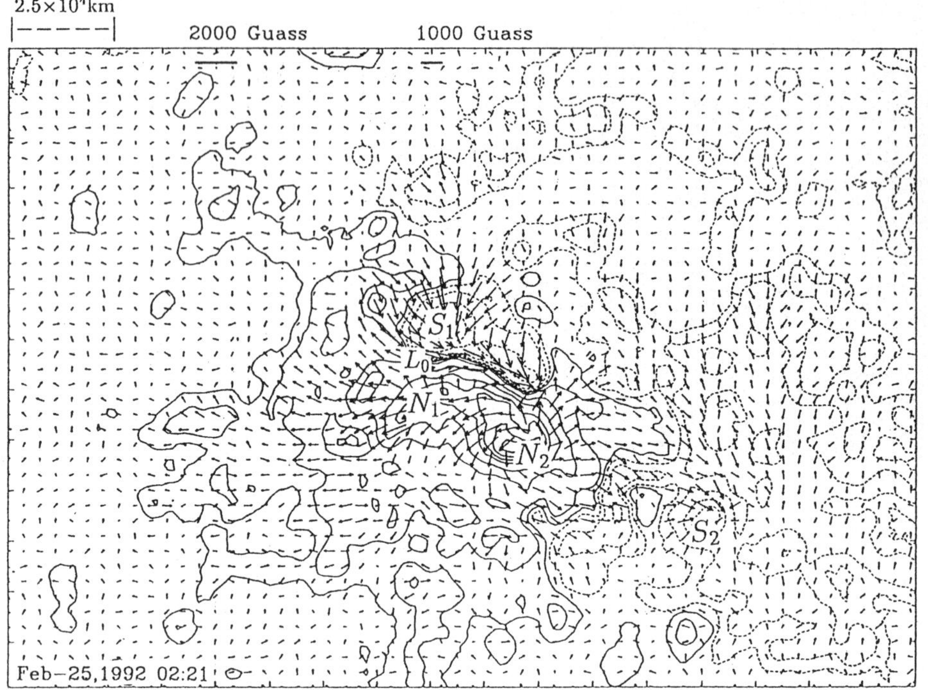

Figure 2. A photospheric vector magnetogram active region on 25 February. The arrows show the transverse component of the magnetic field and the solid (dash) contour indicates the positive (negative) longitudinal magnetic field distribution of 40, 160, 640, 1280, 1920, 2240, 2560G. The north is top and the east is at left.

23:18:40 UT on Feb. 24, 1992, it probably is due to the background emission of the solar disk and the perspective effect of the soft X-ray features. From a series of soft X-ray images, we can find that the soft X-ray flare on February 24-25, 1992 is caused by the eruption of the bright X-ray feature L_1.

In normally, it is belive that the reconnected magnetic field in the post flare stage above the photosphere can be approximated by the potential magnetic field (Svestka, et al, 1987). It can be extrapolated by the photospheric longitudinal magnetogram on Feb. 25, 1992. As this magnetogram is rotated at the position of that on Feb. 21, 1992, we can find that the distribution of the magnetic field lines almost consists with that of post soft X-ray flare loops of Feb. 21, 1992 in Figure 4. The both foot points of the flare loops is just located in the large-scale magnetic poles of opposite polarity. This also means that the change of the basic topology of the magnetic field is not significant, thus the spatial configuration of the soft

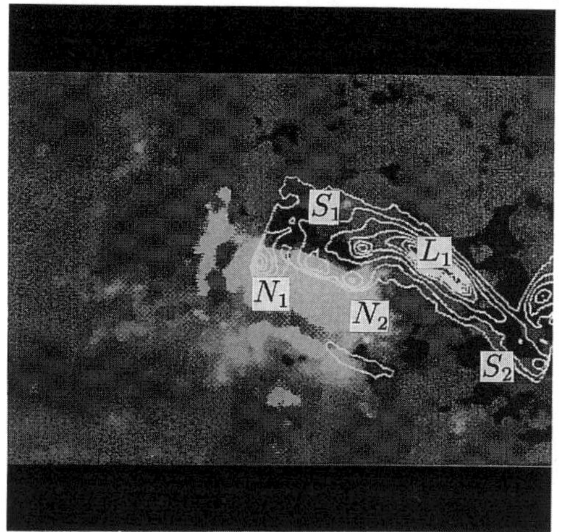

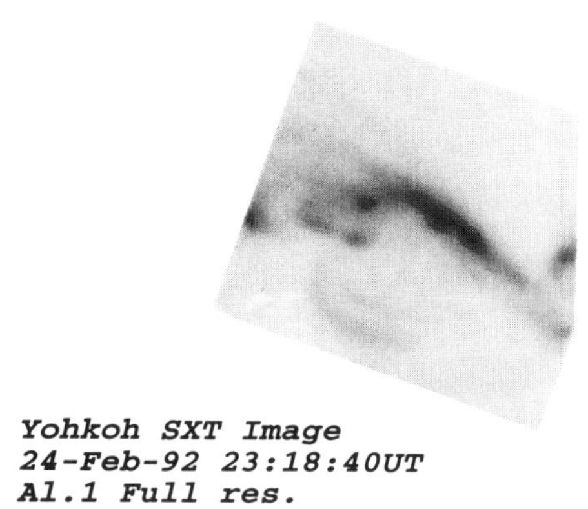

Figure 3. The soft X-ray image (contours) is overlapped on the photospheric magnetogram (left) on 25 February 1992. In the magnetogram, the white (black) shows positive (negative) polarity. In the soft X-ray image (right), the high intensity areas are dark.

X-ray flare on Feb. 21, 1992 can be approximately inferred by that of its homologous flare on Feb. 25, 1992.

If the features in the soft X-ray images reflect the distribution of the coronal magnetic field in active region (NOAA 7070), we can infer the possible process of the magnetic reconnection during the soft X-ray flares in the active region in Figure 1. The reconnection hypothesis of magnetic field is following: (a) The initial reconnection of the magnetic lines of force occurs near the magnetic neutral line due to the interaction between the newly emerging magnetic flux and former magnetic field of opposite polarity. (b) The reconnected magnetic lines of force eject up and push the overlying magnetic field connected with the large-scale photospheric magnetic structures of opposite polarities in the active region. (c) The reconnection of the large-scale magnetic field occurs under the erupting magnetic lines of force and the cusp-shaped loop structures form. (d) The closed magnetic rings occur high above the reconnection region of the magnetic lines of force and move away. (e) After the LDE flare, the non-potential magnetic energy

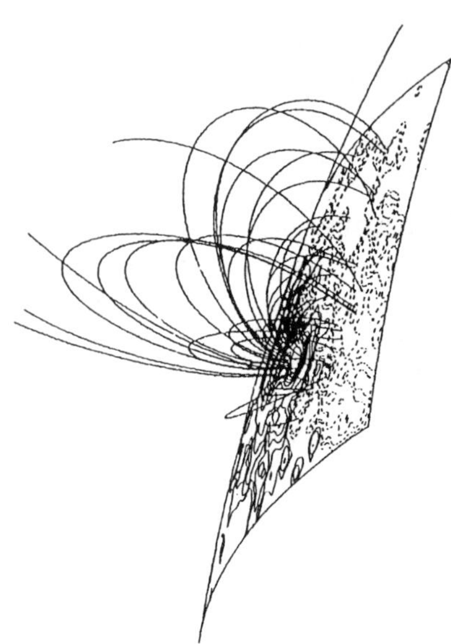

Figure 4. The photospheric longitudinal magnetogram of 25 Feb. 1992 is rotated to the east limb of the sun. The magnetic field lines are extrapolated under the approximation of the potential magnetic field. The north is top and the east is at left.

stores continuously in the active region for the trigger of next homologous flare due to the emergence of new magnetic flux.

References

Carmichael, H. (1996) in *proc. AAS-NASA Symp. on the Physics of Solar Flares*, ed. W.H. Hess (Washington, NASA), NASA-Sp 50, 451.

Hirayama, T. (1974) *Solar Phys.*, **34**, 323.

Kopp, R.A. and Pneuman, G.W. (1976) *Solar Phys.*, **50**, 85.

Kurokawa, H., Kawai, G., Tsuneta, S. and Ogawara, Y. (1994) *X-ray Solar Physics from Yohkoh*, ed. Y. Uchida et al., University Academy Press. Inc., 59.

Shibata, K., Masuda, S., Shimojo, M., Hara, H., Yokoyama, T., Tsuneta, s., Kosugi, T. and Ogawara, Y. (1995) *Astrophys. J.*, **451**, L83.

Strurrock, P.A. (1966) *Nature*, **211**, 695.

Svestka, Z., Fontenla, J., Machado, M., Martin, S., Neidig, D. and Poletto, G. (1987) *Solar Phys.*, **108**, 237.

Tandberg-Hanssen E. and Emslie A.G.: (1988) *The Physics of Solar Flares*, Cambridge Uni. Press.

Tsuneta, S. and Lemen, J. (1993) *Physics of Solar and Stellar Coronae*, ed. J.F. Linsky and S. Serio, 113

Zhang, H. (1995) *Astron. Astrophys.*, **304**, 541.

Zhang, H. (1996) *Astrophys. J.*, **471**, 1049.

List of Participants

Acton, L. W. (Montana State U.)
Akioka, M. (CRLHI)
Akita, K. (Osaka Gakuin U.)
Akiyama, S. (Tokai U.)
Anwar, B. (CRLHI)
Aschwanden, M. (U. Maryland)
Bentley, R. D. (MSSL)
Browning, P. K. (UMIST)
Canfield, R. C. (Montana State U.)
Enome, S. (VSOP Office/NAOJ)
Farnik, F. (Astron. Inst. Czech)
Fujiki, K. (NRO/NAOJ)
Fujisaki, K. (SUT)
Gary, D. E. (Caltech)
Grandpierre, A. (Konkoly Obs.)
Guhathakurta, M. (NASA/GSFC)
Hanaoka, Y. (NRO/NAOJ)
Hara, H. (NAOJ)
Harra-Murnion, L. K. (MSSL)
Hayashi, K. (U. Tokyo)
Hayashi, M. (Chiba U.)
Hayashi, T. (NIFS)
Hiei, E. (Meisei U.)
Hirayama, T. (Meisei U.)
Hirotani, K. (NAOJ)
Hori, K. (Tohoku U.)
Hudson, H. S. (SPRC)
Ichimoto, K. (NAOJ)
Inoue, H. (ISAS)
Ishida, M. (ISAS)
Ishii, T. (Kyoto U.)
Iwata, F. (ISAS)
Kawakami, S. (OMS)
Khan, J. I. (MSSL)
Kitai, R. (Hida Obs.)
Koide, S. (Toyama U.)
Koshiishi, H. (NASDA)
Kosugi, T. (NAOJ)
Koutchmy, S. (IAP)
Kozai, Y.
Kubo, S. (Meisei U.)
Kudoh, T. (NAOJ)
Kurokawa, H. (Hida Obs.)
Kusano, K. (Hiroshima U.)
Longcope, D. W. (Montana State U.)
Magara, T. (Kyoto U.)
Makino, F. (ISAS)
Makishima, K. (U. Tokyo)
Martens, P. C. (ESA/NASA/GSFC)
Marubashi, K. (CRL)
Masuda, S. (STE Lab.)
Matsumoto, R. (Chiba U.)
Matsuzaki, T. (Chiba U.)
McLeod, A. (CRL)
McTiernan, J. (UCB)
Mitsuda, M. (ISAS)
Morita, S. (SUT)
Murakami, T. (ISAS)
Nagase, F. (ISAS)
Nagata, S (NAOJ)
Nakagawa, Y. (Ibaraki U.)
Nakajima H. (NRO/NAOJ)
Nakakubo, K. (Tokyo Gakugei U.)
Nishio, M. (NRO/NAOJ)
Nitta, N. (LMSAL)
Oda, M. (TUIS)
Ogawara, Y. (ISAS)
Ohyama, M. (Nagoya U./NAOJ)
Okubo, A. (Chiba U.)

Peres G. (Inst. Obs. Astron. Palermo)
Phillips, A. (MSSL)
Phillips, K. J. H. (RAL)
Saita, N. (SPU)
Sakao, T. (NAOJ)
Sakurai, T. (NAOJ)
Sanchez-Ibarra, A. (U. Sonora)
Sato, J. (NAOJ)
Savy, S. K. (ISAS)
Schmieder, B. (Obs. Paris)
Setiahadi, B. (Watukosek Solar Obs.)
Shibasaki, K. (NRO/NAOJ)
Shibata, K. (NAOJ)
Shimizu, T. (NAOJ)
Shimojo, M. (NAOJ)
Shin, J. (NAOJ)
Shoji, M. (Kyoto U.)
Slater, G. (LMATC)
Somov, B. (Moscow State U.)
Sterling, A. C. (CPI/ISAS)
Suematsu, Y. (NAOJ)
Suzuki, Y. (NIFS)
Takahashi, T. (ISAS)
Takakura, T. (U. Tokyo)
Takano, T. (NRO/NAOJ)
Tonooka, H. (Chiba U.)
Tsuneta, S. (NAOJ)
Uchida, Y. (SUT)
UeNo, S. (Kyoto U.)
Watanabe, Ta. (Ibaraki U.)
Watanabe, Te. (NAOJ)
Watari, S. (CRL)
Yaji, K. (Kawabe Cosmic Park)
Yamaguchi, A. (NAOJ)
Yamamoto, M. (Ibaraki U.)
Yashiro, S. (U. Tokyo)
Yokoyama, T. (NAOJ)
Yoshida, T. (U. Tokyo)
Yoshimura, K. (Kyoto U.)
Zarro, D. (NASA/GSFC)
Zhang, H. Y. (Yunnan Obs.)
Zhang, H. Q. (Beijing Obs./NAOJ)